THE KLUWER INTERNATIONAL SERIES
IN ENGINEERING AND COMPUTER SCIENCE

STABILITY OF FINITE AND INFINITE DIMENSIONAL SYSTEMS

STABILITY OF FINITE AND INFINITE DIMENSIONAL SYSTEMS

by

Michael I. Gil'
Ben Gurion University
Beer Sheva, ISRAEL

KLUWER ACADEMIC PUBLISHERS
Boston / Dordrecht / London

Distributors for North, Central and South America:
Kluwer Academic Publishers
101 Philip Drive
Assinippi Park
Norwell, Massachusetts 02061 USA

Distributors for all other countries:
Kluwer Academic Publishers
Distribution Centre
Post Office Box 322
3300 AH Dordrecht, THE NETHERLANDS

Library of Congress Cataloging-in-Publication Data

A C.I.P. Catalogue record for this book is available
from the Library of Congress.

The publisher offers discounts on this book when ordered in bulk quantities. For more information contact: Sales Department, Kluwer Academic Publishers, 101 Philip Drive, Assinippi Park, Norwell, MA 02061

Printed on acid-free paper.

Printed in the United States of America

CONTENTS

PREFACE

1. This book deals with the stability of nonautonomous linear and nonlinear systems with concentrated and distributed parameters, as well as time-delay systems. Explicit conditions for the asymptotic stability and the input-output stability are considered. Applications to feedback stabilization are discussed. Moreover, the solution estimates established here provide the bounds for regions of attraction of steady states and estimates for the time of transitional processes.

The main methodology presented in this publication is based on a combined usage of recent norm estimates for matrix-valued and operator-valued functions with the following well-known methods and results:

a) the method of characteristic exponents (the first Lyapunov method);

b) the multiplicative representations of solutions;

c) the freezing method;

d) the positivity of impulse (Green) functions.

A significant part of this book is devoted to a solution to the problem connected with the Aizerman conjecture.

2. The problem of stability analysis of various systems, particularly control systems, continues to attract the attention of many specialists despite its long history. It is still one of the most burning problems of control theory, because of the absence of its complete solution. The problem of the synthesis of a stable system is closely connected with the problem of stability analysis. Any progress in the problem of analysis implies success in the problem of synthesis of stable systems. The basic method for the stability analysis of nonlinear systems is the Lyapunov functions (functionals) one. By this method many very strong results are obtained. Here we do not consider the Lyapunov functions (functionals) method because several excellent books cover this topic. It should be noted that finding Lyapunov's functions is often connected with serious mathematical difficulties, especially in regard to nonstationary systems.

To the contrary, the stability conditions presented in the book are mainly formulated in terms of the eigenvalues of auxiliary matrices. This fact allows us to apply the well-known stability criteria for polynomials and matrices (for example the Hurwitz criterion) to the stability analysis of time-varying (linear and nonlinear) systems.

The other powerful method for stability analysis is the method of characteristic exponents. A generalization of that method to distributed parameter systems is connected with the theory of differential equations in a Banach space. The characteristic exponents method is considered in this book, as well as the freezing method introduced by V. M. Alekseev for the stability analysis of linear finite-dimensional systems. The extension of the freezing method to time-delay and distributed parameter systems is also presented.

3. In 1949 M. A. Aizerman conjectured that a single input-single output system is absolutely stable in the Hurwitz angle. This hypothesis caused great interest among the specialists. Counter-examples were set up that demonstrated it was not, in general, true. The problem of finding a class of systems that satisfy Aizerman's hypothesis arose. One of the most powerful results in this direction was obtained in 1966 by N. Truchan who showed that Aizerman's hypothesis is satisfied by systems having linear parts in the form of single loop circuits with up to five stable aperiodic links connected in tandem. In 1981 the author showed that any system satisfies the Aizerman hypothesis if its impulse function is a non-negative one. That result includes Truchan's. Similar results are derived for multivariable and distributed parameter systems. They are also presented here.

4. The aim of this book is to provide new tools for specialists in control system theory, stability theory of ordinary and partial differential equations, and differential-delay equations.

This is the first book that:

i) gives a systematic exposition of the approach to stability analysis which is based on estimates for matrix-valued and operator-valued functions, allowing us to investigate various classes of finite and infinite dimensional systems from the unified viewpoint;

ii) contains solutions to the problems connected with the Aizerman and generalized Aizerman conjectures;

iii) presents fundamental results by A.Yu. Levin for the stability of nonautonomous systems having variable real characteristic roots,

iv) includes the freezing method for time-delay systems and distributed parameter systems.

The book is intended not only for specialists in stability theory, but for anyone interested in various applications who has had at least a first year graduate level course in analysis.

I was very fortunate to have fruitful discussions with Professors M.A. Aizerman, V.B. Kolmanovskii, M.A. Krasnoselskii, A.D. Myshkis, A. Pokrovskii, and A.A. Voronov, to whom I am very grateful for their interest in my investigations.

I wish to express my thanks to the editorial staff of Kluwer Academic Publishers for their cooperation during the preparation of this book for publication.

INTRODUCTION

The topic of this book is the stability analysis of finite and infinite dimensional systems.

The book consists of 17 chapters. Chapters 1-7 are devoted to finite dimensional systems. In Chapters 8-12 retarded and neutral type systems are considered. Chapters 13-16 deal with the stability of systems governed by linear and nonlinear evolution equations in a Banach space. Chapter 17 contains proofs of auxiliary results.

0.1 Finite dimensional systems

In Chapter 1 some preliminary results are collected. They are systematically used in the next chapters.

In Chapter 2 we review estimates for norms and absolute values of matrix-valued functions and impulse (Green) functions. They are the main tool of our investigations.

Let $||.||_{C^n}$ be the Euclidean norm, and $A = (a_{jk})$ be an $n \times n$-matrix. *The following quantity plays an essential role in the book:*

$$g(A) = (N^2(A) - \sum_{k=1}^{n} |\lambda_k(A)|^2)^{1/2},$$

where $N(A)$ is the Frobenius norm of A, and $\lambda_k(A)$ $(k = 1, ..., n)$ are the eigenvalues of A taken with their multiplicities. The relations

$$g^2(A) \leq N^2(A) - |Trace A^2| \text{ and } g^2(A) \leq N^2(A - A^*)/2$$

are true. Here A^* is the adjoint matrix. Put

$$\alpha(A) = \max_{k=1,\dots,n} Re\ \lambda_k(A).$$

The estimate

$$||exp(At)||_{C^n} \leq \exp[\alpha(A)t] \sum_{k=0}^{n-1} \frac{g^k(A)t^k}{(k!)^{3/2}} \ (t \geq 0) \tag{1.1}$$

is often applied in the book. If A is a normal matrix, then $g(A) = 0$, and $||exp(At)||_{C^n} \leq \exp[\alpha(A)t]$ $(t \geq 0)$.

In Chapter 3 we consider the system

$$\dot{x}(t) = A(t)x(t) \ (\dot{x}(t) = dx(t)/dt, \ t \geq 0) \tag{1.2}$$

with a variable matrix $A(t)$. To illustrate a typical result assume that the matrix $A(t)$ satisfies the conditions

$$v = \sup_{t \geq 0} g(A(t)) < \infty, \ \rho \equiv -\inf_{t \geq 0} \alpha(A(t)) > 0, \tag{1.3}$$

and

$$\|A(t) - A(s)\|_{C^n} \leq q_0 |t - s| \ \ (t, s \geq 0), \tag{1.4}$$

where q_0 is a positive constant. Denote by $z(q_0, v)$ the extreme right-hand (positive) root of the algebraic equation $z^{n+1} = q_0 P(z)$, where

$$P(z) = \sum_{k=0}^{n-1} \frac{v^k (k+1)}{\sqrt{k!}} z^{n-k-1}.$$

By using the freezing method and estimate (1.1), the following result is proved:

Theorem 0.1.1 *Let conditions (1.3) and (1.4) hold. In addition, let the matrix $A(t) + z(q_0, v) I$ be a Hurwitz one for all $t \geq 0$. Then equation (1.2) is exponentially stable.*

Here and below I denotes the unit matrix. Due to a simple estimate for $z(q_0, v)$ that theorem gives

Corollary 0.1.2 *Let the conditions (1.3), (1.4) and $q_0 P(1) < 1$ hold. In addition, let the matrix $A(t) + \sqrt[n]{q_0 P(1)}\, I$ be a Hurwitz one for all $t \geq 0$. Then equation (1.2) is exponentially stable.*

Stability conditions are also derived for systems with piecewise constant matrices, triangular systems and systems which are close to triangular ones.

In Chapter 4 we continue to investigate linear multivariable finite dimensional systems. In particular, the multiplicative representation is introduced for solutions of ordinary linear differential systems. That representation is used for constructing of majorants and minorants for solutions. Moreover, by the multiplicative representation, the Lozinskii and Wazewski inequalities are proved. One of the central results of Chapter 4 is the stability criterion for single input-single output nonautonomous systems with real variable characteristic roots by A. Yu. Levin. In addition, in Chapter 4 the input-output stability and feedback stabilization of linear systems are explored.

In Chapter 5, nonlinear finite dimensional systems with leading autonomous linear parts are considered. Section 5.1 is devoted to the famous Aizerman conjecture. Let us consider in a real Euclidean space $\mathbf{R}^n$ the equation

$$\dot{y} = Ay + b\, f(s, t) \ \ (s = cy, \ t \geq 0), \tag{1.5}$$

where A is a constant $n \times n$-matrix, b is a column, c is a row, f maps $\mathbf{R}^1 \times [0.\infty)$ into $\mathbf{R}^1$ with the property

$$0 \leq f(s, t)/s \leq q \ \ (s \in \mathbf{R}^1, \ s \neq 0, \ t \geq 0). \tag{1.6}$$

In 1949 M. A. Aizerman conjectured the following conjecture: under the condition $f(s,t) \equiv f(s)$, for the absolute stability of the zero solution of equation (1.5) in the class of nonlinearities (1.6) it is necessary and sufficient that the linear equation

$$\dot{y} = Ay + q_1 bcy$$

be asymptotically stable for any $q_1 \in [0, q]$. This conjecture is not, in general, true. Therefore, the following problem arose: to find the class of systems that satisfy Aizerman's hypothesis. To formulate the relevant result let us introduce the transfer function of the linear part of system (1.5)

$$W(\lambda) = c(\lambda I - A)^{-1}b = P^{-1}(\lambda)L(\lambda).$$

Here $P(\lambda)$ and $L(\lambda)$ are polynomials. Besides, let

$$K(t) \equiv \frac{1}{2\pi i}\int_{d-i\infty}^{d+i\infty} exp[\lambda t]c(\lambda I - A)^{-1}b \, d\lambda \; (d = const).$$

That is, $K(t)$ is the corresponding impulse function. In Section 5.1 it is proved that under the condition $K(t) \geq 0$ $(t \geq 0)$, the zero solution of system (1.5) is absolutely stable in the class of nonlinearities (1.6) if and only if the polynomial $P(\lambda) - qL(\lambda)$ is Hurwitzian. Clearly, that result singles out one of the classes of linear parts of systems that satisfy the Aizerman conjecture. Moreover, in Chapter 5 the generalized Aizerman conjecture for multivariable systems, and the Aizerman conjecture in the input-output setting are considered. We also investigate systems with autonomous leading linear parts and continuous nonlinearities. In particular, a bound for the region of attraction is established.

Chapter 6 is devoted to the system

$$dx/dt = A(t)x + F(x,t) \; (t \geq 0), \tag{1.7}$$

where $A(t)$ is a variable $n \times n$-matrix and F maps $\mathbf{C}^n \times [0,\infty)$ into $\mathbf{C}^n$ with the property

$$||F(h,t)||_{C^n} \leq \nu ||h||_{C^n} \text{ for all } h \in \Omega(r) \text{ and } t \geq 0.$$

Here $\nu = const \geq 0$, and $\Omega(r) = \{h \in C^n : ||h||_{C^n} \leq r\}$ $(r \leq \infty)$. By estimate (1.1) and the first Lyapunov method, conditions for the asymptotic stability of system (1.7) are established. These conditions are applied to the problem of the global feedback stabilization of nonlinear systems.

Chapter 7 deals with nonlinear systems without linear leading (principle) parts of the type

$$\dot{x}(t) = B(x(t),t)x(t) \;\; (t \geq 0), \tag{1.8}$$

where

$$B(h,t) = (b_{jk}(h,t))_{j,k=1}^{n}$$

is an $n \times n$-matrix for every $h \in \mathbf{C}^n$ and $t \geq 0$. In particular, the freezing method for the system (1.8) is developed. Also nonlinear dissipative systems, nonlinear triangular systems and their perturbations are investigated. Moreover, a generalization of the above mentioned Levin stability criterion for nonlinear single input-single output systems with real variable characteristic roots is presented.

0.2 Retarded and neutral type systems

In Chapters 8, 9, and 10 systems with delay are investigated. Our main tool in these chapters is the following estimate for the resolvent of an $n \times n$-matrix A:

$$\|(A - \lambda I)^{-1}\|_{C^n} \leq \sum_{k=0}^{n-1} \frac{g^k(A)}{\sqrt{k!}\rho^{k+1}(A, \lambda)} \text{ for all regular } \lambda, \tag{2.1}$$

where $\rho(A, \lambda)$ is the distance between the spectrum of A and a complex point λ. In particular, in Chapter 8 by that inequality, norm estimates are established for the Green functions of linear autonomous retarded systems.

In Chapter 9, the freezing method for ordinary differential systems is extended to the retarded system

$$\dot{x}(t) = A_0(t)x(t) + A_1(t)x(t - h_1) + ... + A_m(t)x(t - h_m) \quad (t \geq 0),$$

where $h_k : 0 = h_0 < h_1 < ... < h_m$ are positive constants, and $A_k(t)$ are variable $n \times n$-matrices with the properties

$$\|A_k(t) - A_k(s)\|_{C^n} \leq q_k \, |t - s| \; (q_k = const \geq 0; \; t, s \geq 0; \; k = 1, ..., m).$$

The main object of Chapter 10 is the system

$$\dot{x}(t) = \int_0^{\eta} dR(\tau)x(t - \tau) + F(t, x_t) \;\; (t \geq 0, \; x_t(\theta) \equiv x(\theta + t), \; -\eta \leq \theta \leq 0), \tag{2.2}$$

where $R(\tau)$ is an $n \times n$-variable matrix defined on a finite segment $[0, \eta]$, and having a bounded variation, and F continuously maps $[0, \infty) \times C([-\eta, 0], \mathbf{C}^n)$ into $\mathbf{C}^n$. By estimate (2.1) stability conditions and estimates for the Lyapunov exponents of solutions of system (2.2) are obtained under certain assumptions. Moreover, input-output stability and the feedback stabilization of nonlinear retarded systems are discussed.

In Chapter 11 linear neutral type functional differential systems are considered. Namely, let $R(\tau)$ and $W(\tau)$ be $n \times n$-matrix-valued functions defined on a finite segment $[0, \eta]$ and having bounded variations. Consider the neutral type functional differential system

$$\dot{x}(t) - \int_0^{\eta} dW(\tau)\dot{x}(t - \tau) - \int_0^{\eta} dR(\tau)x(t - \tau) =$$

$$\sum_{k=1}^{m_1} A_k(t)\dot{x}(t-l_k(t)) + \sum_{j=1}^{m} B_j(t)x(t-h_j(t)) \ (t \geq 0),$$

where $A_k(t), B_j(t)$ are uniformly bounded variable matrices, and

$$l_k(t), h_j(t) \ (k = 1, ..., m_1; \ l = 1, ..., m)$$

are scalar-valued positive functions. Stability conditions and solution estimates for that system are derived.

Chapter 12 is devoted to the nonlinear neutral type functional differential system

$$\dot{x}(t) - \int_0^\eta dW(\tau)\dot{x}(t-\tau) = \int_0^\eta dR(\tau)x(t-\tau) + F(t, x_t, \dot{x}_t)$$

$$(t \geq 0, \ \dot{x}_t(\theta) = \dot{x}(t+\theta), \ -\eta \leq \theta \leq 0),$$

where the function

$$F : [0, \infty) \times (C([-\eta, \infty), \mathbf{C}^n))^2 \to \mathbf{C}^n$$

supplies the solvability and has the following property: there are constants $q_0, q_1 \geq 0$ such that

$$\|F(., u_t, \dot{u}_t)\|_{L^2([0,\infty),\mathbf{C}^n)} \equiv [\int_0^\infty \|F(t, u_t, \dot{u}_t)\|^2_{C^n} dt]^{1/2} \leq$$

$$q_0\|u\|_{L^2([-\eta,\infty),\mathbf{C}^n)} + q_1\|\dot{u}\|_{L^2([-\eta,\infty),\mathbf{C}^n)}$$

for all $u \in L^2([-\eta, \infty), \mathbf{C}^n)$ with $\dot{u} \in L^2([-\eta, \infty), \mathbf{C}^n)$. Stability conditions are derived for these systems.

0.3 Evolution equations and distributed systems

In Chapter 13 the well-known results about sectorial operators and analytical semigroups are collected. In addition, norm estimates for semigroups of a class are established by a combined usage of estimates for operator-valued functions with properties of operators acting in a Hilbert space.

Let $A(t)$ be a variable closed linear operator in a Banach space X with a dense constant domain $D(A(t)) \equiv D \subseteq X$. In Chapter 14 the equation

$$du/dt = A(t)u \quad (t \geq 0). \tag{3.1}$$

is investigated. In that chapter, in particular, the above mentioned freezing method is extended to a class of linear time-variant evolution equations. An extension of the freezing method to evolution equations is based on norm estimations for relevant semigroups. However obtaining these estimations is usually difficult. For this reason we restrict ourselves to a class of equations in a Hilbert space with operators having the form of pencils of selfadjoint operators. The coefficients of the pencils are variable matrices. This class contains various parabolic and integrodifferential systems. The stability conditions derived by the freezing method are particularly formulated in terms of the Hurwitzness of auxiliary matrix-valued functions.

Further, let operator $A(t)$ be continuously differentiable in t on D. That is, the function $A(t)v$ is strongly continuously differentiable for any $v \in D$. Assume that for all $t \geq 0$ and small enough $\delta > 0$, the operator $I_X - \delta A(t)$ is invertible. Here I_X is the unit operator in X. Additionally, for some partiating of a segment $[0,t]$: $0 = t_0^{(n)} < t_1^{(n)} < ... < t_n^{(n)} = t$ let us denote

$$U_{n,k} = (I_X - A(t_n^{(n)})\delta_n)^{-1}(I_X - A(t_{n-1}^{(n)})\delta_{n-1})^{-1}...(I_X - A(t_{k+1}^{(n)})\delta_{k+1})^{-1},$$

for $k < n$, and $U_{n,n} = I_X$. Here $\delta_k = \delta_k^{(n)} = t_k^{(n)} - t_{k-1}^{(n)}$, $(k = 1, ..., n)$. Then with $u(0) = u_0 \in D$ the relation

$$u(t) = \lim U_{n0}u_0 \text{ as } \max_k \delta_k^{(n)} \to 0$$

is valid in the sense of the norm of space X. This relation is called *the multiplicative representation* of the solution $u(t)$ of equation (3.1). By using the multiplicative representation, the above mentioned estimates of Wazewski and Lozinskii (from the theory of ordinary differential equations) are generalized to equation (3.1).

In Chapter 15 the typical object is the equation

$$du/dt = Au + F(u,t) \; (t \geq 0), \tag{3.2}$$

where A is a linear constant operator in X with a dense domain $D(A)$, and $F(A)$ maps $D(A) \times [0, \infty)$ into X. Let $Y \supseteq D(A)$ be a subspace of X, and $\Omega_r(Y) = \{h \in Y : ||h||_Y \leq r\}$ with a positive number $r \leq \infty$. It is assumed that

$$||F(h,t)||_X \leq q||h||_Y \; (t \geq 0, \; h \in \Omega_r(Y)) \tag{3.3}$$

with a constant q. By the Lyapunov exponent method and the estimates for semigroups presented in Chapter 13, stability conditions and a bound for the region of attraction of the zero solution of equation (3.2) are derived. In addition, semilinear distributed parameter systems with positive spectral impulse functions are considered. It is shown that under some restrictions such systems satisfy the generalized Aizerman hypothesis.

In Chapter 16 we consider the semilinear equation of the form (1.7) with a variable operator $A(t)$ and a nonlinearity F satisfying the condition of the type (3.3).

Chapter 1

PRELIMINARIES

In this chapter we present bits of information from the theory of finite dimensional systems, which will be systematically used in the next chapters.

Sections 1.1 deals with the notions of vector and matrix norms.

Stability definitions of ordinary differential equations are presented in Sections 1.2.

In Sections 1.3 some classes of matrices are introduced.

Basic properties of eigenvalues of matrices are reviewed in Section 1.4.

A definition and examples of matrix-valued functions are presented in Section 1.5.

Section 1.6 deals with some standard facts on evolution operators of linear differential equations.

In Section 1.7 some integral inequalities are collected.

In Section 1.8 we have compiled simple facts on perturbations of evolution operators.

Section 1.9 contains a brief summary of some classical results from the Lyapunov exponents theory.

In Section 1.10 relations between contour integrals and divided differences are discussed.

In Section 1.11 a class of algebraic equations is considered.

1.1 Vector and matrix norms

Let $\mathbf{C}^n$ be an n-dimensional complex Euclidean space, and $R_+ = [0, \infty)$.

Recall that a function $\nu : \mathbf{C}^n \to R_+$ is said to be *a norm on* $\mathbf{C}^n$ (or a vector norm) if ν satisfies the following conditions:

1. $\nu(x) = 0$ iff $x = 0$,
2. $\nu(\alpha x) = |\alpha|\nu(x)$,
3. $\nu(x + y) \leq \nu(x) + \nu(y)$ for all $x, y \in \mathbf{C}^n$, $\alpha \in \mathbf{C}$.

Usually, a norm is denoted by the symbol $||.||$. That is, $\nu(x) = ||x||$. The following important properties follow immediately from the definition:

$$||x - y|| \geq ||x|| - ||y||, \text{ and } ||x|| = || - x||.$$

There are an infinite number of norms on $\mathbf{C}^n$. However, three of these are most commonly used in practice. Namely,

$$||x||_1 = \sum_{k=1}^{n} |x_k|, \ ||x||_\infty = \max_{k=1,\ldots,n} |x_k|,$$

and

$$||x||_{C^n} = \sqrt{\sum_{k=1}^{n} |x_k|^2} \ (x = (x_k) \in \mathbf{C}^n).$$

The norm $||x||_{C^n}$ is called also *the Euclidean norm.*

Throughout this chapter $||x||$ *means an arbitrary norm of a vector* x.

We will use the following matrix norms: the operator norm and the Frobenius (Hilbert-Schmidt) norm. The operator norm of a matrix (a linear operator in $\mathbf{C}^n$) A is

$$||A|| = \sup_{x \in C^n} \frac{||Ax||}{||x||}.$$

The relations

$$||A|| > 0 \ (A \neq 0); \ ||\lambda A|| = |\lambda| ||A|| \ (\lambda \in \mathbf{C}),$$

$$||AB|| \leq ||A|| ||B||, \text{ and } ||A + B|| \leq ||A|| + ||B||$$

are valid for all matrices A and B. The Frobenius norm of A is

$$N(A) = \sqrt{\sum_{j,k=1}^{n} |a_{jk}|^2}.$$

Here a_{jk} are the entries of matrix A in some orthonormed basis. The Frobenius norm does not depend on the choice of an orthonormed basis. The relations

$$N(A) > 0 \ (A \neq 0); \ N(\lambda A) = |\lambda| N(A) \ (\lambda \in \mathbf{C}),$$

$$N(AB) \leq N(A)N(B) \text{ and } N(A + B) \leq N(A) + N(B)$$

are true for all matrices A and B.

1.2 Definitions of stability

Consider in $\mathbf{C}^n$ the differential equation

$$\dot{x}(t) = f(t, x(t)) \ (t \geq 0; \ \dot{x}(t) \equiv dx/dt), \tag{2.1}$$

where $f : R_+ \times \mathbf{C}^n \to \mathbf{C}^n$ is a continuous vector-valued function. *A solution* of (2.1) is a differentiable function $x : R_+ \to \mathbf{C}^n$ satisfying that equation for all $t \geq 0$.

It is further assumed that the function f is of such a nature that equation (2.1) has a unique solution over R_+ corresponding to each initial condition $x(0) = x_0$. This is in the case, for example, if f satisfies the Lipshitz condition (see for instance (Vidyasagar, 1993, Theorem 2.4.25)).

A point $x_1 \in \mathbf{C}^n$ is said to be *an equilibrium point* of the system (2.1) if

$$f(t, x_1) \equiv 0 \text{ for all } t \geq 0.$$

In other words, if the system starts at an equilibrium point, it stays there. The converse is also true. Throughout this chapter we shall assume that 0 is an equilibrium point of the system (2.1). This assumption does not result in any loss of generality, because if x_1 is an equilibrium point of (2.1), then 0 is an equilibrium point of the system

$$\dot{z}(t) = f_1(t, z(t)),$$

where

$$f_1(t, z(t)) = f(t, z(t) + x_1).$$

Everywhere below in this section, $x(t)$ is a solution of (2.1).

Definition 1.2.1 *The equilibrium point 0 is said to be stable (in the sense of Lyapunov) if, for every $t_0 \geq 0$ and $\epsilon > 0$, there exists $\delta(t_0) > 0$, such that the condition*

$$||x(t_0)|| \leq \delta(t_0)$$

implies

$$||x(t)|| \leq \epsilon \text{ for all } t \geq t_0. \tag{2.2}$$

It is uniformly stable if, for each $\epsilon > 0$, there exists $\delta > 0$ independent of t_0, such that the condition $||x(t_0)|| \leq \delta$ implies inequality (2.2).

The equilibrium is *unstable*, if it is not stable.

Because all norms on $\mathbf{C}^n$ are topologically equivalent, it follows that the stability status of an equilibrium does not depend on the particular norm.

Definition 1.2.2 *The equilibrium 0 is attractive if, for each $t_0 \in R_+$, there is an $\eta(t_0) > 0$ such that*

$$||x(t_0)|| \leq \eta(t_0) \text{ implies } x(t) \to 0 \text{ as } t \to \infty.$$

The equilibrium 0 is uniformly attractive if there is an $\eta > 0$ independent of t_0, such that

$$||x(t_0)|| \leq \eta \text{ implies } x(t) \to 0 \text{ as } t \to \infty.$$

Also, the set $B(\eta)$, defined by

$$B(\eta) = \{x \in \mathbf{C}^n : ||x|| \leq \eta\}$$

is called the region of attraction (the stability domain) for the equilibrium point 0.

Note that, in order for an equilibrium to be attractive, a necessary (though not sufficient) condition is that it be isolated, i.e., that there exists a neighborhood of the equilibrium that does not contain any other equilibria. This is in contrast to the property of stability, which can apply even to equilibria that are not isolated.

Definition 1.2.3 *The equilibrium point* 0 *is said to be (uniformly) asymptotically stable if it is stable in the sense of Lyapunov and (uniformly) attractive.*

The equilibrium point 0 *is said to be globally (uniformly) asymptotically stable if it is (uniformly) asymptotically stable and the region of attraction* $B(\eta) = \mathbf{C}^n$.

As shown by Vinograd (1957) (see also (Vidyasagar, 1993, p. 141)), attractivity and stability are really independent properties, i.e., an equilibrium can be attractive without being stable.

Definition 1.2.4 *The equilibrium point* 0 *is exponentially stable if for any* $t_0 \geq 0$ *there exist constants* $r, a, b > 0$ *such that*

$$||x(t)|| \leq a||x(t_0)||e^{-b(t-t_0)} \ (t \geq t_0), \ \textit{if} \ ||x(t_0)|| \leq r. \tag{2.3}$$

It is uniformly exponentially stable if constants $r, a, b > 0$ *in (2.3) are independent of* t_0. *The equilibrium point* 0 *is globally exponentially stable if inequality (2.3) holds with* $r = \infty$.

Consider the linear equation

$$\dot{u}(t) = A(t)u(t) \ (t \geq 0) \tag{2.4}$$

with a variable $n \times n$-matrix $A(t)$. Since the zero is the unique equilibrium point of a linear equation, *we will say that (2.4) is stable (uniformly stable, asymptotically stable, exponentially stable)* if the zero solution of (2.4) is stable (uniformly stable, asymptotically stable, exponentially stable).

For linear equations, the notions of the global asymptotic (exponential) stability and asymptotic (exponential) stability coincide.

In addition, *we will say that (2.4) is uniformly asymptotically (exponentially) stable* if the zero solution of (2.4) is uniformly asymptotically (exponentially) stable.

1.3 Classes of matrices

For an $n \times n$-matrix (a linear operator in $\mathbf{C}^n$) A, by A^* the conjugate transposed matrix is denoted. That is, if a_{jk} are entries of A, then $\overline{a}_{kj}$ $(j, k = 1, ..., n)$ are entries of A^*. In other words

$$(Ax, y) = (x, A^*y) \ (x, y \in \mathbf{C}).$$

The symbol $(.,.) = (.,.)_{C^n}$ means the scalar product in $\mathbf{C}^n$. We use I to denote the unit matrix in $\mathbf{C}^n$.

Definition 1.3.1 *A matrix $A = (a_{jk})_{j,k=1}^n$ is*

1. *symmetric (Hermitian) if $A^* = A$;*
2. *positive definite (positive semidefinite, negative definite, negative semidefinite) if it is Hermitian and $(Ah, h) > (\geq, <, \leq)\ 0$ for any nonzero $h \in \mathbf{C}^n$;*
3. *unitary if $A^*A = AA^* = I$;*
4. *normal if $AA^* = A^*A$;*
5. *upper triangular if $a_{jk} = 0$ for $j > k$; i.e., if it is zero below its diagonal;*
6. *lower triangular if $a_{jk} = 0$ for $j < k$; i.e., if it is zero above its diagonal;*
7. *diagonal if it is upper and lower triangular; i.e. its nonzero elements are on its diagonal;*
8. *nilpotent if $A^n = 0$.*

Let A be an arbitrary matrix. Then the matrices

$$A_I = (A - A^*)/2i \text{ and } A_R = (A + A^*)/2$$

are the imaginary Hermitian component and the real Hermitian one of A, respectively. A matrix A *is dissipative* if its real Hermitian component is negative semidefinite. By A^{-1} the matrix inverse to A is denoted: $AA^{-1} = A^{-1}A = I$.

1.4 Eigenvalues of matrices

Let A be an arbitrary matrix. Then if for some $\lambda \in \mathbf{C}$, the equation

$$Ah = \lambda h$$

has a nontrivial solution, λ is an eigenvalue of A and h is its eigenvector. An eigenvalue $\lambda(A)$ has the (algebraic) multiplicity r if

$$dim(\cup_{k=1}^n ker(A - \lambda I)^k) = r.$$

Let $\lambda_k(A)$ $(k = 1, \ldots, n)$ be eigenvalues of A, including with their multiplicities. Then the set

$$\sigma(A) = \{\lambda_k(A)\}_{k=1}^n$$

is the spectrum of A.

All eigenvalues of a Hermitian matrix A are real. If, in addition, A is positive (negative) definite, then all its eigenvalues are positive (negative). Furthermore,

$$r_s(A) = \max_{k=1,\ldots,n} |\lambda_k(A)|$$

is the spectral radius of A. Denote

$$\alpha(A) = \max_{k=1,\ldots,n} Re\lambda_k(A), \ \beta(A) = \min_{k=1,\ldots,n} Re\lambda_k(A).$$

A matrix A *is said to be a Hurwitz one* if all its eigenvalues lie in the open left half-plane, i.e., $\alpha(A) < 0$.

A complex number λ is *a regular point of A* if it does not belong to the spectrum of A, i.e., if $\lambda \neq \lambda_k(A)$ for any $k = 1, \ldots, n$.

The trace of A is denoted by $Tr\ A$:

$$Trace\ A = Tr\ A = \sum_{k=1}^{n} \lambda_k(A).$$

Recall that

$$Tr(AB) = Tr(BA) \text{ and } Tr(A+B) = Tr(A) + Tr(B)$$

for all matrices A and B. In addition, $det(A)$ means the determinant of A:

$$det(A) = \prod_{k=1}^{n} \lambda_k(A).$$

A polynomial

$$p(\lambda) = det(\lambda I - A) = \prod_{k=1}^{n} (\lambda - \lambda_k(A))$$

is said to be the characteristic polynomial of A. All the eigenvalues of A are the roots of its characteristic polynomial. The algebraic multiplicity of an eigenvalue of A coincides with the multiplicity of the corresponding root of the characteristic polynomial. A polynomial $P(\lambda)$ is said to be *a Hurwitz one* if all its roots lie in the open left half-plane. Thus, the characteristic polynomial of a Hurwitz matrix is a Hurwitz polynomial.

1.5 Matrix-valued functions

Let A be a matrix and let $f(\lambda)$ be a scalar-valued function which is analytical on a neighborhood D of $\sigma(A)$. We define the function $f(A)$ of A by the generalized integral formula of Cauchy

$$f(A) = -\frac{1}{2\pi i} \int_{\Gamma} f(\lambda) R_\lambda(A) d\lambda, \tag{5.1}$$

where $\Gamma \subset D$ is a closed smooth contour surrounding $\sigma(A)$, and

$$R_\lambda(A) = (A - \lambda I)^{-1}$$

is the resolvent of A. If an analytic function $f(\lambda)$ is represented in some domain by the Taylor series

$$f(\lambda) = \sum_{k=0}^{\infty} c_k \lambda^k,$$

and the series

$$\sum_{k=0}^{\infty} c_k A^k$$

converges in the norm of space $\mathbf{C}^n$, then

$$f(A) = \sum_{k=0}^{\infty} c_k A^k.$$

In particular, for any matrix A,

$$e^A = \sum_{k=0}^{\infty} \frac{A^k}{k!}.$$

Example 1.5.1 *Let A be a diagonal matrix:*

$$A = \begin{pmatrix} a_1 & 0 & \dots & 0 \\ 0 & a_2 & \dots & 0 \\ . & \dots & . & . \\ 0 & \dots & 0 & a_n \end{pmatrix}.$$

Then

$$f(A) = \begin{pmatrix} f(a_1) & 0 & \dots & 0 \\ 0 & f(a_2) & \dots & 0 \\ . & \dots & . & . \\ 0 & \dots & 0 & f(a_n) \end{pmatrix}.$$

Example 1.5.2 *If a matrix J is an $n \times n$-Jordan block:*

$$J = \begin{pmatrix} \lambda_0 & 1 & 0 & \dots & 0 \\ 0 & \lambda_0 & 1 & \dots & 0 \\ . & . & . & \dots & . \\ . & . & . & \dots & . \\ . & . & . & \dots & . \\ 0 & 0 & \dots & \lambda_0 & 1 \\ 0 & 0 & \dots & 0 & \lambda_0 \end{pmatrix},$$

then

$$f(J) = \begin{pmatrix} f(\lambda_0) & \frac{f'(\lambda_0)}{1!} & \dots & \frac{f^{(n-1)}(\lambda_0)}{(n-1)!} \\ 0 & f(\lambda_0) & \dots & \\ . & . & \dots & . \\ . & . & \dots & . \\ . & . & \dots & . \\ 0 & \dots & f(\lambda_0) & \frac{f'(\lambda_0)}{1!} \\ 0 & \dots & 0 & f(\lambda_0) \end{pmatrix}.$$

Let us consider the equation

$$\dot{x}(t) = Ax(t) \ (t \geq 0)$$

with an arbitrary constant matrix A. For any solution $x(t)$ of that equation the representation

$$x(t) = e^{At}x(0) \ (t \geq 0) \tag{5.2}$$

is valid. Now consider the equation

$$\dot{x}(t) = Ax(t) + f(t) \ (t \geq 0) \tag{5.3}$$

with a piecewise continuous function $f(t) : R_+ \to \mathbf{C}^n$.

Lemma 1.5.3 *A solution $x(t)$ of equation (5.3) admits the representation*

$$x(t) = e^{tA}x(0) + \int_0^t e^{(t-s)A}f(s)ds \ (t \geq 0). \tag{5.4}$$

Relations (5.2) and (5.4) can be proved by direct differentiation (see also (Gantmaher, 1967, Section 5.6)).

1.6 Evolution operators

Let $u(t)$ be a solution of the equation

$$\dot{u}(t) = A(t)u(t) \ (t \geq 0) \tag{6.1}$$

with a variable $n \times n$-matrix $A(t)$. Then a linear operator (matrix) $U(t, s)$: $\mathbf{C}^n \to \mathbf{C}^n$ defined by the equality

$$U(t, s)u(s) = u(t) \ (t, s \geq 0)$$

is called *the evolution operator* of equation (6.1). The evolution operator has the following properties:

$$U(t, t) = I \ ,$$

$$\frac{d}{dt}U(t, s) = A(t)U(t, s). \tag{6.2}$$

Moreover,

$$\frac{d}{ds}U(t, s) = -U(t, s)A(s) \tag{6.3}$$

and

$$U(t, s)U(s, \tau) = U(t, \tau) \ (t, s, \tau \geq 0).$$

The latter property implies $U(s, t) = U^{-1}(t, s)$. If $U(t, s)$ is the evolution operator of equation (6.1), then $U(s, t) = U^{-1}(t, s)$ is the evolution operator of the equation

$$\dot{u}(t) = -A(t)u(t) \ (t \geq 0). \tag{6.4}$$

The operator $U(t) = U(t, 0)$ is called *the Cauchy operator* for equation (6.1). Since

$$U(t, s) = U(t, 0)U(0, s), \text{ and } U(0, s) = U^{-1}(s, 0),$$

the relation

$$U(t, s) = U(t)U^{-1}(s) \quad (t, s \geq 0)$$

is valid. If $A(t) \equiv A$ is a constant matrix, then

$$U(t) = e^{tA} \text{ and } U(t, s) = U(t - s) = e^{(t-s)A}.$$

Furthermore, a solution of the non-homogeneous equation

$$\dot{u}(t) = A(t)u(t) + f(t) \ (t \geq 0) \tag{6.5}$$

with a given piecewise continuous function $f : R_+ \to \mathbf{C}^n$ can be obtained in the form

$$u(t) = U(t, s)u(s) + \int_s^t U(t, \tau)f(\tau)d\tau \ (t, s \geq 0). \tag{6.6}$$

It is simple to check that if $U(t, s)$ is the evolution operator of (6.1), then for any complex number α, $U_\alpha(t, s) = e^{\alpha(t-s)}U(t, s)$ is the evolution operator of the equation

$$\dot{u}(t) = (A(t) + \alpha I)u(t) \ (t \geq 0). \tag{6.7}$$

The Cauchy operator can be represented in the form

$$U(t) = I + \int_0^t A(t_1)dt_1 + \sum_{k=2}^{\infty} \int_0^t \int_0^{t_k} \ldots \int_0^{t_2} A(t_k)\, A(t_{k-1}) \ldots A(t_1) dt_k\, dt_{k-1} \ldots dt_1.$$

Lemma 1.6.1 *A necessary and sufficient condition for the stability of equation (6.1) is the uniform boundedness of its Cauchy operator:*

$$\sup_{t \geq 0} \|U(t)\| < \infty.$$

For the proof see (Daleckii and Krein, 1974, Lemma 3.3.1).

Rephrasing the conditions of the lemma, we will say that a necessary and sufficient condition for the stability of equation (6.1) is the existence of a constant $N > 0$ such that any solution $u(t)$ of equation (6.1) satisfies the estimate

$$\|u(t)\| \leq N\|u(0)\| \ (t \geq 0).$$

The least value N_0 of this constant is clearly given by the equality

$$N_0 = \sup_{t \geq 0} \|U(t)\|.$$

Linear equation (6.1) is uniformly stable if there exists a constant $N > 0$ such that any solution $u(t)$ of this equation satisfies the following estimate for all $t \geq s \geq 0$:

$$\|u(t)\| \leq N\|u(s)\|.$$

Since $u(t) = U(t,s)u(s)$, it can be shown that the uniform stability of the equation is equivalent to the condition

$$\sup_{t \geq s \geq 0} \|U(t,s)\| < \infty. \tag{6.8}$$

Note that if $A(t) \equiv A$ is a constant matrix, then the stability condition

$$\sup_{t \geq 0} \|e^{tA}\| < \infty$$

coincides with the uniform stability condition.

1.7 Integral inequalities

Let $h = (h_k)_{k=1}^n$ and $w = (w_k)_{k=1}^n$ be real vectors. We will write

$$h \leq w \text{ if } w_k \leq h_k \ (k = 1, ..., n).$$

The vector w is non-negative, if

$$w_k \geq 0 \ (k = 1, ..., n).$$

A vector-valued function is non-negative if its values are non-negative vectors. In the sequel we will often use the following result.

Lemma 1.7.1 *Let a non-negative continuous vector-valued function $\phi : R_+ \to \mathbf{R}^n$ satisfy the inequality*

$$\phi(t) \leq f(t) + \int_0^t K(t,s)\phi(s)ds \ (t \geq 0)$$

with a non-negative continuous vector-valued function $f : R_+ \to \mathbf{R}^n$ and a continuous non-negative matrix-valued kernel $K(t,s)$. Then $\phi(t) \leq \psi(t)$, where $\psi(t)$ is a solution of the integral equation

$$\psi(t) = f(t) + \int_0^t K(t,s)\psi(s)ds.$$

Similarly, the inequality

$$\phi(t) \geq f(t) + \int_0^t K(t,s)\phi(s)ds \ (t \geq 0)$$

implies $\phi(t) \geq \psi(t)$ $(t \geq 0)$.

Proof: The lemma is due to an obvious application of the well-known Theorem 1.9.3 (Daleckii and Krein, 1974) to the space $C(R_+, \mathbf{R}^n)$ of real continuous vector-valued functions defined on R_+, since the integral operator

$$(Ku)(t) = \int_0^t K(t,s)u(s)ds$$

is a Volterra operator and, as is well known (Gohberg and Krein, 1970), its spectral radius equals zero. □

Corollary 1.7.2 *(The Gronwall inequality). Suppose a positive scalar-valued function ϕ subordinates the inequality*

$$\phi(t) \leq c + \int_0^t h(s)\phi(s)ds \ (c = const > 0, \ t \geq 0),$$

where $h(t)$ is a continuous positive scalar-valued function. Then

$$\phi(t) \leq c \, exp \, [\int_0^t h(s)ds] \ (t \geq 0).$$

More about the Gronwall inequality can be found, for instance, in (Reissig et al., 1974), (Vidyasagar, 1993, p. 236).

Corollary 1.7.3 *Suppose a positive scalar-valued function ϕ subordinates the inequality*

$$\phi(t) \leq c \, exp \, [\int_0^t p(\tau)d\tau] + \int_0^t exp \, [\int_s^t p(\tau)d\tau]h(s)\phi(s)ds \ (t \geq 0),$$

where $h(t)$ and $p(t)$ are continuous scalar-valued functions. Moreover, $h(t)$ is positive. Then

$$\phi(t) \leq c \, exp \, [\int_0^t (p(s) + h(s))ds] \ (t \geq 0).$$

To establish that result it is sufficient to set

$$\phi_1(t) = \phi(t) \, exp \, [-\int_0^t p(s)ds]$$

and to apply Corollary 1.7.2.

1.8 Perturbations of evolution operators

Lemma 1.8.1 *Let $U(t,s)$ be the evolution operator of equation (6.1) and $U_1(t,s)$ be the evolution operator of the equation*

$$\dot{u}(t) = A_1(t)u(t) \ (t \geq 0) \tag{8.1}$$

with a variable $n \times n$-matrix $A_1(t) \neq A(t)$. Then the following relation is true:

$$U(t,s) - U_1(t,s) = \int_s^t U(t,\tau)(A(\tau) - A_1(\tau))U_1(\tau,s)d\tau \ (t,s \geq 0).$$

Proof: Subtract (8.1) from (6.1) and put $\delta(t) = u(t) - u_1(t)$, where $u(t)$ is a solution of (6.1), and $u_1(t)$ is a solution of (8.1) with the initial conditions $u_1(s) = u(s) = h$. Here h is a given vector. We get

$$\dot{\delta}(t) = A(t)\delta(t) + (A(t) - A_1(t))u_1(t).$$

Taking into account that $\delta(s) = 0$ and applying formula (6.6), we arrive at the equality

$$\delta(t) = \int_s^t U(t,\tau)(A(\tau) - A_1(\tau))u_1(\tau)d\tau.$$

But

$$u_1(\tau) = U_1(t,s)u(s) \text{ and } \delta(t) = (U(t,s) - U_1(t,s))u(s).$$

This proves the result, since $u(s) = h$ is an arbitrary vector. □

Furthermore, Lemma 1.8.1 implies,

$$\|U(t,s) - U_1(t,s)\| \leq \int_s^t \|U(t,\tau)\| \, \|A(\tau) - A_1(\tau)\| \, \|U_1(\tau,s)\| d\tau \ (0 \leq s < t).$$

Hence,

$$\|U_1(t,s)\| \leq \|U(t,s)\| + \int_s^t \|U(t,\tau)\| \, \|A(\tau) - A_1(\tau)\| \, \|U_1(\tau,s)\| d\tau$$
$$(0 \leq s < t). \tag{8.2}$$

Let us suppose that

$$\|U(t,s)\| \leq \phi(t,s) \ (0 \leq s < t), \tag{8.3}$$

where $\phi(t,s) : R_+ \times R_+ \to R_+$ is a continuous function. Then it can be written

$$\|U_1(t,s)\| \leq \phi(t,s) + \int_s^t \phi(t,\tau)\|A(\tau) - A_1(\tau)\| \, \|U_1(\tau,s)\| d\tau.$$

Lemma 1.7.1 and the latter inequality yield

Lemma 1.8.2 *Let $U(t,s)$ and $U_1(t,s)$ be the evolution operators of the equations (6.1) and (8.1), respectively. Let inequality (8.3) hold. Then $\|U_1(t,s)\| \leq \eta_s(t)$, where $\eta_s(t)$ is the solution of the equation*

$$\eta_s(t) = \phi(t,s) + \int_s^t \phi(t,\tau)\|A(\tau) - A_1(\tau)\|\eta_s(\tau)d\tau \ (0 \leq s < t).$$

Corollary 1.8.3 *Let the evolution operator $U(t,s)$ of equation (6.1) satisfy the inequality*

$$\|U(t,s)\| \leq Ne^{-\nu(t-s)} \ (t, s \geq 0) \tag{8.4}$$

with positive constants N and ν. Then the evolution operator $U_1(t,s)$ of equation (8.1) subordinates the inequality

$$||U_1(t,s)|| \le Ne^{-\nu(t-s)}e^{N\int_s^t ||A(\tau)-A_1(\tau)||d\tau} \quad (s \le t). \tag{8.5}$$

Moreover,

$$||U(t,s) - U_1(t,s)|| \le Ne^{-\nu(t-s)}(e^{N\int_s^t ||A(\tau)-A_1(\tau)||d\tau} - 1). \tag{8.6}$$

Indeed, to prove inequality (8.5) we need, according to Lemma 1.8.2, only to check that the right part of (8.5) is a solution of the equation

$$\eta(t) = Ne^{-\nu(t-s)} + N\int_s^t e^{-\nu(t-\tau)}||A(\tau) - A_1(\tau)||\eta(\tau)d\tau \ (s \le t).$$

The checking of inequality (8.6) is left to the reader (see also (Daleckii and Krein, 1974, Section 3.2)). The latter corollary and relation (6.8) imply

Corollary 1.8.4 *Let equation (6.1) be uniformly exponentially stable and let*

$$\int_0^\infty ||A(\tau) - A_1(\tau)||d\tau < \infty.$$

Then equation (8.1) is also uniformly exponentially stable.

When $A_1(t) = 0$ using (8.6), we easily get:

Corollary 1.8.5 *Under condition (8.4) the evolution operator $U(t,s)$ of equation (6.1) satisfies the inequality*

$$||U(t,s) - I|| \le Ne^{-\nu(t-s)}(e^{\int_s^t ||A(\tau)||d\tau} - 1) \ \ (s \le t).$$

Lemma 1.8.6 *Let $U(t,s)$ and $U_1(t,s)$ be the evolution operators of equations (6.1) and (8.1), respectively. For some positive $T \le \infty$, let the condition*

$$J \equiv \sup_{0\le t\le T} \int_0^t ||U(t,s)|| ||A(s) - A_1(s)||ds < 1$$

be fulfilled. Then the inequality

$$\sup_{0\le t\le T} ||U_1(t,0)|| \le (1-J)^{-1} \sup_{0\le t\le T} ||U(t,0)||$$

holds.

Proof: According to (8.2),

$$\sup_{0\le t\le T} ||U_1(t,0)|| \le \sup_{0\le t\le T} ||U(t,0)|| +$$

$$\sup_{0\le t\le T}\|U_1(\tau,0)\|\int_0^t \|U(t,\tau)\|\|A(\tau)-A_1(\tau)\|d\tau \le$$

$$\sup_{0\le t\le T}\|U(t,0)\| + \sup_{0\le \tau\le T}\|U_1(\tau,0)\|J.$$

Now the result is due to the condition $J<1$. □

Lemma 1.8.7 *Let $U(t,s)$ and $U_1(t,s)$ be the evolution operators of equations (6.1) and (8.1), respectively. Let the condition*

$$\|U(t,s)\| \le c\, exp[\int_s^t a(\tau)]d\tau \ (0\le s\le t)$$

be fulfilled with a continuous scalar-valued function $a(t)$ and a positive constant c. Then the inequality

$$\|U_1(t,s)\| \le c\, exp[\int_s^t (a(\tau)+c\|A(\tau)-A_1(\tau)\|)d\tau]$$

holds.

Proof: Thanks to Lemma 1.8.2, $\|U_1(t,0)\|\le \eta(t)$, where $\eta(t)$ is the solution of the equation

$$\eta(t) = c\, exp[\int_s^t a(\tau)d\tau] + c\int_s^t exp[\int_\tau^t a(\tau_1)d\tau_1]\|A(\tau)-A_1(\tau)\|\eta(\tau)d\tau.$$

Put

$$z(t) = \eta(t)exp[-\int_0^t a(\tau)d\tau].$$

Then

$$z(t) = c + c\int_s^t \|A(\tau)-A_1(\tau)\|z(\tau)d\tau.$$

Now the result is due to the Gronwall lemma. □

1.9 Lyapunov exponents

Again consider the equation

$$\dot u = A(t)u \ (t\ge 0) \tag{9.1}$$

with a variable matrix

$$A(t) = (a_{jk}(t))_{j,k=1}^n.$$

The (upper) Lyapunov exponent of a solution $u(t)$ of equation (9.1) is

$$\kappa(u) = \overline{\lim}_{t\to\infty} \frac{ln\|u(t)\|}{t}.$$

If $A(t)$ is bounded:

$$\sup_{t\geq 0} \|A(t)\| = M < \infty$$

or, more generally, integrally bounded

$$\sup_{t\geq 0} \int_t^{t+1} \|A(s)\|ds = M < \infty,$$

then its solutions are subject to the estimate

$$\|u(t)\| \leq \|u(0)\|\, e^{Mt}\ (t \geq 0).$$

This implies that the (upper) Lyapunov exponent of a solution $u(t)$ of equation (9.1) is finite.

The set Σ of Lyapunov exponents of all possible solutions of equation (9.1) is called *the (upper) Lyapunov spectrum* of equation (9.1). Since equation (9.1) has n linearly independent solutions, the Lyapunov spectrum Σ consists of no more that n exponents. *The upper Lyapunov exponent of equation (9.1)* refers to the quantity

$$\kappa_L = \max_{\kappa\in\Sigma} \kappa.$$

Theorem 1.9.1 *The Lyapunov exponent of equation (9.1) coincides with the upper Lyapunov exponents of the Cauchy operator $U(t)$ of this equation:*

$$\kappa_L = \overline{\lim}_{t\to\infty} \frac{ln\|U(t)\|}{t}.$$

For the proof see (Daleckii and Krein, 1974, Theorem 3.4.1). With the notation

$$\overline{p} = \overline{\lim}_{t\to\infty} \frac{1}{t}\int_0^t p(s)ds,$$

the estimates

$$\overline{\beta}(A_R) \leq \kappa_j \leq \overline{\alpha}(A_R)\ (\kappa_j \in \Sigma)$$

can be found in (Izobov, 1974). Here $\beta(A_R(t))$ is the smallest eigenvalue and $\alpha(A_R(t))$ is the largest one of the matrix $A_R(t) = (A(t) + A^*(t))/2$.

Lozinskii (1958) introduced the logarithm norms

$$l(t) \equiv \overline{\lim}_{h\to -0} \frac{1}{h}[\|I + hA(t)\| - 1]$$

and

$$L(t) \equiv \overline{\lim}_{h\to +0} \frac{1}{h}[\|I + hA(t)\| - 1].$$

By the logarithm norms he derived the estimates $\bar{l} \le \kappa_j \le \bar{L}$ $(\kappa_j \in \Sigma)$. In particular, these estimates imply

$$\overline{\min_i [a_{ii} - \sum_{k \ne i} |a_{ik}|]} \le \kappa_j \le \overline{\max_i [a_{ii} + \sum_{k \ne i} |a_{ik}|]}$$

and

$$\overline{\min_k [a_{kk} - \sum_{i \ne k} |a_{ik}|]} \le \kappa_j \le \overline{\max_k [a_{kk} + \sum_{i \ne k} |a_{ik}|]} \ (j = 1, ..., n).$$

1.10 Estimates for contour integrals

Lemma 1.10.1 *Let D be the closed convex hull of points $x_0, x_1, ..., x_n \in \mathbf{C}$ and let a scalar-valued function f be regular on a neighborhood D_1 of D. In addition, let $\Gamma \subset D_1$ be a Jordan closed contour surrounding the points $x_0, x_1, ..., x_n$. Then the inequality*

$$\left|\frac{1}{2\pi i} \int_\Gamma \frac{f(\lambda) d\lambda}{(\lambda - x_0)...(\lambda - x_n)}\right| \le \frac{1}{n!} \sup_{\lambda \in D} |f^{(n)}(\lambda)|$$

is valid.

Proof: First, let all the points be distinct: $x_j \ne x_k$ for $j \ne k$ $(j, k = 0, ..., n)$, and let $D_f(x_0, x_1, ..., x_n)$ be a divided difference of the scalar-valued function f at points $x_0, x_1, ..., x_n$ of the complex plane. If f is regular on a neighborhood of the closed convex hull D of the points $x_0, x_1, ..., x_n$, then the divided difference admits the representation

$$D_f(x_0, x_1, ..., x_n) = \frac{1}{2\pi i} \int_\Gamma \frac{f(\lambda) d\lambda}{(\lambda - x_0)...(\lambda - x_n)} \tag{10.1}$$

(see (Gelfond, 1967, formula (54)). But, on the other hand, the following estimate is well known:

$$| D_f(x_0, x_1, ..., x_n) | \le \frac{1}{n!} \sup_{\lambda \in D} |f^{(n)}(\lambda)|$$

(Gelfond, 1967, formula (49)). Combining that inequality with relation (10.1), we arrive at the required result. If $x_j = x_k$ for some $j \ne k$, then the claimed inequality can be obtained by small perturbations and the previous reasonings. □

Lemma 1.10.2 *Let $x_0 \le x_1 \le ... \le x_n$ be real points and let a function f be regular on a neighborhood D_1 of the segment $[x_0, x_n]$. In addition, let $\Gamma \subset D_1$ be* *a Jordan closed contour surrounding* $[x_0, x_n]$. *Then there is a point* $\eta \in [x_0, x_n]$, *such that the equality*

$$\frac{1}{2\pi i}\int_\Gamma \frac{f(\lambda)d\lambda}{(\lambda - x_0)...(\lambda - x_n)} = \frac{1}{n!}f^{(n)}(\eta)$$

is true.

Proof: First suppose that all the points are distinct: $x_0 < x_1 < ... < x_n$. Then the divided difference $D_f(x_0, x_1, ..., x_n)$ of f in the points $x_0, x_1, ..., x_n$ admits the representation

$$D_f(x_0, x_1, ..., x_n) = \frac{1}{n!}f^{(n)}(\eta)$$

with some point $\eta \in [x_0, x_n]$ (Gelfond, 1967, formula (43)), (Ostrowski, 1973, p.5). Combining that equality with representation (10.1), we arrive at the required result. If $x_j = x_k$ for some $j \neq k$, then the claimed inequality can be obtained by small perturbations and the previous reasonings. □

1.11 Algebraic equations

Let us consider the algebraic equation

$$z^n = P(z), \tag{11.1}$$

where $P(z)$ is the polynomial

$$P(z) = \sum_{j=0}^{n-1} c_j z^{n-j-1}$$

with non-negative coefficients c_j $(j = 0, ..., n)$.

Lemma 1.11.1 *The extreme right-hand root* z_0 *of equation (11.1) is non-negative and the following estimates:*

$$z_0 \leq [P(1)]^{1/n} \ \ \text{if} \ \ P(1) \leq 1, \tag{11.2}$$

and

$$z_0 \leq P(1) \ \ \text{if} \ \ P(1) \geq 1 \tag{11.3}$$

are valid.

For the proof see (Gil', 1995, Lemma 4.3.2).

Setting in (11.1) $z = ax$ with a positive constant a, we obtain

$$x^n = \sum_{j=0}^{n-1} c_j a^{-j-1} x^{n-j-1}. \tag{11.4}$$

If

$$a = max_{j=0,\dots,n-1} \sqrt[j+1]{2c_j},$$

then

$$\sum_{j=0}^{n-1} c_j a^{-j-1} \le \sum_{j=0}^{n-1} 2^{-j-1} = 1 - 2^{-n+1} < 1. \tag{11.5}$$

Let x_0 be the extreme right-hand root of equation (11.4), then by (11.2) $x_0 < 1$. Since $z_0 = ax_0$, we have derived

Corollary 1.11.2 *The extreme right-hand root z_0 of equation (11.1) is non-negative. Moreover,*

$$z_0 \le \max_{j=0,\dots,n-1} \sqrt[j+1]{2c_j}.$$

1.12 Notes

This book presupposes a knowledge of basic matrix theory, for which there are good introductory texts. The books (Gantmaher, 1967), (Marcus and Minc, 1964), (Bellman, 1970) are classical. For more details about the notions presented in Sections 1.1, 1.3 and 1.4 also see (Collatz, 1967) and (Stewart and Sun, 1990).

The stability definitions presented in Section 1.2 are particularly taken from the books (Reissig et al., 1974), (Bellman, 1953), and (Vidyasagar, 1993).

The material of Sections 1.6-1.8 is based on Sections 3.1 and 3.2 of the book (Daleckii and Krein, 1974).

Section 1.9 is based on the survey (Izobov, 1974) (see also (Bylov et al., 1966)).

Lemma 1.10.1 was proved in (Gil',1995).

Estimates for roots of algebraic equations similar to estimate (11.5) can be found in the book (Ostrovski, 1973, p. 277).

References for Chapter 1

Bellman, R.E. (1953). *Stability Theory of Differential Equations.* McGraw-Hill, New York.

Bellman, R.E. (1970). *Introduction to Matrix Analysis.* McGraw-Hill, New York.

Bylov, B. F., Grobman, B. M., Nemyckii V. V. and Vinograd R.E. (1966). *The Theory of Lyapunov Exponents.* Nauka, Moscow. In Russian.

Collatz, L. (1966). *Functional Analysis and Numerical Mathematics.* Academic Press, New York-London.

Daleckii, Yu L. and Krein, M. G. (1974). *Stability of Solutions of Differential Equations in Banach Space*, Amer. Math. Soc., Providence, R. I.

Gantmaher, F. R. (1967). *Theory of Matrices.* Nauka, Moscow. In Russian.

Gelfond, A. O. (1967). *Calculations of Finite Diferences.* Nauka, Moscow, In Russian.

Gil', M. I. (1995). *Norm Estimations for Operator-valued Functions and Applications.* Marcel Dekker, Inc, New York.

Gohberg, I. C. and Krein, M. G. (1970). *Theory and Applications of Volterra Operators in Hilbert Space.* Trans. Mathem. Monographs, vol. 24, Amer. Math. Soc., R. I.

Izobov, N. A. (1974). Linear systems of ordinary differential equations. *Itogi Nauki i Tekhniki. Mat. Analis, 12*: 71-146. In Russian.

Lozinskii, S.M. (1958). Error estimates for numerical solutions of differential equations, *Izvestia VUZ. Matematika, N 5*, 52-90. In Russian.

Marcus, M. and Minc, H. (1964). *A Survey of Matrix Theory and Matrix Inequalities.* Allyn and Bacon, Boston .

Ostrowski, A. M. (1973). *Solution of Equations in Euclidean and Banach spaces.* Ac. Press, New York - London.

Reissig, R., Sansone, G. and Conti, R. (1974). *Nonlinear Differential Equations of High Order.* Noordhoff International Publishing, Leiden.

Stewart, G. W. and Sun Ji-guang (1990). *Matrix Perturbation Theory*, Academic Press, New York.

Vidyasagar, M. (1993). *Nonlinear System Analysis*, second edition. Prentice-Hall, Upper Saddle River, New Jersey.

Vinograd R.E. (1957). The inadequace of the method of characteristic exponents for the study of nonlinear differential equations, *Math. Sbornik, 41*, 431-438. In Russian.

Chapter 2

ESTIMATES FOR MATRIX-VALUED FUNCTIONS

In the present chapter we summarize estimates for norms and absolute values of matrix-valued functions and for impulse (Green) functions. These estimates are the main tool of our investigations.

In Section 2.1 norm estimates for matrix-valued functions are presented.

Section 2.2 deals with estimates for absolute values of matrix-valued functions.

In Section 2.3 estimates for impulse functions are derived.

Section 2.4 contains positivity conditions for impulse functions.

Section 2.5 deals with applications of estimates for the norm of the matrix exponent to the Lyapunov equation.

Throughout the present chapter $\|.\|$ *means the Euclidean norm* (see Section 1.1).

2.1 Norm estimates for matrix-valued functions

2.1.1 Notation

Let $A = (a_{jk})$ be an $n \times n$-matrix. *The following quantity plays a key role in the sequel:*

$$g(A) = (N^2(A) - \sum_{k=1}^{n} |\lambda_k(A)|^2)^{1/2}. \tag{1.1}$$

Recall that $N(A)$ is the Frobenius norm of A, and $\lambda_k(A)$ $(k = 1, ..., n)$ are the eigenvalues taken with their multiplicities. Since

$$\sum_{k=1}^{n} |\lambda_k(A)|^2 \geq |TrA^2|,$$

we get

$$g^2(A) \leq N^2(A) - |TrA^2|. \tag{1.2}$$

Moreover, if the eigenvalues of A are real, then $|Tr\ A^2| = Tr\ A^2$ and therefore,

$$g^2(A) = N^2(A) - TrA^2.$$

It is easy to see that

$$N(A) \leq \sqrt{n}||A||.$$

Hence,

$$g(A) \leq [n||A||^2 - \sum_{k=1}^{n} |\lambda_k(A)|^2]^{1/2} \leq \sqrt{n}||A||.$$

Let us mention some properties of $g(A)$ (for the proof see Appendix 1, Section 17.4 below, or (Gil', 1995, Corollaries 1.3.7 and 1.3.8)).

Proposition 2.1.1 *For any matrix A, the relations*

$$g^2(A) \leq \frac{1}{2} N^2(A^* - A) \tag{1.3}$$

and

$$g(e^{i\tau} A + zI) = g(A) \tag{1.4}$$

are valid with every $\tau \in \mathbf{R}, z \in \mathbf{C}$.

If A is a normal matrix: $AA^* = A^*A$, then

$$N(A) = [\sum_{k=1}^{n} |\lambda_k(A)|^2]^{1/2}$$

and therefore, $g(A) = 0$.

Example 2.1.2 *Consider the matrix*

$$A = \begin{pmatrix} a_{11} & a_{12} \\ a_{21} & a_{22} \end{pmatrix}$$

where a_{jk} $(j, k = 1, 2)$ are real numbers.

First, consider the case of *nonreal eigenvalues*: $\lambda_2(A) = \overline{\lambda}_1(A)$. It can be written

$$det(A) = \lambda_1(A)\overline{\lambda}_1(A) = |\lambda_1(A)|^2$$

and

$$|\lambda_1(A)|^2 + |\lambda_2(A)|^2 = 2|\lambda_1(A)|^2 = 2det(A) = 2[a_{11}a_{22} - a_{21}a_{12}].$$

Thus,

$$g^2(A) = N^2(A) - |\lambda_1(A)|^2 - |\lambda_2(A)|^2 =$$

$$a_{11}^2 + a_{12}^2 + a_{21}^2 + a_{22}^2 - 2[a_{11}a_{22} - a_{21}a_{12}].$$

Hence,

$$g(A) = \sqrt{(a_{11} - a_{22})^2 + (a_{21} + a_{12})^2}. \tag{1.5}$$

Let $n = 2$ and a matrix A have real entries again, but now the *eigenvalues of A are real.* Then

$$|\lambda_1(A)|^2 + |\lambda_2(A)|^2 = TraceA^2.$$

Obviously,

$$A^2 = \begin{pmatrix} a_{11}^2 + a_{12}a_{21} & a_{11}a_{12} + a_{12}a_{22} \\ a_{21}a_{11} + a_{21}a_{22} & a_{22}^2 + a_{21}a_{12} \end{pmatrix}.$$

We thus get the relation

$$|\lambda_1(A)|^2 + |\lambda_2(A)|^2 = a_{11}^2 + 2a_{12}a_{21} + a_{22}^2.$$

Consequently,

$$g^2(A) = N^2(A) - |\lambda_1(A)|^2 - |\lambda_2(A)|^2 =$$

$$a_{11}^2 + a_{12}^2 + a_{21}^2 + a_{22}^2 - (a_{11}^2 + 2a_{12}a_{21} + a_{22}^2).$$

Hence,

$$g(A) = |a_{12} - a_{21}|. \tag{1.6}$$

Example 2.1.3 *Let A be an upper-triangular matrix:*

$$A = \begin{pmatrix} a_{11} & a_{12} & \dots & a_{1n} \\ 0 & a_{21} & \dots & a_{2n} \\ . & \dots & . & . \\ 0 & \dots & 0 & a_{nn} \end{pmatrix}.$$

Then

$$g(A) = \sqrt{\sum_{k=1}^{n}\sum_{j=1}^{k-1} |a_{jk}|^2}, \tag{1.7}$$

since eigenvalues of a triangular matrix are its diagonal elements.

Example 2.1.4 *Consider the matrix*

$$A = \begin{pmatrix} -a_1 & \dots & -a_{n-1} & -a_n \\ 1 & \dots & 0 & 0 \\ . & \dots & . & . \\ 0 & \dots & 1 & 0 \end{pmatrix}.$$

with real positive numbers a_k. Such matrices play a key role in the theory of scalar ordinary differential equations. Take into account that

$$A^2 = \begin{pmatrix} a_1^2 - a_2 & \dots & a_1 a_{n-1} & a_1 a_n \\ -a_1 & \dots & -a_{n-1} & -a_n \\ 0 & \dots & 0 & 0 \\ . & \dots & . & . \\ 0 & \dots & 0 & 0 \end{pmatrix}.$$

We thus obtain,

$$Trace\ A^2 = a_1^2 - 2a_2.$$

Therefore

$$g^2(A) \leq N^2(A) - |Trace A^2| =$$

$$\sum_{k=1}^{n} a_k^2 + n - 1 - a_1^2 + 2a_2.$$

Hence

$$g(A) \leq \sqrt{\sum_{k=3}^{n} a_k^2 + n - 2 + (1 + a_2)^2}. \tag{1.8}$$

In particular, if $n = 2$, then

$$g(A) \leq |1 + a_2|. \tag{1.9}$$

2.1.2 Estimates for norms

For a natural $n > 1$ introduce the numbers

$$\gamma_{n,p} = \sqrt{\frac{C_{n-1}^p}{(n-1)^p}}\ (p = 1, ..., n-1) \text{ and } \gamma_{n,0} = 1.$$

Here

$$C_{n-1}^p = \frac{(n-1)!}{(n-p-1)!p!}$$

are binomial coefficients. Evidently, for $n > 2$

$$\gamma_{n,p}^2 = \frac{(n-2)(n-3)\dots(n-p)}{(n-1)^{p-1}p!} \leq \frac{1}{p!}\ (p = 1, 2, ..., n-1). \tag{1.10}$$

Theorem 2.1.5 *Let A be a linear operator in a complex Euclidean space $\mathbf{C}^n$, and let f be a function regular on a neighborhood of the closed convex hull $co(A)$ of the eigenvalues of A. Then the inequality*

$$||f(A)|| \leq \sum_{k=0}^{n-1} \sup_{\lambda \in co(A)} |f^{(k)}(\lambda)| g^k(A) \frac{\gamma_{n,k}}{k!}$$

is true.

For the proof of that theorem see Appendix 1, Section 17.1 below. More exact estimates for $||f(A)||$ can be found in the book (Gil', 1995, Section 1.2) .

Corollary 2.1.6 *Let A be an $n \times n$-matrix. Then*

$$||exp(At)|| \leq exp[\alpha(A)t] \sum_{k=0}^{n-1} g^k(A) t^k \frac{\gamma_{n,k}}{k!} \ (t \geq 0),$$

where $\alpha(A) = \max_{k=1,\dots,n} Re\lambda_k(A)$.

Corollary 2.1.6 and inequality (1.10) imply

Corollary 2.1.7 *Let A be an $n \times n$-matrix. Then*

$$||exp(At)|| \leq \exp[\alpha(A)t] \sum_{k=0}^{n-1} \frac{g^k(A)t^k}{(k!)^{3/2}} \ \textit{for all } t \geq 0.$$

If A is a normal matrix, then $g(A) = 0$, and thus $||exp(At)|| \leq \exp[\alpha(A)t]$, $t \geq 0$.

Let $\beta(A) = \min_{k=1,\dots,n} Re\lambda_k(A)$.

Corollary 2.1.8 *Let A be an $n \times n$-matrix. Then for any $h \in \mathbf{C}^n$, the inequality*

$$||exp(At)h|| \geq e^{[\beta(A)t]} \ [\sum_{k=0}^{n-1} g^k(A) t^k (k!)^{-1} \gamma_{n,k}]^{-1} \ ||h|| \ (t \geq 0)$$

is valid. Therefore, in the accordance with (1.10),

$$||exp(At)h|| \geq e^{[\beta(A)t]} [\sum_{k=0}^{n-1} g^k(A) (k!)^{-3/2} \ t^k]^{-1} \ ||h|| \ (t \geq 0).$$

Indeed, since

$$\max Re\sigma(-A) = -\min Re\sigma(A) = -\beta(A),$$

and $g(-A) = g(A)$, Corollary 2.1.6 yields

$$||exp(-At)v|| \leq ||v|| e^{-\beta(A)t} \sum_{k=0}^{n-1} g^k(A) t^k \frac{\gamma_{n,k}}{k!} \ (t \geq 0)$$

for any $v \in \mathbf{C}^n$. Taking into account that the operator $exp(-At)$ is the inverse one to $exp(At)$ and putting $exp(-At)v = h$, we arrive at the assertion.

In the sequel we will often use the following notation:

$$p(t, A) = e^{\alpha(A)t} \sum_{k=0}^{n-1} \frac{g^k(A)t^k}{(k!)^{3/2}}.$$

Due to Corollary 2.1.7 $\|exp(At)\| \leq p(t, A)$ $(t \geq 0)$.

Lemma 2.1.9 *Let A be a Hurwitz matrix. Then*

$$\sup_{t\geq 0} \|exp(At)\| \leq \chi(A), \quad \textit{where } \chi(A) = \max_{t\geq 0} p(t, A).$$

The proof is obvious.

Lemma 2.1.10 *Let A be a Hurwitz matrix. Then*

$$\int_0^\infty \|exp(At)\|dt \leq \Gamma(A), \quad \textit{where } \Gamma(A) = \sum_{k=0}^{n-1} \frac{g^k(A)}{|\alpha(A)|^{k+1}(k!)^{1/2}}.$$

To prove this lemma observe that

$$\sum_{k=0}^{n-1} \int_0^\infty e^{\alpha(A)t} \frac{g^k(A)t^k}{(k!)^{3/2}} dt = \Gamma(A),$$

and take into account Corollary 2.1.7.

2.2 Estimates for absolute values

2.2.1 Statement of the result

Let $A = (a_{lj})_{j,l=1}^n$ be a matrix, and let $m_{lj}(\lambda)$ *be the cofactor of the element* $\delta_{lj}\lambda - a_{lj}$ *of the matrix* $\lambda I - A$. Here δ_{lj} is the Kronecker symbol: $\delta_{lj} = 0$ if $j \neq l$, and $\delta_{jj} = 1$. Assume that after division by the common zeros, the equality

$$\frac{m_{lj}(\lambda)}{det(I\lambda - A)} = \frac{\mu_{lj}(\lambda)}{d_{lj}(\lambda)} \quad (j, l = 1, ..., n) \tag{2.1}$$

holds, where $d_{lj}(\lambda)$ and $\mu_{lj}(\lambda)$ are polynomials which have no common zeros. That is, $\mu_{lj}(\lambda)/d_{lj}(\lambda)$ are the entries of the matrix $(I\lambda - A)$. Furthermore, the degree of $d_{lj}(\lambda)$ is equal to $n_{lj} \leq n$. Obviously, if $m_{lj}(\lambda)$ and $det(I\lambda - A)$ have no common zeros, then $n_{lj} = n$ and $m_{lj}(\lambda) = \mu_{lj}(\lambda)$. Put for $k = 0, ..., n-1$,

$$b_{jl}(k) = \frac{1}{k!(n_{lj} - 1 - k)!} \sup_{\lambda \in co(A)} |\mu_{lj}^{(n_{lj}-k-1)}(\lambda)| .$$

Since $(n_{lj} - j)! = 0$ for $j > n_{lj}$, we have

$$b_{jl}(k) = 0 \text{ for } k = n_{lj}, ..., n-1 \text{ if } n_{lj} < n - 1. \tag{2.2}$$

Theorem 2.2.1 *Let A be an $n \times n$-matrix, and let f be a function regular on a neighborhood of the closed convex hull $co(A)$ of the eigenvalues of A. Then the entries $f_{lj}(A)$ of the matrix $f(A)$ satisfy the inequalities*

$$|f_{lj}(A)| \leq \sum_{k=0}^{n-1} \max_{\lambda \in co(A)} |f^{(k)}(\lambda)| b_{lj}(k) \ (l, j = 1, ..., n).$$

The proof is presented in the next subsection.

Let $|A|$ denote the matrix of the absolute values of the elements of A and let $|h|$ have the similar interpretation for the vector h. Inequalities between vectors and between matrices are interpreted as inequalities between corresponding components. Now Theorem 2.2.1 can be rewritten in the vector form.

Theorem 2.2.2 *Let A be an $n \times n$-matrix, and let f be a function regular on a neighborhood of the closed convex hull $co(A)$ of the eigenvalues of A. Then*

$$|f(A)| \leq \sum_{k=0}^{n-1} \max_{\lambda \in co(A)} |f^{(k)}(\lambda)| B_k,$$

where $B_k = (b_{lj}(k))_{l,j=1}^{n}$ $(k = 0, 1, ..., n-1)$.

Corollary 2.2.3 *For any $n \times n$-matrix A we have the inequality*

$$|e^{At}| \leq exp[\alpha(A)t] \sum_{k=0}^{n-1} t^k B_k \ (t \geq 0).$$

I.e. the entries $e_{lj}(At)$ of e^{At} satisfy the inequality

$$|e_{lj}(A)t| \leq exp[\alpha(At)] \sum_{k=0}^{n-1} t^k b_{lj}(k) \ (t \geq 0; \ l, j = 1, ..., n).$$

2.2.2 Proofs

Denote by $co_{lj}(A)$ the closed convex hull of the roots of the polynomial $d_{lj}(\lambda)$ defined by (2.1).

Lemma 2.2.4 *Let A be an $n \times n$-matrix, and let f be a function regular on a neighborhood of the closed convex hull $co(A)$ of the eigenvalues of A. Then the entries $f_{jl}(A)$ of $f(A)$ satisfy the inequalities*

$$|f_{jl}(A)| \leq \frac{1}{(n_{lj}-1)!} \sup_{\lambda \in co_{lj}(A)} \left| \frac{d^{n_{lj}-1}(f(\lambda)\mu_{lj}(\lambda))}{d\lambda^{n_{lj}-1}} \right| \ (l, j = 1, ..., n).$$

Proof: We begin with the representation

$$f(A) = \frac{1}{2\pi i}\int_\Gamma f(\lambda)(I\lambda - A)^{-1}d\lambda,$$

where Γ is a smooth contour containing all eigenvalues of A. It follows from the rule for inversion of matrices that the entries $r_{jl}(\lambda)$ of the matrix $(I\lambda - A)^{-1}$ are

$$r_{jl}(\lambda) = \frac{m_{lj}(\lambda)}{det(I\lambda - A)} = \frac{\mu_{lj}(\lambda)}{d_{lj}(\lambda)} \ (l, j = 1, ..., n).$$

Hence,

$$f_{jl}(A) = \frac{1}{2\pi i}\int_\Gamma f(\lambda)\frac{\mu_{lj}(\lambda)}{d_{lj}(\lambda)}d\lambda = \frac{1}{2\pi i}\int_\Gamma f(\lambda)\frac{\mu_{lj}(\lambda)}{(\lambda-\omega_1)...(\lambda-\omega_{n_{lj}})}d\lambda,$$

where $\omega_1, ..., \omega_{n_{lj}}$ are the roots of $d_{lj}(\lambda)$ which are simultaneously the eigenvalues of A taken with their multiplicities. Now by virtue of Lemma 1.10.1 we conclude that the required assertion is valid. □

Proof of Theorem 2.2.1: It can be written

$$\left|\frac{d^{n_{lj}-1}(f(\lambda)\mu_{lj}(\lambda))}{d\lambda^{n_{lj}-1}}\right| \leq \sum_{k=0}^{n_{lj}-1} C^k_{n_{lj}-1}|\mu_{lj}^{(n_{lj}-k-1)}(\lambda)||f^{(k)}(\lambda)|.$$

Recall that C^k_m are the binomial coefficients. Hence,

$$\sup_{\lambda\in co(A)}\left|\frac{d^{n_{lj}-1}(f(\lambda)\mu_{lj}(\lambda))}{d\lambda^{n_{lj}-1}}\right| \leq (n_{lj}-1)!\sum_{k=0}^{n_{lj}-1} b_{ij}(k)\sup_{\lambda\in co(A)}|f^{(k)}(\lambda)|.$$

Now Lemma 2.2.4 and (2.2) yield the required result, since

$$co_{lj}(A) \subseteq co(A) \ (l, j = 1, ..., n). \ \square$$

2.3 Impulse functions

2.3.1 Estimates for the impulse function

Let

$$P(\lambda) = \lambda^n + a_1\lambda^{n-1} + ... + a_n \tag{3.1}$$

be polynomial of the degree n, and $L(\lambda)$ be a polynomial of the degree $m < n$. Assume that these polynomials are real and have no common zeros. Consider the equation

$$P(D)y = L(D)f(t) \ \ (t \geq 0, \ D = d/dt), \tag{3.2}$$

where $f : [0, \infty) \to \mathbf{C}$ is a given sufficiently smooth function. Denote by $W(\lambda)$ *the transfer function* of the equation. I.e.,

$$W(\lambda) = P^{-1}(\lambda)L(\lambda) \ (\lambda \in \mathbf{C}). \tag{3.3}$$

Let $\lambda_1, ..., \lambda_n$ be roots of the polynomial $P(\lambda)$, $co(P)$ be the closed convex hull of the numbers $\lambda_1, ..., \lambda_n$, and

$$\alpha = \max_{1 \le k \le n} Re\lambda_k.$$

The impulse (Green) function of equation (3.2) is defined by the relation

$$K(t) = \frac{1}{2\pi i}\int_{d-i\infty}^{d+i\infty} exp[\lambda t]W(\lambda)d\lambda =$$

$$\frac{1}{2\pi i}\int_{d-i\infty}^{d+i\infty} exp[\lambda t]P^{-1}(\lambda)L(\lambda)d\lambda. \tag{3.4}$$

Here $d = constant > \alpha$.

Lemma 2.3.1 *The inequality*

$$|K(t)| \le e^{\alpha t}\sum_{k=0}^{m}\frac{t^{n-k-1}}{(n-1-k)!k!}\theta_k(L) \ (t \ge 0)$$

is true with the notation

$$\theta_k(L) = \max_{\lambda \in co(P)} |L^{(k)}(\lambda)| \ (k = 0, ..., m).$$

Proof: Deforming the contour in (3.4), we can write down

$$K(t) = \frac{1}{2\pi i}\int_\Gamma \frac{L(\lambda)e^{t\lambda}}{\prod_{k=1}^{n}(\lambda - \lambda_k)}d\lambda, \tag{3.5}$$

where Γ is a smooth contour encompassing all the roots of $P(\lambda)$. Due to Lemma 1.10.1,

$$|K(t)| \le \frac{1}{(n-1)!}\sup_{\lambda \in co(P)} |\frac{d^{n-1}}{d\lambda^{n-1}}e^{\lambda t}L(\lambda)|.$$

This clearly forces

$$|K(t)| \le \frac{1}{(n-1)!}\sum_{k=0}^{n-1} C_{n-1}^k t^{n-k-1} \sup_{\lambda \in co(P)} |e^{\lambda t}L^{(k)}(\lambda)| \le$$

$$\frac{e^{\alpha t}}{(n-1)!}\sum_{k=0}^{n-1} C_{n-1}^k t^{n-k-1} \sup_{\lambda \in co(P)} |L^{(k)}(\lambda)|.$$

Taking into account that $L^{(k)}(\lambda) = 0$ for $k > m$, we arrive at the result. □

2.3.2 Estimates for derivatives of impulse functions

In this subsection it is assumed that $P(\lambda)$ *is a Hurwitz polynomial* defined by (3.1), and $L(\lambda) \equiv 1$. That is, we consider the equation

$$P(D)y = f(t) \quad (t \geq 0). \tag{3.6}$$

The corresponding impulse function has the form

$$K(t) = \frac{1}{2\pi}\int_{-\infty}^{\infty} e^{ti\omega}\frac{d\omega}{P(i\omega)}. \tag{3.7}$$

Lemma 2.3.2 *The impulse function of equation (3.6) satisfies the inequalities*

$$|K^{(j)}(t)| \leq e^{\alpha t}\eta_j(t) \quad (t \geq 0) \ (j = 0, 1, ..., n-1),$$

where

$$\eta_j(t) = \sum_{k=0}^{j} \frac{j!\Lambda^{j-k}t^{n-k-1}}{(j-k)!(n-k-1)!k!},$$

and

$$\Lambda = \max_{k=1,\dots,n} |\lambda_k|.$$

Proof: Deforming the contour in (3.7) we can write down

$$K^{(j)}(t) = \frac{1}{2\pi i}\int_{\Gamma} \frac{\lambda^j e^{t\lambda}}{\prod_{k=1}^{n}(\lambda - \lambda_k)}d\lambda,$$

where the contour Γ is the same as in the previous section. Due to Lemma 1.10.1

$$|K^{(j)}(t)| \leq \frac{1}{(n-1)!}\sup_{\lambda \in co(P)} |\frac{d^{n-1}}{d\lambda^{n-1}}e^{\lambda t}\lambda^j|.$$

Recall that $co(P)$ is the closed convex hull of the roots $\lambda_1, ..., \lambda_n$ of polynomial $P(\lambda)$. This clearly forces

$$|K^{(j)}(t)| \leq \sum_{k=0}^{j} \frac{j!t^{n-k-1}}{k!(n-k-1)!(j-k)!}\sup_{\lambda \in co(P)} |e^{\lambda t}\lambda^{j-k}| \leq$$

$$e^{\alpha t}\sum_{k=0}^{j} \frac{j!t^{n-k-1}}{k!(n-k-1)!(j-k)!}\sup_{\lambda \in co(P)} |\lambda^{j-k}|.$$

Since $\sup_{\lambda \in co(P)} |\lambda| = \Lambda$, this is precisely the assertion of the lemma. □

Lemma 2.3.3 *For a solution $z(t)$ of the equation*

$$P(D)z = 0,$$

the following estimate is true:

$$|z^{(m)}(t)| \leq e^{\alpha t} \sum_{i=0}^{n-1} |z^{(n-i-1)}(0)| \sum_{s=0}^{i} a_s \eta_{i-s+m}(t)$$

$$(a_0 = 1,\ t \geq 0;\ m = 1, ..., n-1).$$

Proof: By the Laplace transformation (Döetsch, 1961, formula (14.6)), we have

$$z(t) = \sum_{j=0}^{n-1} z^{(j)}(0) \sum_{k=1}^{n-j} a_{k-1} K^{(n-k-j)}(t).$$

This implies

$$z^{(m)}(t) = \sum_{j=0}^{n-1} z^{(j)}(0) \sum_{k=1}^{n-j} a_{k-1} K^{(n-k-j+m)}(t) =$$

$$\sum_{i=0}^{n-1} z^{(n-i-1)}(0) \sum_{k=1}^{i+1} a_{k-1} K^{(i+1-k+m)}(t) =$$

$$\sum_{i=0}^{n-1} z^{(n-i-1)}(0) \sum_{s=0}^{i} a_s K^{(i-s+m)}(t).$$

Now the required result follows from Lemma 2.3.2. □

2.4 Positivity conditions for impulse functions

2.4.1 Statements of the results

Again, let $L(\lambda)$ and $P(\lambda)$ be polynomials with real coefficients, which have no common zeros; the degrees of $L(\lambda)$ and $P(\lambda)$ are equal to m and $n > m$, respectively. In addition, $P(\lambda)$ is a Hurwitz polynomial defined by (3.1). In this section we consider conditions guaranteeing the positivity of scalar-valued impulse functions represented by formula (3.4).

Furthermore we denote by $\lambda_1, ..., \lambda_\nu$ the real zeros of $P(\lambda)$ arranged in the non-decreasing order and counting with their multiplicities: $\lambda_1 \leq ... \leq \lambda_\nu$. The nonreal complex conjugate zeros of $P(\lambda)$ are denoted by

$$w_1, \overline{w}_1, ..., w_\mu, \overline{w}_\mu.$$

Therefore $2\mu + \nu = n$. It is assumed that

$$Rew_j < \lambda_1 \ (j = 1, ..., \mu).$$

Set

$$N(t) = \frac{1}{(\nu-1)!} \min_{\lambda_1 \le \lambda \le \lambda_\nu} \frac{d^{\nu-1}}{d\lambda^{\nu-1}} \exp(\lambda t) \prod_{j=1}^{\mu} |\lambda - w_j|^{-2} L(\lambda),$$

and

$$M(t) = \frac{1}{(2\mu-1)!} \max_{\lambda \in \gamma} \left| \frac{d^{2\mu-1}}{d\lambda^{2\mu-1}} \exp(\lambda t) \prod_{j=1}^{\nu} (\lambda - \lambda_j)^{-1} L(\lambda) \right|,$$

where γ is the convex hull of the nonreal zeros of $P(\lambda)$.

Theorem 2.4.1 *Let the inequality*

$$N(t) \ge M(t) \ for \ all \ t \ge 0$$

be fulfilled. Then

$$K(t) \equiv \frac{1}{2\pi} \int_{-\infty}^{\infty} exp(t\ i\omega) \frac{L(i\omega)}{P(i\omega)} d\omega \ge 0 \ (t \ge 0). \tag{4.1}$$

The proofs are presented in the next subsection

Let all the roots of polynomial $P(\lambda)$ be real. Then $M(t) = 0$, and

$$N(t) = \frac{1}{(n-1)!} \min_{\lambda_1 \le \lambda \le \lambda_n} \frac{d^{n-1}}{d\lambda^{n-1}} exp[\lambda t] L(\lambda).$$

Since

$$\frac{d^{n-1}}{d\lambda^{n-1}} exp[\lambda t] L(\lambda) = exp[\lambda t] \sum_{k=0}^{n-1} C_{n-1}^{k} t^{n-k-1} L^{(k)}(t),$$

and

$$exp[\lambda t] \ge 0 \text{ with every real } \lambda,$$

Theorem 2.4.1 yields

Corollary 2.4.2 *Let all the roots of polynomial $P(\lambda)$ be real and let the inequalities*

$$L^{(k)}(t) \ge 0 \ (k = 0, 1, ..., m;\ t \ge 0) \tag{4.2}$$

be fulfilled. Then $K(t) \ge 0$ $(t \ge 0)$.

Corollary 2.4.3 *Let all the roots of polynomial $P(\lambda)$ be real. Then the relation*

$$\frac{1}{2\pi} \int_{-\infty}^{\infty} exp(t\ i\omega) P^{-1}(i\omega) d\omega \ge 0 \ (t \ge 0)$$

is true.

In fact, in the considered case $L(\lambda) \equiv 1$, and therefore condition (4.2) is obviously fulfilled.

We will also prove the following

Lemma 2.4.4 *Let a real polynomial $P_3(\lambda) = \lambda^3 + b_1\lambda^2 + b_2\lambda + b_3$ have a pair of complex conjugate roots: $-\gamma \pm i\omega$, and a real root $-z$ $(z, \gamma, \omega > 0)$. Then the function*

$$K_3(t) \equiv \frac{1}{2\pi i}\int_{-i\infty}^{i\infty} exp(t\lambda)P_3^{-1}(\lambda)d\lambda \tag{4.3}$$

is positive provided the inequality $\gamma \geq z$ holds.

Since the Laplace transform of a convolution is a product of the Laplace transforms, this lemma yields

Corollary 2.4.5 *Let a polynomial $P(\lambda)$ of the degree $n \geq 3$, for each of its pairs of nonreal conjugate zeros (if they exist), have a real root lying to the right of this pair. (For each nonreal conjugate pair there is a corresponding real root lying to the right). Then the inequality (4.1) is valid.*

2.4.2 Proofs

Proof of Lemma 2.4.1: Let Γ_0 be a smooth contour containing all the real zeros of $P(\lambda)$, and not containing the nonreal zeros, and let Γ_1 be a smooth contour containing all the nonreal zeros of $P(\lambda)$, and not containing the real zeros. We have $J(t) = J_0(t) + J_1(t)$, where

$$J_k(t) \equiv \frac{1}{2\pi i}\int_{\Gamma_k} exp(t\lambda)P^{-1}(\lambda)L(\lambda)d\lambda \ (k = 0, 1).$$

Clearly,

$$P(\lambda) \equiv \prod_{j=1}^{\mu}(\lambda - w_j)(\lambda - \overline{w}_j)\prod_{m=1}^{\nu}(\lambda - \lambda_m).$$

By virtue of Lemma 1.10.2, $J_0(t) \geq N(t)$. By Lemma 1.10.1, $|J_1(t)| \leq M(t)$. Therefore

$$K(t) = J_0(t) + J_1(t) \geq J_0(t) - |J_1(t)| \geq N(t) - M(t) \geq 0$$

as claimed. □

Proof of Lemma 2.4.4: First, assume that $\gamma > z$. Taking into account that $K_3(t)$ is a solution of the equation

$$P_3(D)x(t) = 0,$$

with the initial condition $K(0) = K'(0) = 0$, $K"(0) = 1$ (Döetsch, 1961, Section 12), we get

$$K(t) = c[e^{-zt} - e^{-\gamma t}(cos(\omega t) - bsin(\omega t))] \ (c = const > 0)$$

with

$$b = (\gamma - z)\omega^{-1} > 0.$$

Consider the equation $K_3(t) = 0$. Multiplying it by $e^{\gamma t}$ and substituting $s = \omega t$, we have the equation

$$e^{bs} = \cos s - b \sin s.$$

By virtue of the Taylor series that equation can be rewritten as

$$e^{bs} = 1 + bs + g(s) = \cos s - b \sin s \quad (g(s) > 0).$$

Since

$$|\cos s| \leq 1, |\sin s| \leq s \ (s \geq 0),$$

one can assert that this equation has no zeros for $s > 0$. Therefore, $K_3(t)$ has no zeros. This finishes the proof in the case $\gamma > z$, since $K_3(t) > 0$ for all sufficiently large t. In the case $\gamma = z$ we arrive at the result by a small perturbation. □

2.5 The Lyapunov equation

Recall the following famous Lyapunov theorem (Daleckii and Krein, 1974, p. 33) (see also (Vidyasagar, 1993, p. 198)).

Theorem 2.5.1 *In order for the eigenvalues of a matrix A to lie in the interior of the left half-plane, it is necessary and sufficient that there exists a positive definite Hermitian matrix W, such that the matrix $WA + A^*W$ is a negative Hermitian one. Moreover, if the eigenvalues of A lie in the interior of the left half-plane, then for any positive definite Hermitian matrix H there exists a positive definite Hermitian matrix W_H such that*

$$W_H A + A^* W_H = -2H. \tag{5.1}$$

In addition,

$$W_H = 2\int_0^\infty e^{A^* t} H e^{At} dt. \tag{5.2}$$

Equation (5.1) is called *the Lyapunov equation.* In many applications, it is important to know the norm of the solution W_H of the Lyapunov equation. We are beginning with

Lemma 2.5.2 *Let A be an $n \times n$-Hurwitz matrix. Then the solution W_H of the Lyapunov equation (5.1) subordinates the inequality*

$$\|W_H\| \leq \|H\| \sum_{j,k=0}^{n-1} \frac{g^{j+k}(A)(k+j)!}{2^{j+k}|\alpha(A)|^{j+k+1}(j!\, k!)^{3/2}}.$$

Proof: Take into account that $g(A) = g(A^*)$, and $\alpha(A) = \alpha(A^*)$. Then by virtue of Corollary 2.1.7 and equality (5.2),

$$||W_H|| \le 2\int_0^\infty ||e^{A^*t}|| ||H|| ||e^{At}||dt \le$$

$$2||H|| \int_0^\infty \exp[2\alpha(A)t]\Big(\sum_{k=0}^{n-1} \frac{g^k(A)t^k}{(k!)^{3/2}}\Big)^2 dt.$$

The integration gives

$$||W_H|| \le 2||H|| \int_0^\infty \exp[2\alpha(A)t] \sum_{j,k=0}^{n-1} \frac{(g(A)t)^{k+j}}{(j!\ k!)^{3/2}} dt =$$

$$||H|| \sum_{j,k=0}^{n-1} 2\frac{(k+j)!g^{j+k}(A)}{(2|\alpha(A)|)^{j+k+1}(j!\ k!)^{3/2}}$$

as claimed. □

Lemma 2.5.3 *Let A be an $n \times n$-Hurwitz matrix. Then the solution W_H of the Lyapunov equation (5.1) satisfies the inequality*

$$(W_H h, h) \ge 2\beta(H) \int_0^\infty e^{2\beta(A)t} \Big[\sum_{k=0}^{n-1} g^k(A)t^k (k!)^{-3/2}\Big]^{-2} dt\ ||h||^2 \quad (h \in \mathbf{C}^n),$$

where

$$\beta(A) = \min_k Re\lambda_k(A), \beta(H) = \min_k Re\lambda_k(H).$$

Proof: Take into account that

$$g(A) = g(A^*) \text{ and } \beta(A) = \beta(A^*).$$

Then by Corollary 2.1.8,

$$(e^{A*t}He^{At}h, h) = (He^{At}h, e^{At}h) \ge$$

$$\beta(H)(e^{At}h, e^{At}h) \ge$$

$$\beta(H)e^{2\beta(A)t}\Big[\sum_{k=0}^{n-1} g^k(A)t^k(k!)^{-3/2}\Big]^{-2} ||h||^2 \ (t \ge 0).$$

Now equality (5.2) yields

$$(W_H h, h) = 2\int_0^\infty (e^{A*t}He^{At}h, h)dt \ge$$

$$2\beta(H)\int_0^\infty e^{2\beta(A)t}\,[\sum_{k=0}^{n-1} g^k(A)t^k(k!)^{-3/2}]^{-2}dt\|h\|^2$$

as claimed. □

Let us consider the Lyapunov equation with $H = I$:

$$WA + A^*W = -2I. \tag{5.3}$$

Furthermore, let $u(t)$ be a solution of the equation

$$\dot{u}(t) = Au(t), \tag{5.4}$$

and let W be a solution of equation (5.3). Multiplying equation (5.4) by W and doing the scalar product, we get

$$(W\dot{u}(t), u(t)) = (WAu(t), u(t)).$$

Since

$$\frac{d}{dt}(Wu(t), u(t)) = (W\dot{u}(t), u(t)) + (u(t), W\dot{u}(t)),$$

it can be written

$$\frac{d}{dt}(Wu(t), u(t)) = (WAu(t), u(t)) + (u(t), WAu(t)) = ((WA + A^*W)u(t), u(t)).$$

Now (5.3) yields

$$\frac{d}{dt}(Wu(t), u(t)) = -2(u(t), u(t)) = -2(W^{-1}Wu(t), u(t)) \leq$$

$$-2b(W)(Wu(t), u(t)),$$

where

$$b(W) = \inf_{h \in \mathbf{C}^n} \frac{(W^{-1}h, h)}{(h, h)} = \inf_{w \in \mathbf{C}^n} \frac{(w, w)}{(Ww, w)} = \|W\|^{-1}.$$

Solving this inequality, we get

$$(Wu(t), u(t)) = \|W^{1/2}u(t)\|^2 \leq e^{-t2\|W\|^{-1}}\|W^{1/2}u(0)\|^2.$$

Since $u(t) = e^{At}u(0)$, this inequality means that

$$\|W^{1/2}e^{At}h\| \leq \|W^{1/2}h\|e^{-t\|W\|^{-1}} \quad (t \geq 0) \tag{5.5}$$

for every $h \in \mathbf{C}^n$. We have obtained

Lemma 2.5.4 *Let A be a Hurwitz matrix, and W be a solution of equation (5.3). Then inequality (5.5) is true for any $h \in \mathbf{C}^n$.*

This lemma and Lemma 2.5.2 yield

Corollary 2.5.5 *Let A be a Hurwitz matrix, and W be a solution of equation (5.3). Then with the notation*

$$\psi(A) = \sum_{j,k=0}^{n-1} \frac{g^{j+k}(A)(k+j)!}{2^{j+k}|\alpha(A)|^{j+k+1}(j!\,k!)^{3/2}},$$

the inequality

$$\|W^{1/2}e^{At}\| \leq \sqrt{\psi(A)}e^{-t\psi^{-1}(A)} \ (t \geq 0)$$

is true.

2.6 Notes

The quantity $g(A)$ was introduced by Henrici (1962) and independently by Gil' (1983). I.M. Gel'fand and G.E. Shilov (1958) established the estimate

$$\|f(A)\| \leq \sum_{k=0}^{n-1} \sup_{\lambda \in co(A)} |f^{(k)}(\lambda)|(2\|A\|)^k.$$

About other estimations for the matrix exponent see (Coppel, 1978).

Theorem 2.1.5 was derived in the paper (Gil', 1983) (see also (Gil', 1995)).

The material of Section 2.2 is taken from the paper (Gil', 1987).

Section 2.3 is based on the paper (Gil' and Shargorodsky, 1988).

Theorem 2.4.1 was proved in (Gil', 1994). Corollary 2.4.3 was established in the book (Krasnosel'skii et al., 1970) (see also (Krasnosel'skii et al., 1989)).

Lemma 2.5.2 is new.

References for Chapter 2

Coppel, W.A. (1978). *Dichotomies in Stability Theory.* Lectures Notes in Mathematics 629, Springer-Verlag, New York.

Daleckii, Yu L. and Krein, M. G. (1974). *Stability of Solutions of Differential Equations in Banach Space.* Amer. Math. Soc., Providence, R. I.

Döetsch, G. (1961). *Anleitung zum Praktishen Gebrauch der Laplace-transformation.* Oldenburg, Munchen.

Gel'fand, I.M. and Shilov, G.E. (1958). *Some Questions of Theory of Differential Equations.* Nauka, Moscow. In Russian.

Gil', M. I. (1983). On an estimate for resolvents of nonselfadjoint operators which are "near" to selfadjoint and to unitary ones, *Mathematical Notes, 33,*

81-84.

Gil', M.I. (1987). Two-sides bounds for solutions of linear ordinary differential equations, *Differential Equations, 23*, 2031-2036.

Gil', M. I. (1994). Class of absolutely stable multivariable systems, *International Journal of Systems Sciences 25(3)*, 613-617.

Gil', M. I. (1995). *Norm Estimations for Operator-valued Functions and Applications.* Marcel Dekker, Inc, New York.

Gil', M. I. and Shargorodsky, L. L. (1988). Boundedness conditions for solutions and existence of limit cycles, *Izvestiya VUZ, Matematika, 32, No. 5*, 6-10 (Russian). English translation in *Soviet Math., 32, No. 5* , 1-6.

Henrici, P. (1962). Bounds for iterates, inverses, spectral variation and field values of nonnormal matrices. *Numerishe Mathematik, 4*, 24-39.

Krasnosel'skii, M. A., Burd, Sh., and Yu. Kolesov. (1970). *Nonlinear Almost Periodic Oscillations*, Nauka, Moscow. In Russian.

Krasnosel'skii, M. A., Lifshits, J., and A. Sobolev. (1989). *Positive Linear Systems. The Method of Positive Operators.* Heldermann Verlag, Berlin.

Vidyasagar, M. (1993). *Nonlinear System Analysis*, second edition. Prentice Hall, Upper Saddle River, New Jersey.

Chapter 3

LINEAR FINITE DIMENSIONAL SYSTEMS

In the present chapter stability conditions for linear finite dimensional systems are proposed. In many cases these conditions are formulated in terms of the eigenvalues of auxiliary matrices. This means that one can apply the well known stability criteria for linear time-invariant systems (for example, Hurwitz's criterion) to the investigation of nonautonomous systems.

Section 3.1 deals with general linear systems.

In Section 3.2 systems with the Lipschitz property are analyzed by the freezing method.

Section 3.3 is devoted to systems with piecewise constant matrices.

In Sections 3.4 and 3.5, triangular systems and systems which are "close" to triangular ones are investigated.

In Section 3.6 systems satisfying the matrix Lipschitz condition are discussed.

Throughout the present chapter $||.||$ *means the Euclidean norm* (see Section 1.1).

3.1 General systems

Recall that the quantity $g(A)$ is defined in Section 2.1. In addition, Corollary 2.1.7 gives for any $n \times n$-matrix A the inequality:

$$||exp[At]|| \leq p(t, A) \; (t \geq 0), \tag{1.1}$$

where

$$p(t,A) \equiv exp[\alpha(A)t] \sum_{k=0}^{n-1} \frac{g^k(A)t^k}{(k!)^{3/2}} \tag{1.2}$$

with $\alpha(A) = \max_k Re\lambda_k(A)$. Consider in $\mathbf{C}^n$ the equation

$$\dot{x}(t) = A(t)x(t) \ (t \geq 0), \tag{1.3}$$

where $A(t)$ is a variable piecewise-continuous $n \times n$ - matrix. Denote

$$q(t,s) \equiv ||A(t) - A(s)|| \ (t, s \geq 0),$$

and assume that

$$\chi \equiv \sup_{t\geq 0} p(t, A(t)) = \sup_{t \geq 0} exp[\alpha(A(t))t] \sum_{k=0}^{n-1} \frac{g^k(A(t))t^k}{(k!)^{3/2}} < \infty.$$

Theorem 3.1.1 *Let the inequality*

$$\zeta \equiv \sup_{t\geq 0} \int_0^t p(t-t_1, A(t))q(t,t_1)dt_1 < 1$$

hold. Then equation (1.3) is uniformly stable. Moreover, the estimate

$$||x(t)|| \leq \frac{\chi ||x(s)||}{1-\zeta} \ (t > s \geq 0)$$

is true for any solution $x(t)$ of (1.3).

Proof: Exploiting inequality (1.1), we get

$$||exp[A(\tau)t]|| \leq p(t, A(\tau)) \ (t, \ \tau \geq 0). \tag{1.4}$$

Equation (1.3) can be rewritten in the form

$$dx/dt - A(\tau)x = [A(t) - A(\tau)]x$$

with an arbitrary fixed $\tau \geq 0$. This equation is equivalent to the following one:

$$x(t) = exp[A(\tau)t]x(0) + \int_0^t exp[A(\tau)(t-t_1)][A(t_1) - A(\tau)]x(t_1)dt_1. \tag{1.5}$$

According to (1.4), it can be written as

$$||x(t)|| \leq p(t, A(\tau))||x(0)|| + \int_0^t p(t-t_1, A(\tau))q(\tau, t_1)||x(t_1)||dt_1.$$

With $\tau = t$ this relation gives

$$\sup_{0 \leq t \leq T} ||x(t)|| \leq \chi ||x(0)|| + \sup_{0\leq t \leq T} ||x(t)|| \zeta$$

for any positive finite T. By the condition $\zeta < 1$ we arrive at the inequality

$$\sup_{0 \le t \le T} \|x(t)\| \le \chi \|x(0)\| (1-\zeta)^{-1}.$$

Since the right-hand part of the latter inequality does not depend on T, we get

$$\sup_{t \ge 0} \|x(t)\| \le \chi \|x(0)\| (1-\zeta)^{-1}.$$

This is the required inequality for $s = 0$. It can be similarly proved for any $s \le t$. □

3.2 The freezing method for linear systems

3.2.1 Statement of results

Again consider the system

$$\dot{x}(t) = A(t)x(t) \ (t \ge 0), \tag{2.1}$$

assuming now that matrix $A(t)$ satisfies the following conditions:

$$v \equiv \sup_{t \ge 0} g(A(t)) < \infty, \text{ and } \rho \equiv -\inf_{t \ge 0} \alpha(A(t)) > 0, \tag{2.2}$$

and

$$\|A(t) - A(s)\| \le q_0 |t-s| \text{ for all } t, s \ge 0, \tag{2.3}$$

where q_0 is a positive constant. Denote

$$\zeta_1 = \sum_{k=0}^{n-1} \frac{(k+1)v^k}{\sqrt{k!}\rho^{2+k}}.$$

Recall that χ is defined in Section 3.1. Obviously, under conditions (2.2) $\chi < \infty$.

Theorem 3.2.1 *Let the conditions (2.2), (2.3) and $q_0 \zeta_1 < 1$ hold. Then the evolution operator $U(t,s)$ of equation (2.1) subordinates the inequality*

$$\|U(t,s)\| \le \chi (1 - q_0 \zeta_1)^{-1}$$

for all $t, s \ge 0$. Moreover, equation (2.1) is exponentially stable.

The proofs are presented in the next subsection.

Furthermore, by $z(q_0, v)$ denote the extreme right-hand root of the algebraic equation

$$z^{n+1} = q_0 P(z), \tag{2.4}$$

where

$$P(z) = \sum_{k=0}^{n-1} \frac{(k+1)v^k}{\sqrt{k!}} z^{n-k-1}.$$

Recall that I is the unit matrix.

Theorem 3.2.2 *Let conditions (2.2) and (2.3) hold. In addition, let the matrix $A(t) + z(q_0, v)I$ be a Hurwitz one for all $t \geq 0$. Then equation (2.1) is exponentially stable.*

It is simple to show that $z(q_0, v)$ is a unique positive root. Setting $z = vy$ in (2.4) and applying Lemma 1.11.1, we can assert that with the notation

$$w_n = \sum_{k=0}^{n-1} \frac{k+1}{\sqrt{k!}},$$

under the condition

$$q_0 w_n \geq v^2, \tag{2.5}$$

the following estimate is true:

$$z(q_0, v) \leq v^{1-\frac{1}{2(n+1)}} [q_0 w_n]^{1/(n+1)}. \tag{2.6}$$

Theorem 3.2.2 and bound (2.6) entail the following results.

Corollary 3.2.3 *Let the conditions (2.2), (2.3) and (2.5) hold. If, in addition, the matrix*

$$A(t) + q_0 v^{1-\frac{1}{2(n+1)}} [q_0 w_n]^{1/(n+1)} I$$

is a Hurwitz one for all $t \geq 0$, then (2.1) is exponentially stable.

We will also prove

Theorem 3.2.4 *Under condition (2.2), (2.3), any solution $x(t)$ of equation (2.1) satisfies the inequalities*

$$a^{-1} exp[(\beta_0 - z(q_0, v))t] \leq \frac{\|x(t)\|}{\|x(0)\|} \leq a\, exp[(-\rho + z(q_0, v))t] \ (t \geq 0),$$

where a is a positive constant, and

$$\beta_0 = \inf_{t \geq 0} \min_k Re\lambda_k(A(t)).$$

In other words, the Lyapunov exponents κ_j of equation (2.1) subordinate the estimates

$$\beta_0 - z(q_0, v) \leq \kappa_j \leq -\rho + z(q_0, v) \ (j = 1, ..., n).$$

Lemma 3.2.5 *Let conditions (2.2) and (2.3) be fulfilled. In addition, let the inequality*

$$q_0 < \frac{\rho^{n+1}}{P(\rho)} \tag{2.7}$$

hold. Then $q_0\zeta_1 < 1$, and the matrix $A(t) + z(q_0, v)I$ is a Hurwitz one.

Now Theorem 3.2.2 implies

Corollary 3.2.6 *Let conditions (2.2), (2.3) and (2.7) hold. Then equation (2.1) is exponentially stable.*

3.2.2 Proofs

Set

$$p_0(t) \equiv \exp[-\rho t] \sum_{k=0}^{n-1} \frac{t^k v^k}{(k!)^{3/2}}, \tag{2.8}$$

and

$$\zeta_0 \equiv \int_0^\infty p_0(t)dt.$$

Simple calculations show that

$$\zeta_0 = \sum_{k=0}^{n-1} \frac{v^k}{\sqrt{k!}\rho^{1+k}}, \tag{2.9}$$

and

$$\zeta_1 = \int_0^\infty t p_0(t)dt.$$

Lemma 3.2.7 *Let the conditions (2.2), (2.3) and $q_0\zeta_1 < 1$ be fulfilled. Then the evolution operator $U(t,s)$ of (2.1) satisfies the estimate*

$$\int_0^t ||U(t,w)||dw \leq \zeta_0(1 - q_0\zeta_1)^{-1} \text{ for all } t \geq 0.$$

Moreover, equation (2.1) is exponentially stable.

Proof: Again rewrite equation (2.1) in the form

$$dx/dt - A(\tau)x = [A(t) - A(\tau)]x$$

with a fixed $\tau \geq 0$. Hence

$$x(t) = exp[A(\tau)t]x(w) + \int_w^t exp[A(\tau)(t-t_1)][A(t_1) - A(\tau)]x(t_1)dt_1$$

$$(0 \leq w \leq t < \infty).$$

Using relation (2.3), we get

$$\|x(t)\| \leq \|e^{A(\tau)(t-w)}x(w)\| + \int_w^t \|e^{A(\tau)(t-t_1)}\| \, q_0|t_1 - \tau|\|x(t_1)\|dt_1. \quad (2.10)$$

Consequently, due to (1.1) and (2.8),

$$\|x(t)\| \leq p_0(t-w)\|x(w)\| + \int_w^t p_0(t-t_1)q_0|t_1 - \tau|\|x(t_1)\|dt_1.$$

When $\tau = t$ this relation implies the inequality

$$y(t,w) \leq p_0(t-w) + q_0 \int_0^t p_0(t-t_1)(t-t_1)y(t_1,w)dt_1$$

with

$$y(t,w) = \|x(t)\|/\|x(w)\|. \quad (2.11)$$

The integration with respect to w yields

$$\int_0^t y(t,w)dw \leq \zeta_0 + q_0 \int_0^t \int_w^t p_0(t-t_1)(t-t_1)y(t_1,w)dt_1 dw, \quad (2.12)$$

because

$$\int_0^t p_0(t-w)dw \leq \int_0^\infty p_0(s)ds = \zeta_0.$$

Obviously,

$$\int_0^t \int_w^t p_0(t-u)(t-u)y(u,w)du\,dw =$$

$$\int_0^t \int_0^u p_0(t-u)(t-u)y(u,w)dudw.$$

Hence,

$$\int_0^t \int_w^t p_0(t-u)(t-u)y(u,w)dudw \leq \int_0^t p_0(t-u)(t-u)du \max_{u \leq t} \int_0^u y(u,w)dw.$$

Furthermore, it is easy to see that

$$\int_0^t p_0(t-u)(t-u)du \leq \int_0^\infty p_0(t)tdt = \zeta_1. \quad (2.13)$$

Now, due to (2.12) and (2.13) we can write down

$$\int_0^t y(t,w)dw \leq \zeta_0 + q_0 \max_{u \leq t} \int_0^u y(u,w)dw \int_0^t p_0(t-u)(t-u)du \leq$$

$$\zeta_0 + q_0\zeta_1 \max_{u \leq t} \int_0^u y(u,w)dw.$$

For any fixed positive T, this implies

$$\max_{0 \leq t \leq T} \int_0^t y(t,w)dw \leq \zeta_0 + q_0\zeta_1 \max_{0 \leq u \leq T} \int_0^u y(u,w)dw.$$

Thus, due to the inequality $q_0\zeta_1 < 1$, it can be written

$$\max_{0 \leq t \leq T} \int_0^t y(t,w)dw \leq (1 - q_0\zeta_1)^{-1}\zeta_0.$$

The right hand part of the latter inequality does not depend on T. Therefore,

$$\max_{t \geq 0} \int_0^t y(t,w)dw \leq (1 - \zeta_1)^{-1}\zeta_0.$$

According to (2.11), this result proves the lemma, since $x(t)$ is an arbitrary solution. □

Proof of Theorem 3.2.1: Take into account that

$$p(t, A(\tau)) \leq p_0(t) \ (t, \tau \geq 0) \text{ and } \chi = \sup_{t \geq 0} p(t, A(t)).$$

Then due to (1.4) and (2.10) it can be written

$$||x(t)|| \leq \chi ||x(w)|| + \int_w^t q_0 p_0(t-t_1)|t_1 - \tau| ||x(t_1)|| dt_1.$$

Thus, with $\tau = t$, we get

$$||x(t)|| \leq \chi ||x(w)|| + \int_w^t q_0 p_0(t-t_1)|t_1 - t| ||x(t_1)|| dt_1 \ (0 \leq w \leq t).$$

Furthermore, for a fixed $T > 0$, put

$$y_T = \sup_{0 \leq t \leq T} ||x(t)|| \, ||x(w)||^{-1}.$$

Then

$$y_T \leq \chi + q_0 \, y_T \sup_{t \leq T} \int_0^t p_0(t-u)(t-u)du.$$

Consequently, due to (2.13)

$$y_T \leq \chi + q_0\, y_T\, \zeta_1.$$

But $q_0\zeta_1 < 1$, and therefore

$$y_T \leq \chi(1 - q_0\zeta_1)^{-1}.$$

Since T is arbitrary, we get the required estimate which implies the stability in the sense of Lyapunov. To prove the exponential stability it is sufficient to make in (2.1) the substitution $x(t) = x_1(t)e^{-t\epsilon}$ with a positive ϵ and to apply our reasoning to the perturbed system. For a sufficiently small ϵ the perturbed system will be stable in the sense of Lyapunov. This implies that (2.1) is exponentially stable. The proof is complete. □

Proof of Theorem 3.2.2: Under the hypothesis of the theorem we have the inequality $\rho > z(q_0, v)$. Dividing (2.4) by z^{n+1}, and taking into account that $z(q_0, v)$ is the extreme right-hand root of (2.4), we can write down

$$1 > q_0 \sum_{k=0}^{n-1} \frac{(k+1)v^k}{(k!)^{1/2}\rho^{k+2}} \equiv q_0\zeta_1.$$

Now, Theorem 3.2.1 entails the result. □

Proof of Theorem 3.2.4: First, let us prove the upper estimate.

Substituting the equality $x(t) = x_c(t)exp(-tc)$ with a real number c in equation (2.1), we get the equation

$$dx_c/dt = (A(t) + Ic)x_c. \tag{2.14}$$

Take $c = \rho - z(q_0, v) - \epsilon$ with an $\epsilon > 0$. Then we can affirm that the matrix $A(t) + I(z(q_0, v) + c)$ is a Hurwitz one. Proposition 2.1.1 implies

$$g(A(t) + Ic) = g(A(t)).$$

Therefore, due to Theorem 3.2.2 equation (2.14) is stable, i.e. there is a constant a such that

$$\|x_c(t)\| \leq a\|x_c(0)\| \text{ for all } t \geq 0.$$

This ensures the upper estimate, since $x(t) = x_c(t)\exp(-tc)$, and ϵ is arbitrary.

Now let us prove the lower estimate. Take into account that the Cauchy operator $U(t)$ of equation (2.1) has the following property: its inverse operator $U^{-1}(t)$ is the Cauchy operator of the equation

$$\dot{u} = -A(t)u$$

(see Section 1.6). But

$$\sup_{t\geq 0} \max_k Re\lambda_k(-A(t)) = -\inf_{t\geq 0} \min_k Re\lambda_k(A(t)).$$

Due to the proven upper estimate

$$||U^{-1}(t)h|| \leq a||h||exp\,[(-\beta_0 + z(q_0, v))t],$$

since $g(-A) = g(A)$. Putting $w = U^{-1}(t)h$, we arrive at the relation

$$||U(t)w|| \geq a^{-1}||exp\,[(\beta_0 - z(q_0, v))t]\,||w||\ (w \in \mathbf{C}^n).$$

That inequality gives the lower estimate. □

Proof of Lemma 3.2.5: Take into account that by (2.4),

$$q_0 = \frac{z^{n+1}(q_0, v)}{P(z(q_0, v))}.$$

But the function $x^{n+1}P^{-1}(x)$ increases as $x > 0$ increases. Hence by (2.7), $\rho > z(q_0, v)$. This fact yields the stated result. □

3.2.3 Example

Let us consider the equation

$$\ddot{y} + p(t)\dot{y} + w(t)y = 0\ (t \geq 0). \tag{2.15}$$

Here $p = p(t)$ and $w = w(t)$ are positive scalar-valued functions with the property

$$|p(t) - p(s)| + |w(t) - w(s)| \leq q_0|t - s| \text{ for all } t, s \geq 0.$$

In addition, it is assumed that $\sup_{t\geq 0} w(t) < \infty$. According to Example 2.1.4

$$g(A(t)) \leq v \equiv 1 + \sup_{t\geq 0} w(t)\ (t \geq 0).$$

Let us suppose that

$$\rho = -\sup_{t\geq 0} \alpha(A(t)) = \inf Re\,[p(t)/2 - \sqrt{p^2(t)/4 - w(t)}] > 0.$$

Then we can write $\zeta_1 = \rho^{-1} + 2v\rho^{-2}$. So if $\zeta_1 q_0 < 1$, then equation (2.15) is exponentially stable due to Theorem 3.2.1 .

3.3 Systems with piecewise constant matrices

Again, consider in $\mathbf{C}^n$ the equation

$$\dot{x}(t) = A(t)x(t)\ (t \geq 0), \tag{3.1}$$

where the $n \times n$-matrix $A(t)$ is piecewise constant. Namely,

$$A(t) = A_k \ (t_k \le t < t_{k+1};\ k = 0, 1, 2, ...) \tag{3.2}$$

with constant matrices A_k $(k = 0, 1, 2, ...)$, and

$$0 = t_0 < t_1 < t_2 < ... < t_k < ... .$$

Let $U(t, s)$ be the evolution operator of (3.1). Since

$$U(t, s)U(s, \tau) = U(t, \tau) \ (t, s, \tau \ge 0),$$

it can be written as

$$U(t_{k+1}, t_j) = U(t_{k+1}, t_k) ... U(t_{j+1}, t_j) \ (k > j).$$

But $A(t)$ is constant on each segment $[t_j, t_{j+1})$. We thus get

$$U\ (t_{k+1}, t_j) = e^{\delta_k A_k} e^{\delta_{k-1} A_{k-1}} ... e^{\delta_j A_j} \ (k > j \ge 0) \tag{3.3}$$

with $\delta_k = t_{k+1} - t_k$. Thanks to Corollary 2.1.7 this relation implies

$$\|U\ (t_{k+1}, t_j)\| \le \|e^{\delta_k A_k}\| \| e^{\delta_{k-1} A_{k-1}}\| ... \|e^{\delta_j A_j}\| \le$$

$$e^{\delta_k \alpha(A_k)} \gamma(A_k, \delta_k) e^{\delta_{k-1} \alpha(A_{k-1})} \gamma(A_{k-1}, \delta_{k-1}) ... e^{\delta_j \alpha(A_j)} \gamma(A_j, \delta_j) \tag{3.4}$$

with the notation

$$\gamma(A, t) = \sum_{k=0}^{n-1} t^k \frac{g^k(A)}{(k!)^{3/2}}.$$

Consequently,

$$\|U\ (t_{k+1}, t_j)\| \le exp\,[\sum_{m=j}^{k} \delta_m \alpha(A_m)] \prod_{m=j}^{k} \gamma(A_m, \delta_m). \tag{3.5}$$

Let the relations

$$\hat{\alpha} \equiv \overline{\lim}_{n \to \infty} t_n^{-1} \sum_{k=1}^{n} \alpha(A_k) \delta_k < \infty \tag{3.6}$$

and

$$\psi \equiv \overline{\lim}_{n \to \infty} t_n^{-1} \sum_{k=1}^{n} ln\ \gamma(A_k, \delta_k) < \infty \tag{3.7}$$

be fulfilled. Then inequality (3.5) implies

$$\overline{\lim}_{t \to \infty} t^{-1} ln \|U(t, 0)\| \le \hat{\alpha} + \psi.$$

We thus have derived

Theorem 3.3.1 *Under conditions (3.6), (3.7), let the inequality $\hat{\alpha} + \psi < 0$ be fulfilled. Then system (3.1) is asymptotically stable.*

Corollary 3.3.2 *Let*

$$\sup_{k=1,2,\ldots} \alpha(A_k) + ln\ \gamma(A_k, \delta_k) < 0.$$

Then system (3.1) is asymptotically stable.

3.4 Triangular systems

Let us consider the triangular system

$$\dot{u}_j(t) = \sum_{k=1}^{j} a_{jk}(t) u_k(t) \ (t \geq 0;\ j = 1, ..., n) \tag{4.1}$$

with piecewise continuous real functions $a_{jk}(t)$ $(1 \leq k \leq j \leq n)$. Then we have

$$u_j(t) = e^{\int_0^t a_{jj}(s)ds} u_j(0) + \int_0^t e^{\int_\tau^t a_{jj}(s)ds} \sum_{k=1}^{j-1} a_{jk}(\tau) u_k(\tau) d\tau.$$

Hence we arrive at the inequality

$$|u_j(t)| \leq e^{\int_0^t \alpha(s)ds} |u_j(0)| + \int_0^t e^{\int_\tau^t \alpha(s)ds} \sum_{k=1}^{j-1} |a_{jk}(\tau)||u_k(\tau)| d\tau$$

with the notation

$$\alpha(t) = \max_{j=1,\ldots,n} a_{jj}(t). \tag{4.2}$$

Furthermore, putting

$$e^{-\int_0^t \alpha(s)ds} |u_j(t)| = w_j(t), \tag{4.3}$$

we get

$$|w_j(t)| \leq |u_j(0)| + \int_0^t \sum_{k=1}^{j-1} |a_{jk}(\tau)||w_k(\tau)| d\tau \ (j = 1, ..., n). \tag{4.4}$$

Denote by $V(t)$ the lower-triangular matrix with the entries

$$v_{jk}(t) = |a_{jk}(t)| \text{ for } j > k, \text{ and } v_{jk}(t) = 0 \text{ for } j \leq k \ (j, k = 1, ...n),$$

then we can write down inequalities (4.4) in the vector form

$$|w(t)| \leq |u(0)| + \int_0^t V(\tau)|w(\tau)| d\tau.$$

Here $w(t) = (w_1(t), ..., w_n(t))$. For any continuous vector-valued function $h(t)$ defined on the positive half-line set

$$(Jh)(t) = \int_0^t V(\tau)h(\tau)d\tau.$$

But $V(t)$ is a lower-triangular $n \times n$-matrix with the zero diagonal. Therefore

$$V(\tau_1)V(\tau_2)...V(\tau_n) = 0 \ (\tau_1, ..., \tau_n \geq 0).$$

Consequently $J^n = 0$, and

$$(I - J)^{-1} = \sum_{k=0}^{\infty} J^k = \sum_{k=0}^{n-1} J^k. \tag{4.5}$$

It can be written as

$$|w(t)| \leq (I - J)^{-1}|u(0)| = \sum_{k=0}^{n-1} J^k |u(0)|.$$

But according to Example 2.1.3

$$||V(t)|| \leq N(V(t)) = g(A(t)) \equiv \sqrt{\sum_{1 \leq k < j \leq n} a_{jk}^2(t)}. \tag{4.6}$$

The latter inequality yields the relation

$$||(J^k h)(t)|| \leq \int_0^t ... \int_0^{t_{k-1}} ||V(t_1)... V(t_k)||dt_k...dt_1 \leq$$

$$\int_0^t ... \int_0^{t_{k-1}} g(A(t_1))... g(A(t_k))dt_k...dt_1 = \frac{1}{k!}[\int_0^t g(A(t_1))dt_1]^k.$$

This inequality and (4.4) imply

$$||w(t)|| \leq ||(I - J)^{-1}|| \ ||u(0)|| \leq ||u(0)|| \sum_{k=0}^{n-1} \frac{1}{k!}[\int_0^t g(A(t_1))dt_1]^k.$$

According to (4.3), we arrive at the inequality

$$||u(t)|| \leq ||u(0)|| e^{\int_0^t \alpha(s)ds} \sum_{k=0}^{n-1} \frac{1}{k!}[\int_0^t g(A(t_1))dt_1]^k.$$

Instead of 0 take an arbitrary real τ. We thus have derived

Theorem 3.4.1 *With the notation (4.2), (4.6) the inequality*

$$\|u(t)\| \leq \|u(\tau)\| e^{\int_\tau^t \alpha(s)ds} \sum_{k=0}^{n-1} \frac{1}{k!} [\int_\tau^t g(A(t_1))dt_1]^k \ (0 \leq \tau \leq t < \infty)$$

is true for any solution $u(t)$ of system (4.1).

Corollary 3.4.2 *Let the relation*

$$\overline{\lim}_{t\to\infty}\{\int_0^t \alpha(s)ds + \ln [\sum_{k=0}^{n-1} \frac{1}{k!} (\int_0^t g(A(t_1))dt_1)^k]\} < \infty$$

hold. Then system (4.1) is stable.

Corollary 3.4.3 *Let the relation*

$$\overline{\lim}_{t\to\infty} t^{-1}\{\int_0^t \alpha(s)ds + \ln [\sum_{k=0}^{n-1} \frac{1}{k!} (\int_0^t g(A(t_1))dt_1)^k]\} < 0$$

hold. Then system (4.1) is asymptotically stable.

3.5 Perturbations of triangular systems

Let $A(t) = (a_{jk}(t))_{j,k}^n$ be a real piecewise continuous $n \times n$-matrix. Denote the corresponding lower-triangular matrix by $T(t) = (T_{jk}(t))$. That is,

$$T_{jk}(t) = a_{jk}(t) \text{ if } j \geq k \text{ and } T_{jk}(t) = 0 \text{ if } j < k.$$

We will investigate the system

$$\dot{u} = A(t)u \tag{5.1}$$

as a perturbation of the triangular system

$$\dot{w}(t) = T(t)w. \tag{5.2}$$

Due to Theorem 3.4.1, a solution $w(t)$ of equation (5.2) is subject to the inequality

$$\|w(t)\| \leq e^{\int_\tau^t \alpha(s)ds} \sum_{k=0}^{n-1} (k!)^{-1} [\int_\tau^t g(T(t_1))dt_1]^k \|w(\tau)\|. \tag{5.3}$$

Here $\alpha(t)$ is defined by (4.2), and

$$g(T(t)) \equiv \sqrt{\sum_{1\leq k<j\leq n} a_{jk}^2(t)}.$$

Put $q(t) = ||A(t) - T(t)||$. Clearly,

$$q(t) \leq \max_j \sum_{k=j+1}^{n} |a_{jk}(t)|.$$

Employing Lemma 1.8.6 and inequality (5.3), we get

Theorem 3.5.1 *Let the condition*

$$J_A \equiv \sup_{t\geq 0} \int_0^t q(\tau) e^{\int_\tau^t \alpha(s)ds} \sum_{k=0}^{n-1} (k!)^{-1} [\int_\tau^t g(T(t_1))dt_1]^k d\tau < 1$$

be fulfilled. Then equation (5.1) is stable. Moreover, the Cauchy operator $U(t)$ of equation (5.1) satisfies the inequality

$$||U(t)|| \leq (1 - J_A)^{-1} \sup_{t\geq 0} e^{\int_0^t \alpha(s)ds} \sum_{k=0}^{n-1} (k!)^{-1} [\int_0^t g(T(t_1))dt_1]^k.$$

3.6 Systems with the matrix Lipschitz property

3.6.1 Stability criterion

In the present section we investigate the system

$$\frac{dx_j}{dt} = \sum_{k=1}^{n} a_{jk}(t) x_k \ (t \geq 0; \ j = 1, ..., n) \tag{6.1}$$

under the conditions

$$|a_{jk}(t) - a_{jk}(s)| \leq q_{jk}|t - s| \ (t, s \geq 0; \ j, k = 1, ..., n),$$

where q_{jk} are constants. We will write these conditions in the form

$$|A(t) - A(s)| \leq Q|t - s|, \ (t, s \geq 0) \tag{6.2}$$

where

$$A(t) = (a_{jk}(t))_{j,k=1}^n, \text{ and } Q = (q_{jk})_{j,k=1}^n$$

is a constant matrix. Inequality (6.2) permits us to obtain estimates for each coordinate of a solution and to take into account information about each element of the matrix more completely than under the customary condition

$$||A(t) - A(s)|| \leq q|t - s| \ (t, s \geq 0) \tag{6.3}$$

with a constant number q, although applications of condition (6.2) require more calculations than condition (6.3).

Throughout the present section it is assumed that matrix $A(t)$ is uniformly bounded on $[0,\infty)$.

Furthermore, let $m_{ij}(\lambda,t)$ be the cofactor of the element $\delta_{ij}\lambda - a_{ij}(t)$ of the matrix $\lambda I - A(t)$. Recall that $\delta_{ij} = 0,\ j \neq i$ and $\delta_{jj} = 1$. Assume that after division by the common zeros we obtain the equality

$$\frac{m_{ij}(\lambda,t)}{det(I\lambda - A(t))} = \frac{\mu_{ij}(\lambda,t)}{d_{ij}(\lambda,t)},$$

where $d_{ij}(\lambda,t)$ and $\mu_{ij}(\lambda,t)$ are polynomials with respect to λ, and the degree of $d_{ij}(\lambda,t)$ is equal to $n_{ij}(t)$. Let $co(A(t))$ be the convex hull of all eigenvalues of matrix $A(t)$. Put

$$b_{ji}(k,t) = \frac{1}{k!(n_{ij}-1-k)!} \sup_{\lambda\in co(A(t))} |\mu_{ij}^{(n_{ij}-k-1)}(\lambda,t)|,$$

$(n_{ij} \equiv n_{ij}(t);\ k = 0,...,n-1)$ Since $(-m)! = 0$ for any natural m, the equalities

$$b_{ij}(k,t) = 0 \text{ for } k = n_{ij}(t),...,n-1$$

are valid if $n_{ij}(t) < n-1$. Set

$$\bar{b}_{ij}(k) = \sup_{t\geq 0} b_{ij}(k,t)$$

and

$$\overline{B}_k = (\bar{b}_{ij}(k))_{j,i=1}^n .$$

Finally, denote

$$\alpha_0 \equiv \sup_{t\geq 0} \max_{k=1,...,n} Re\lambda_k(A(t)).$$

Theorem 3.6.1 *If $A(t)$ satisfies condition (6.2), then any solution $x(t)$ of equation (6.1) subordinates the inequality*

$$|x(t)| \leq exp[(\alpha_0 + z_0)t]C(t)|x(0)|\ (t \geq 0),$$

where $C(t)$ is a non-negative polynomial matrix, and z_0 is the extreme right-hand (positive) root of the polynomial

$$det(I\lambda^{n+1} - \sum_{k=0}^{n-1} \lambda^{n-k-1}(k+1)!\overline{B}_k Q). \tag{6.4}$$

Below in this section we will derive a simple estimate for z_0. Recall that I is the unit matrix.

Corollary 3.6.2 *Under condition (6.2), let the matrix $A(t)+z_0 I$ be a Hurwitz one for all $t \geq 0$. Then system (6.1) is stable.*

3.6.2 Proof of Theorem 3.6.1

Let $T_1, T_2, ..., T_m$ be arbitrary constant $n \times n$-matrices. Recall that a number ω is called *a characteristic value of the matrix pencil*

$$\Phi(\lambda) = \lambda^m I - \lambda^{m-1} T_1 - ... - T_m \tag{6.5}$$

if there is a non-zero vector h_0 such that $\Phi(\omega)h_0 = 0$ cf. (Markus, 1988).

We will use the following

Lemma 3.6.3 *Each characteristic value of the matrix pencil defined by (6.5) is an eigenvalue of the block matrix*

$$\begin{pmatrix} T_1 & T_2 & \dots & T_{m-1} & T_m \\ I & 0 & \dots & 0 & 0 \\ . & . & \dots & . & . \\ 0 & 0 & \dots & I & 0 \end{pmatrix}.$$

For the proof see (Rodman, 1989, Chapter 1). By virtue of the Frobenius theorem for positive matrices (Gantmaher, 1967), we arrive at the following result.

Corollary 3.6.4 *Let all the matrices T_k $(k = 1, ..., m)$ be non-negative. Then the extreme right characteristic value of the matrix pencil (6.5) is positive.*

Proof of Theorem 3.6.1: Fix $s \geq 0$ and write (6.1) as

$$dx/dt - A(s)x = [A(t) - A(s)]x.$$

This equation is equivalent to the following one:

$$x(t) = e^{A(s)t}x(0) + \int_0^t e^{A(s)(t-\tau)}[A(\tau) - A(s)]x(\tau)d\tau.$$

Hence (6.2) implies that

$$|x(t)| \leq |e^{A(s)t}x(0)| + \int_0^t |e^{A(s)(t-\tau)}|Q|\tau - s||x(\tau)|d\tau.$$

Putting $t = s$ and writing t instead of s, we find that

$$|x(t)| \leq |e^{A(t)t}x(0)| + \int_0^t (t-\tau)|e^{A(t)(t-\tau)}|Q|x(\tau)|d\tau.$$

According to Corollary 2.2.3,

$$|e^{A(\tau)t}| \leq e^{\alpha_0 t}\Psi(t) \ (\tau, t \geq 0),$$

where

$$\Psi(t) = \sum_{k=0}^{n-1} t^k \overline{B}_k.$$

Consequently,

$$|x(t)|e^{-\alpha_0 t} \leq \Psi(t)|x(0)| + \int_0^t \Psi(t-\tau)(t-\tau)Qe^{-\alpha_0\tau}|x(\tau)|d\tau.$$

It follows from Lemma 1.7.1 that

$$|x(t)|e^{-\alpha_0 t} \leq \eta(t), \tag{6.6}$$

where $\eta(t)$ is the solution of the equation

$$\eta(t) = \Psi(t)|x(0)| + \int_0^t \Psi(t-\tau)(t-\tau)Q\eta(\tau)d\tau.$$

To solve the foregoing equation we apply the Laplace transformation. Let λ be the dual variable, $\overline{\eta}(\lambda), \overline{\Psi}(\lambda)$ and $\overline{L}(\lambda)$ be the Laplace transforms of $\eta(t)$, $\Psi(t)$ and $t\Psi(t)$, respectively. Taking the Laplace transform of the latter equation and using the fact that the transform of a convolution is the product of the functions in the convolution, we obtain

$$\overline{\eta}(\lambda) = \overline{\Psi}(\lambda)|x(0)| + \overline{L}(\lambda)Q\overline{\eta}(\lambda).$$

By the inverse Laplace transform,

$$\eta(t) = \frac{1}{2\pi i}\int_{Re\lambda>z_0} e^{t\lambda}(I-\overline{L}(\lambda)Q)^{-1}\overline{\Psi}(\lambda)|x(0)|d\lambda. \tag{6.7}$$

Obviously,

$$\overline{\Psi}(\lambda) = \sum_{k=0}^{n-1}\frac{k!}{\lambda^{k+1}}\overline{B}_k$$

and

$$\overline{L}(\lambda) = \sum_{k=0}^{n-1}\frac{(k+1)!}{\lambda^{k+2}}\overline{B}_kQ.$$

These relations yield

$$\lambda^{n+1}\overline{\Psi}(\lambda) = \sum_{k=0}^{n-1}k!\lambda^{n-k}\overline{B}_k,$$

and

$$\lambda^{n+1}\overline{L}(\lambda) = \sum_{k=0}^{n-1}(k+1)!\lambda^{n-k-1}\overline{B}_kQ.$$

By (6.7) we conclude that

$$\eta(t) = \frac{1}{2\pi i}\int_{Re\lambda>z_0} e^{t\lambda}T^{-1}(\lambda)\sum_{k=0}^{n-1}k!\lambda^{n-k}\overline{B}_k\,d\lambda\,|x_0| \tag{6.8}$$

with the notation

$$T(\lambda) = I\lambda^{n+1} - \sum_{k=0}^{n-1}(k+1)!\lambda^{n-k-1}\overline{B}_k Q.$$

Take into account that

$$T^{-1}(\lambda) = \frac{T_1(\lambda)}{det\, T(\lambda)},$$

where $T_1(\lambda)$ is a polynomial matrix. By virtue of Corollary 3.6.4, z_0 is a positive number. Now according to the residue theorem, (6.8) implies that

$$\eta(t) \le exp(z_0 t)[C_0 + C_1 t + .. + C_m t^{m-1}]|x(0)|,$$

where C_k $(k = 0, ..., m)$ are constant matrices and m is the multiplicity of the root z_0. The assertion of the theorem now follows from the latter inequality and inequality (6.6). □

3.6.3 Estimates for z_0

Lemma 3.6.5 *Let z_1 be the extreme right-hand (positive) characteristic value of the matrix pencil*

$$\lambda^n I - \sum_{k=0}^{n-1} \lambda^{n-k-1} Q_k,$$

where Q_k are non-negative matrices. Then $z_1 \le z_2$, where z_2 is the extreme right-hand root of the algebraic equation

$$\lambda^n = \sum_{k=0}^{n-1} \lambda^{n-k-1} ||Q_k||. \tag{6.9}$$

Moreover, if

$$\sum_{k=0}^{n-1} ||Q_k|| \le 1, \tag{6.10}$$

then

$$z_1^n \le \sum_{k=0}^{n-1} ||Q_k||. \tag{6.11}$$

Proof: Clearly,

$$z_1^n \le \sum_{k=0}^{n-1} z_1^{n-k-1} ||Q_k||.$$

Or

$$1 \le \sum_{k=0}^{n-1} z_1^{-k-1} ||Q_k||.$$

On the other hand, dividing (6.9) by λ^n and taking into account that z_2 is the root of (6.9), we arrive at the equality

$$1 = \sum_{k=0}^{n-1} z_2^{-k-1} \|Q_k\|.$$

The comparison of the two last relations gives the inequality $z_1 \leq z_2$. Now under condition (6.10), inequality (6.11) is due to Lemma 1.11.1. □

Corollary 3.6.6 *Let z_0 be the (positive) extreme right-hand root of the polynomial defined by (6.4), and let*

$$\sum_{k=0}^{n-1} (k+1)! \|\overline{B}_k Q\| \leq 1. \tag{6.12}$$

Then $z_0 \leq c(Q)$, where

$$c(Q) = \sqrt[n+1]{\sum_{k=0}^{n-1} (k+1)! \|\overline{B}_k Q\|}.$$

The latter corollary and Corollary 3.6.2 imply

Corollary 3.6.7 *Under conditions (6.2) and (6.12), let the matrix $A(t) + Ic(Q)$ be a Hurwitz one for all $t \geq 0$. Then system (6.1) is stable.*

3.6.4 Lower estimates

Theorem 3.6.8 *If (6.2) holds and*

$$\beta_0 = \inf_{t \geq 0} \min_{k=1,\ldots,n} Re\lambda_k(A(t)) > -\infty,$$

then any solution of system (6.1) is subject to the estimate

$$|x(t)| \geq exp[(\beta_0 - z_0)t] C^{-1}(t) |x(0)|, \ (t \geq 0),$$

where $C(t)$ is a non-negative polynomial matrix, and z_0 is the extreme right-hand root of the polynomial (6.4).

Proof: Fox a fixed $T > 0$, put in (6.1) $\tau = T - t$, $x(T - \tau) = y(\tau)$, and $-A(T-\tau) = A_1(\tau)$. We obtain the equation

$$dy/d\tau = A_1(\tau) y(\tau) \ (0 \leq \tau \leq T).$$

The eigenvalues of $A_1(\tau)$ are $-\lambda_1(A(T-\tau)), \ldots, -\lambda_n(A(T-\tau))$. Hence,

$$\sup_{0 \leq \tau \leq T} \max_{k=1,\ldots,n} Re\lambda_k(A_1(\tau)) \leq -\beta_0.$$

Moreover, the closed convex hull of the numbers

$$-\lambda_1(A(T-\tau)), ..., -\lambda_n(A(T-\tau))$$

is the mirror image of $co(A(t))$ in the imaginary axis. Moreover, the cofactor of the element $-a_{ij}(T-\tau)-\delta_{ij}\lambda$ of the matrix $A_1(\tau)-\lambda I$ equals $-m_{ij}(-\lambda, T-\tau)$, i.e.

$$\frac{1}{k!(n_{ij}-k-1)!} \sup_{\lambda \in co(-A(T-\tau))} |\mu_{ij}^{(n_{ij}-k-1)}(\lambda, T-\tau)| \leq \bar{b}_{ji}(k)$$

$$(n_{ij} = n_{ij}(t);\ 0 \leq \tau \leq T;\ i, j = 1, ..., n)$$

(see Subsection 3.6.1). Therefore, Theorem 3.6.1 implies that

$$|y(\tau)| \leq exp[(-\beta_0 + z_0)\tau]C(\tau)|y(0)|\ (0 \leq \tau \leq T).$$

Hence, taking into account that $y(0) = x(T)$, and $x(0) = y(T)$, with $T = t$, we obtain the desired inequality. □

3.6.5 Example

Example 3.6.9 *Let us consider the equation*

$$y'' + a(t)y' + y = 0,$$

where the scalar-valued function $a(t)$ satisfies the conditions

$$|a(t) - a(s)| \leq q|t - s|, \text{ and } 0 \leq a(t) \leq 2\ (t, s \geq 0).$$

Reducing this equation to the form (6.1) by the substitution $x_1 = y'$ and $x_2 = y$, we find that

$$\alpha_0 = -\frac{1}{2}\inf_{t\geq 0} a(t),$$

and $co(A(t)) = [\lambda_1(A(t)), \lambda_2(A(t))]$ is the segment joining the eigenvalues

$$\lambda_1(A(t)) = -a(t)/2 + i\sqrt{1 - a^2(t)/4}$$

and

$$\lambda_2(A(t)) = -a(t)/2 - i\sqrt{1 - a^2(t)/4}$$

of the matrix $A(t)$ under consideration. We have

$$Q = \begin{pmatrix} q & 0 \\ 0 & 0 \end{pmatrix}$$

and

$$\overline{B}_k Q = \begin{pmatrix} q\bar{b}_{11}(k) & 0 \\ q\bar{b}_{21}(k) & 0 \end{pmatrix}$$

with $k = 0, 1$. I.e., there is not need to calculate $\bar{b}_{12}(k)$ and $\bar{b}_{22}(k)$. Clearly, $m_{11}(\lambda, t) = -\lambda$ and $m_{12}(\lambda, t) = -1$. Hence $\bar{b}_{21}(1) = 1, \bar{b}_{21}(0) = 0$,

$$\bar{b}_{11}(1) = \max_{\lambda \in co(A(t))} |\lambda| = (a^2(t)/4 + 1 - a^2(t)/4)^{1/2} = 1,$$

and $\bar{b}_{11}(0) = 1$. Thus

$$\Delta(\lambda) = det\,(\lambda^3 I - \lambda \overline{B}_0 Q - 2\overline{B}_1 Q) =$$

$$det \begin{pmatrix} \lambda^3 - \lambda q - 2q & 0 \\ -q & \lambda^3 \end{pmatrix} = \lambda^3(\lambda^3 - \lambda q - 2q).$$

The zero z_0 of $\Delta(\lambda)$ furthest to the right coincides with the zero furthest to the right of the polynomial $\lambda^3 - \lambda q - 2q$. The Cardan formula yields

$$z_0 = (q + \sqrt{q^2 - (q/3)^3}\,)^{1/3} + (q - \sqrt{q^2 - (q/3)^3}\,)^{1/3} \ (q \leq 27).$$

Cubing this expression and making elementary calculations, we find that

$$z_0^3 = q(2 + (q + \sqrt{q^2 - (q/3)^3})^{1/3} + (q - \sqrt{q^2 - (q/3)^3})^{1/3}).$$

It is simple to check that the function

$$(q + \sqrt{q^2 - (q/3)^3})^{1/3} + (q - \sqrt{q^2 - (q/3)^3})^{1/3}$$

monotonically increases as $q < 27$ increases. Hence for an arbitrary fixed $q \leq 27$ we have

$$z_0 \leq \sqrt[3]{2q(1 + q^{1/3})}.$$

Theorem 3.6.1 implies that the upper Lyapunov exponent κ_L of the system satisfies the inequality

$$\kappa_L \leq \alpha_0 + \sqrt[3]{2q(1 + q^{1/3})} \ (q \leq 27). \tag{6.13}$$

3.7 Notes

Note 3.1 There is a much literature on stability theory for ordinary differential equations. Books that provide a broader look at the subjects we cover include (Bellman, 1953), (Coppel, 1965), (Willems, 1970), (Harris and Miles, 1980), (Rugh, 1996).

Note 3.2 The contents of Section 3.1 are adapted from the paper (Gil', 1987). Theorem 3.2.2 is taken from (Gil', 1989). It is based on the freezing method developed in (Bylov et al., 1966). Other results on stability and instability of slowly varying systems can be found in the papers (Skoog and Lau, 1972), (Desoer, 1969). A survey of results on uniform exponential stability under the hypothesis that pointwise eigenvalues of the slowly-varying $A(t)$ have negative real parts is in (Ilchmann et al., 1987).

Note 3.3 Theorem 3.3.1 is a trivial corollary of Theorem 2.1.5. The material of Sections 3.4 and 3.5 is new. For additional results using a diagonal form for $A(t)$, consult (Wu, 1984).

Note 3.4 Contents of Section 3.6 are adapted from (Gil', 1987). R. E. Vinograd (1983) used the freezing method to obtain the following inequality for the upper Lyapunov exponent κ_L of system (2.1):

$$\kappa_L \leq \alpha_0 + cq^{1/n},$$

where

$$c = 2a(C_n/4a^2)^{1/(n+1)}, \ C_n = n(n+1)/2, \ a = \sup_{t\geq 0} \|A(t)\|.$$

In the case Example 3.6.9 Vinograd's inequality yields

$$\kappa_L \leq \alpha_0 + a\sqrt[3]{2e^{-2}q/a^2}.$$

Note that inequality (6.13) is more accurate than the latter one.

References for Chapter 3

Bellman, R.E. (1953). *Stability Theory of Differential Equations.* McGraw-Hill, New York.

Bylov, B. F., Grobman, B. M., Nemyckii V. V. and Vinograd R.E. (1966). *The Theory of Lyapunov Exponents*, Nauka, Moscow. In Russian.

Coppel, W.A. (1965). *Stability and Asymptotic Behavior of Differential Equations.* Heath, Boston.

Desoer, C.A. (1969). Slowly varying systems $\dot{x} = A(t)x$, *IEEE Transactions on Automatic Control, 14*, 780-781.

Gantmaher, F. R. (1967). *Theory of Matrices.* Nauka, Moscow. In Russian.

Gil', M. I. (1985). Sufficient and necessary conditions for stability of linear systems, *Izvestiya of the USSR Ac. Sci., ser. Techn. Cybernetics, 4*, 225-229.

Gil', M.I. (1987). Two-sides bounds for solutions of linear ordinary differential equations, *Differential Equations, 23*, 2031-2036.

Gil', M.I. (1989). The freezing method for nonlinear equations. *Differential Equations, 25*, 912-918.

Harris, C.J. and Miles J.F. (1980). *Stability of Linear Systems*, Academic Press, New York.

Ilchmann A., Owens, D.H, and Pratcel-Wolters, D. (1987). Sufficient conditions for stability of linear time-varying systems, *Systems & Control Letters, 9*, 157-163.

Markus, A.S. (1988). *Introduction to Spectral Theory of Polynomial Operator Pencils.* Trans. of Math. Monographs, v. 71, Providence, R. I.

Rodman, L. (1989). *An Introduction to Operator Polynomials*, Operator Theory. Advances and Appl. v. 38, Birkhäuser, Basel.

Rugh, W.J. (1996). *Linear System Theory.* Prentice Hall, Upper Saddle River, New Jersey.

Skoog, R.A. and Lau, G.Y. (1972). Instability of slowly varying systems, *IEEE Transactions on Automatic Control, 17*, 86-92.

Vinograd, R.E. (1983). An improved estimate in the method of freezing, *Proc. Amer. Soc. 89 (1)*, 125-129.

Willems, J.L. (1970). *Stability Theory of Dynamical Systems.* John Willey, New York.

Wu, M.Y. (1984). Stability of linear time-varying systems, *International Journal of System Sciences, 15*, 137-150.

Chapter 4

LINEAR FINITE DIMENSIONAL SYSTEMS (CONTINUATION)

In the present chapter we continue to investigate linear finite dimensional systems.

In Section 4.1 the multiplicative representation for solutions of ordinary linear differential equations is introduced. In Section 4.2, by virtue of that representation, we derive the Lozinskii and Wazewski inequalities.

In Section 4.3, majorants and minorants for solutions are constructed.

Section 4.4 deals with perturbations of autonomous linear systems.

Section 4.5 is about the second order systems.

In Section 4.6 nonautonomous scalar linear equations with real characteristic roots are considered.

Section 4.7 is devoted to the input-output stability of linear systems.

4.1 The multiplicative representation of solutions

For a bounded Riemann integrable matrix-valued function F defined on a segment [0,t] define *the left multiplicative integral* as the limit of the sequence of the products

$$\prod_{1\leq k\leq N-1}^{\leftarrow} (I + F(t_k^{(N)})\delta_k) \equiv$$

$$(I + F(t_{N-1}^{(N)})\delta_{N-1})(I + F(t_{N-2}^{(N)})\delta_{N-2})...(I + F(t_1^{(N)})\delta_1)$$

as $\max_k |t_{k+1}^{(N)} - t_k^{(N)}|$ tends to zero. Here

$$0 = t_1^{(N)} < t_2^{(N)} < ... < t_N^{(N)} = t.$$

and

$$\delta_k = t_{k+1}^{(N)} - t_k^{(N)} \ (k = 1, ..., N-1).$$

I.e., the arrow over the symbol of the product means that the indexes of the co-factors increase from right to left. *The left multiplicative integral we denote by the symbol*

$$\int_{[0,t]}^{\leftarrow} (I + F(s)ds).$$

Now again consider the equation

$$\dot{x} = A(t)x \ (t \geq 0) \tag{1.1}$$

with a piecewise continuous matrix $A(t)$.

Lemma 4.1.1 *Any solution $x(t)$ of equation (1.1) can be represented by the relation*

$$x(t) = \int_{[0,t]}^{\leftarrow} (I + A(s)ds)x(0). \tag{1.2}$$

Proof: As mentioned in Section 1.6, the Cauchy operator of equation (1.1) can be represented in the form

$$U(t) = I + \int_0^t A(t_1)dt_1 +$$

$$\sum_{k=2}^{\infty} \int_0^t \int_0^{t_k} ... \int_0^{t_2} A(t_k)\ A(t_{k-1})...A(t_1)dt_k\ dt_{k-1}...dt_1.$$

For $N > 2$ put

$$U_N(t) = I + \int_0^t A(t_1)dt_1 +$$

$$\sum_{k=2}^{N} \int_0^t \int_0^{t_k} ... \int_0^{t_2} A(t_k)\ A(t_{k-1})...A(t_1)dt_k\ dt_{k-1}...dt_1.$$

Then, obviously,

$$U_N(t) \to U(t) \text{ as } N \to \infty \tag{1.3}$$

uniformly on each finite segment. Now set

$$W_m = A(t_m^{(N)})\delta_m \ (m \leq N).$$

Clearly,

$$\int_0^t A(t_1)dt_1 = \sum_{k=1}^{N-1} W_k + O(1/N) \tag{1.4}$$

for each fixed $t > 0$. Denote

$$U_{NN} = I + \sum_{k=1}^{N-1} W_k + \sum_{1 \le k_1 < k_2 \le N-1} W_{k_2} W_{k_1} + \dots$$

$$+ \sum_{1 \le k_1 < k_2 < \dots < k_j \le N-1} W_{k_j} \dots W_{k_2} W_{k_1} + \dots + W_{N-1} \dots W_2 W_1.$$

We can write down

$$\overset{\leftarrow}{\prod_{1 \le k \le N-1}} (I + W_k) = I + \sum_{k=1}^{N-1} W_k + \sum_{1 \le k_1 < k_2 \le N-1} W_{k_2} W_{k_1} + \dots$$

$$+ \sum_{1 \le k_1 < k_2 < \dots < k_j \le N-1} W_{k_j} \dots W_{k_2} W_{k_1} + \dots + W_{N-1} \dots W_2 W_1 = U_{NN}.$$

But according to (1.4), for a fixed $t > 0$,

$$U_{NN} \to U_N(t) \text{ as } \max_k \delta_k \to 0.$$

Now the required result follows from (1.3). □

For more details about the multiplicative integral see (Dollard and Friedman, 1979), (Gantmaher, 1959).

4.2 The Lozinskii and Wazewski inequalities

Let $||.||_n$ be an arbitrary norm in $\mathbf{C}^n$. Again consider the system (1.1) with a piecewise continuous matrix $A(t)$.

Theorem 4.2.1 *Let there be a continuous function $\phi(t)$, such that for all sufficiently small $\delta > 0$ the condition*

$$||I + A(t)\delta||_n \le 1 + \phi(t)\delta \;\; (t \ge 0) \tag{2.1}$$

is fulfilled. Then for any solution $x(t)$ of (1.1) the inequality

$$\frac{||x(t)||_n}{||x(s)||_n} \le exp[\int_s^t \phi(s_1)ds_1] \;\; (t \ge s \ge 0) \tag{2.2}$$

is true.

Proof: Condition (2.1) implies

$$\| \prod_{1\le k\le n}^{\leftarrow} (1 + A(t_k^{(n)})\delta_k)\|_n \le \prod_{k=1}^{n} (1 + \phi(t_k^{(n)})\delta_k).$$

Clearly,

$$1 + \phi(t_k^{(n)})\delta_k = e^{\phi(t_k^{(n)})\delta_k} + o(\delta_k) \ (\delta_k \downarrow 0).$$

That is,

$$\| \prod_{1\le k\le n}^{\leftarrow} (I + A(t_k^{(n)})\delta_k)\|_n \le \exp[\sum_{k=1}^{n} \phi(t_k^{(n)})\delta_k + 0(\delta)] \ (\delta \downarrow 0),$$

where $\delta = max\delta_k$. The passage to the limit as $n \to \infty$ and Lemma 4.1.1 give the required estimate. □

Corollary 4.2.2 *(Lozinskii). Let the relation*

$$\overline{\lim}_{\delta\downarrow 0} \frac{\|I + A(t)\delta\|_n - 1}{\delta} = \phi(t) \ (t \ge 0)$$

hold. Then for any solution $x(t)$ of (1.1) the inequality (2.2) is true.

Indeed, under the above conditions we have

$$\|I + A(t)\delta\|_n = 1 + \phi(t)\delta + o(\delta) \ (\delta \downarrow 0).$$

Now, by virtue of Theorem 4.2.1, we easily get the inequality (2.2).

Theorem 4.2.3 *Let there be a continuous function $\psi(t)$, such that for all sufficiently small $\delta > 0$ and $h \in \mathbf{C}^n$, the relation*

$$1 + \psi(t)\delta \le \frac{\|(I + A(t)\delta)h\|_n}{\|h\|_n} \ (t \ge 0)$$

holds. Then for any solution $x(t)$ of (1.1), the inequality

$$exp[\int_s^t \psi(s_1)ds_1] \le \frac{\|x(t)\|_n}{\|x(s)\|_n} \ (t, s \ge 0)$$

is true.

The proof of that theorem is analogous to the proof of Theorem 4.2.1.

Further, let $(.,.)$ be the scalar product in $\mathbf{C}^n$:

$$(h, g) = \sum_{k=1}^{n} h_k \overline{g}_k \ (h = (h_k), g = (g_k) \in \mathbf{C}^n).$$

Recall that $A_R = (A + A^*)/2$ is the real Hermitian component; $\beta(A_R), \alpha(A_R)$ are the smallest and largest eigenvalues of A_R, respectively.

Lemma 4.2.4 *Let $\|.\|_{C^n}$ be the Euclidean norm. Then for any $n \times n$-matrix A and any $h \in \mathbf{C}^n$, the relations*

$$1 + \beta(A_R)\delta + o(\delta) \leq \frac{\|(I + A\delta)h\|_{C^n}}{\|h\|_{C^n}} \leq 1 + \alpha(A_R)\delta + o(\delta) \ (\delta \downarrow 0)$$

are true.

Proof: It is obvious that

$$\|(I + A\delta)h\|_{C^n}^2 = \|h\|_{C^n}^2 + 2(A_R h, h)\delta + (Ah, Ah)\delta^2 \leq 1 + 2\alpha(A_R)\delta + o(\delta)$$

when $\|h\|_{C^n} = 1$. But

$$\sqrt{1 + 2\alpha(A_R)\delta} = 1 + \alpha(A_R)\delta + o(\delta).$$

This clearly forces the upper estimate. The proof of the lower estimate is left to the reader. □

Theorems 4.2.1, 4.2.3 and Lemma 4.2.4 imply

Corollary 4.2.5 *(Wazewski). For any solution $x(t)$ of equation (1.1) and all $t > s \geq 0$, the estimations*

$$exp[\int_s^t \beta(A_R(s_1))ds_1] \leq \frac{\|x(t)\|_{C^n}}{\|x(s)\|_{C^n}} \leq exp[\int_s^t \alpha(A_R(s_1))ds_1]$$

are true.

Let E be a positive definite matrix. Let us introduce the scalar product $(.,.)_E$ by the equality

$$(h, g)_E = (Eh, g) \ (h, g \in \mathbf{C}^n).$$

Denote the corresponding norm by $\|.\|_E$. That is,

$$\|h\|_E = \sqrt{(Eh, h)} = \|E^{1/2}h\|_{C^n}.$$

Applying Theorem 4.2.1 and Lemma 4.2.4, and omitting the evident details, we get

Corollary 4.2.6 *Let the relation*

$$EA^*(t) + A(t)E \leq 0 \ (t \geq 0)$$

hold with a positive definite matrix E. Then equation (1.1) is stable.

4.3 Linear systems with majorants and minorants

Let $A(t)$ $(t \geq 0)$ be a variable matrix with entries $a_{jk}(t)$ $(j,k=1,...,n)$. Then if there is a real matrix $M(t) = (m_{jk}(t))_{j,k=1}^n$ with non-negative out diagonal entries such that

$$|I + A(t)\delta| \leq I + M(t)\delta \ (t \geq 0) \tag{3.1}$$

for all sufficiently small positive δ, we will say that $A(t)$ *has a majorant* $M(t)$. Inequality (3.1) means that

$$|a_{jk}(t)| \leq m_{jk}(t) \text{ for } j \neq k,$$

and

$$|1 + \delta\, a_{kk}(t)| \leq 1 + \delta\, m_{kk}(t) \ (t \geq 0;\ j,k = 1,...,n).$$

Lemma 4.3.1 *Let a piecewise continuous matrix $A(t)$ have a majorant $M(t)$. Then for any solution $x(t)$ of the linear system*

$$\dot{x} = A(t)x, \tag{3.2}$$

the inequality

$$|x(t)| \leq z(t) \ (t \geq 0)$$

is true, where $z(t)$ is a solution of the equation

$$\dot{z} = M(t)z \tag{3.3}$$

with the initial condition $z(0) = |x(0)|$.

Proof: Due to condition (3.1) it can be written

$$|\overleftarrow{\prod_{1\leq k\leq n}} (I + A(t_k^{(n)})\delta_k)u_0| \leq \overleftarrow{\prod_{1\leq k\leq n}} (I + M(t_k^{(n)})\delta_k)|u_0|.$$

But according to Lemma 4.1.1 one can assert that the sequences at the left-hand part and at the right-hand part of this inequality tend to the solutions $x(t)$ and $z(t)$ of equations (3.2) and (3.3), respectively. This finishes the proof. □

Corollary 4.3.2 *Let a piecewise continuous matrix $A(t)$ have a constant majorant M. Then for any solution $x(t)$ of (3.2) the inequality*

$$|x(t)| \leq e^{Mt}|x(0)| \ (t \geq 0)$$

holds.

Now we will obtain a lower solution estimate. Suppose that the matrix $-A(t)$ has a majorant $P(t)$. In other words, *suppose that there is a matrix $P(t)$ with non-negative out diagonal entries such that*

$$|I - A(t)\delta| \leq I + P(t)\delta \ (t \geq 0) \tag{3.4}$$

for all sufficiently small positive δ. Then we will say that $A(t)$ has the minorant $P(t)$.

Further, let $U(t)$ be the Cauchy operator of equation (3.2). Take into account the operator $U^{-1}(t)$ is the Cauchy operator of the equation

$$\dot{x} = -A(t)x. \tag{3.5}$$

By the previous lemma $|U^{-1}(t)| \leq Z(t)$, where $Z(t)$ is the Cauchy operator of the equation $\dot{z} = P(t)z$. Now we easily obtain

$$|U(t)h| \geq Z^{-1}(t)|h| \ (h \in \mathbf{C}^n, t \geq 0).$$

But $Z^{-1}(t)$ is the Cauchy operator of the equation

$$\dot{z} = -P(t)z. \tag{3.6}$$

We thus have derived

Lemma 4.3.3 *Let a piecewise continuous matrix $A(t)$ have a minorant $P(t)$. Then for any solution $x(t)$ of equation (3.2), the inequality*

$$|x(t)| \geq z_-(t) \ (t \geq 0)$$

is valid where $z_-(t)$ is the solution of equation (3.6) with the initial condition $z_-(0) = |x(0)|$.

Corollary 4.3.4 *Let $A(t)$ have a constant minorant P. Then any solution $u(t)$ of (3.2) is subject to the inequality*

$$|u(t)| \geq e^{-Pt}|u(0)| \ (t \geq 0).$$

4.4 Perturbations of autonomous systems

Let us consider in $\mathbf{C}^n$ the equation

$$\dot{x} = Ax + B(t)x, \ t \geq 0, \tag{4.1}$$

where A is a constant $n \times n$-matrix, and $B(t)$ is a variable one with the property

$$\|B(t)\| \leq \nu \text{ for all } t \geq 0. \tag{4.2}$$

Here $\nu = const > 0$. Denote by $z(\nu, A)$ the extreme right-hand (unique positive and simple) zero of the algebraic equation.

$$z^n = \nu \sum_{j=0}^{n-1} \frac{g^j(A)}{\sqrt{j!}} z^{n-j-1}.$$

Recall that

$$\Gamma(A) = \sum_{j=0}^{n-1} \frac{g^j(A)}{|\alpha(A)|^{j+1}\sqrt{j!}},$$

and

$$\chi(A) = \max_{t\geq 0} exp[\alpha(A)t] \sum_{j=0}^{n-1} \frac{g^j(A)t^j}{\sqrt{j!}}.$$

Theorem 4.4.1 *Under condition (4.2), let the matrix $A + z(\nu, A)I$ be a Hurwitz one. Then equation (4.1) is asymptotically stable. Moreover, the inequality $\nu\Gamma(A) < 1$ holds, and any solution $x(t)$ of (4.1) subordinates the estimate*

$$||x(t)|| \leq \frac{\chi(A)||x(0)||}{1 - \nu\Gamma(A)} \ (t \geq 0).$$

This theorem is a particular case of Theorem 5.3.1.

4.5 Second order systems

Consider in $\mathbf{C}^n$, the system

$$\ddot{u} + B(t)\dot{u} + C^2(t)u = 0 \ (t \geq 0). \tag{5.1}$$

Theorem 4.5.1 *Let $B(t)$ be a piecewise continuous matrix, and $C(t)$ be a differentiable positive definite one. If, in addition,*

$$K(t) \equiv -C^{-1}(t)(\dot{C}(t) + B(t)C(t)) \tag{5.2}$$

is a dissipative matrix for all $t \geq 0$, then system (5.1) is stable. Moreover, the inequality

$$||u(t)||^2_{C^n} + ||C^{-1}(t)\dot{u}(t)||^2_{C^n} \leq ||u(0)||^2_{C^n} + ||C^{-1}(0)\dot{u}(0)||^2_{C^n} \ (t \geq 0) \tag{5.3}$$

is true.

Proof: Clearly, system (5.1) is equivalent to the following one

$$\dot{x} = -Bx - C^2y, \ \dot{y} = x,$$

where $B = B(t), C = C(t)$. Put $x(t) = Cz(t)$. Then the latter system takes the form

$$C\dot{z} + \dot{C}z = -BCz - C^2y, \ \dot{y} = Cz.$$

Hence

$$\dot{z} = K(t)z - Cy,\ \dot{y} = Cz.$$

Let $A(t)$ be the matrix of the obtained system. That is,

$$A(t) = \begin{pmatrix} K(t) & -C(t) \\ C(t) & 0 \end{pmatrix}.$$

Then

$$A(t) + A^*(t) = \begin{pmatrix} K(t) + K^*(t) & 0 \\ 0 & 0 \end{pmatrix}.$$

Clearly, if $K(t)$ is dissipative, then $A(t)$ is dissipative, and the required result is due to Corollary 4.2.6. □

Corollary 4.5.2 *Let $B(t)$ be a piecewise continuous matrix, and $C(t)$ be a differentiable positive definite one. If, in addition, $B(t)$ and $C(t)$ commute, and*

$$-C^{-1}(t)\dot{C}(t) - B(t)$$

is a dissipative matrix for all $t \geq 0$, then system (5.1) is stable. Moreover, inequality (5.3) holds.

Example 4.5.3 *Consider the equation*

$$\ddot{u} + b(t)\dot{u} + c^2(t)u = 0 \tag{5.4}$$

with a positive piecewise continuous scalar-valued function $b(t)$ and a positive differentiable one $c(t)$. By the previous corollary, the inequality

$$b(t) + \dot{c}(t)c^{-1}(t) \geq 0\ (t \geq 0)$$

ensures the stability of equation (5.4). Moreover, the inequality

$$u^2(t) + (c^{-1}(t)\dot{u}(t))^2 \leq u^2(0) + (c^{-1}(0)\dot{u}(0))^2\ (t \geq 0)$$

holds for any solution of equation (5.4).

4.6 Scalar linear equations with real characteristic roots

In this section we consider the scalar equation

$$x^{(n)} + p_1(t)x^{(n-1)} + \dots + p_n(t)x = 0\ (t \geq 0) \tag{6.1}$$

with real piecewise continuous coefficients. Introduce the algebraic equation

$$\lambda^n + p_1(t)\lambda^{n-1} + \dots + p_n(t) = 0\ (t \geq 0). \tag{6.2}$$

Lemma 4.6.1 *Suppose that, for a sufficiently large t_0, the roots $\lambda_1(t), ..., \lambda_n(t)$ of (6.2) are real and satisfy the inequalities*

$$\nu_0 \leq \lambda_1(t) \leq \nu_1 \leq \lambda_2(t) \leq ... \leq \nu_{n-1} \leq \lambda_n(t) \leq \nu_n \ (t \geq t_0), \qquad (6.3)$$

where ν_i are some constants $\nu_0 < \nu_1 < ... < \nu_n$. Then there are linearly independent solutions $x_1(t), ..., x_n(t)$ of equation (6.1) with the properties

$$c_j e^{\nu_j t} \leq x_1(t) \leq d_j e^{\nu_j t} \ (t \geq t_0;\ c_j, d_j = const \geq 0;\ j = 1, ..., n). \qquad (6.4)$$

For the proof of that lemma see (Levin, 1969, Section 5).

Corollary 4.6.2 *Let condition (6.3) hold with $\nu_n \leq 0$. Then equation (6.1) is stable.*

4.7 Input-output stability of linear systems

In the present section $||.||_n$ is an arbitrary norm in $\mathbf{C}^n$. Let us consider the system

$$\dot{x} = A(t)x + M(t)u(t), \qquad (7.1)$$

$$y(t) = C(t)x(t) + D(t)u(t) \ \ (t \geq 0), \qquad (7.2)$$

where $x : [0, \infty) \to \mathbf{C}^n$ is the state, $u : [0, \infty) \to \mathbf{C}^m$ $(m \leq n)$ is the input, $y : [0, \infty) \to \mathbf{C}^l$ $(l \leq n)$ is the output. Besides, $A(t), M(t), C(t), D(t)$ are variable matrices.

Definition 4.7.1 *System (7.1), (7.2) is bounded-input, bounded-output stable (BIBO-stable) if there exists a positive constant γ_0, such that for any measurable bounded input, the condition $x(0) = 0$ implies the relation*

$$\sup_{t \geq 0} ||y(t)||_l \leq \gamma_0 \sup_{t \geq 0} ||u(t)||_m.$$

Lemma 4.7.2 *Let the evolution operator $U(t, s)$ of the equation*

$$\dot{x} = A(t)x \ (t \geq 0) \qquad (7.3)$$

satisfy the inequality

$$||U(t, s)||_n \leq ce^{-\nu(t-s)} \ (c, \nu = const > 0) \qquad (7.4)$$

(that is, equation (7.3) is uniformly exponentially stable). In addition, let matrices $M(t), C(t), D(t)$ be uniformly bounded on the positive half-line. Then system (7.1), (7.2) is BIBO-stable

Proof: Equation (7.1) with the zero initial condition is equivalent to the following one:

$$x(t) = \int_0^t U(t,s)M(s)u(s)ds.$$

Hence,

$$\sup_{t\geq 0} \|x(t)\|_n \leq c \sup_{t\geq 0} \|M(t)u(t)\|_n \int_0^\infty e^{-\nu s} ds = c\nu^{-1} \sup_{t\geq 0} \|M(t)u(t)\|_n.$$

That inequality implies the required result. □

Let us consider the converse assertion.

Lemma 4.7.3 *Let system (7.1), (7.2) be BIBO-stable. In addition, let $C(t)$ be an invertible matrix, and let matrices $M(t), C^{-1}(t), D(t)$ be uniformly bounded on the positive half-line. Then for some $c, \nu > 0$, inequality (7.4) is valid.*

Proof: Due to (7.2) we have the relation

$$\sup_{t\geq 0} \|x(t)\|_n \leq \sup_{t\geq 0} \|C^{-1}(t)(y(t) - D(t)u(t))\|_n < \infty.$$

Now (7.4) is due to the well-known Theorem 3.5.1 from (Daleckii and Krein, 1974). □

4.8 Notes

Note 4.1 The multiplicative representation for solutions of linear ordinary differential equations is well-known cf. (Dollard and Friedman, 1979), (Gantmaher, 1959). Theorems 4.2.1, 4.2.3 and 4.3.1 are trivial corollaries of the multiplicative representation. Corollary 4.2.2 was derived by Lozinskii (1958), Corollary 4.2.5 is due to Wazewski (see (Izobov, 1974)). Similar result was established by Winter (1946). **Note 4.2** Theorem 4.4.1 is adapted from (Gil', 1995, Chapter 7). About some other results on perturbations of autonomous systems see for instance (Bellman, 1953, Section 2.5), (Bellman et al., 1985).

Note 4.3 The material of Section 4.6 is based on the paper (Levin, 1969).

Note 4.4 In the book (Rugh, 1996, Chapter 12), the BIBO stability is called the uniform bounded-input, bounded-output stability. A more economical definition is that a system (7.1), (7.2) is bounded-input, bounded-output stable if given $t_0, x(t_0) = 0$ and $u(t)$ satisfying $\|u(t)\|_m \leq \delta$ $(t \geq t_0;\ \delta = const)$, there is a constant $\epsilon > 0$, such that $\|y(t)\|_l \leq \epsilon$ $(t \geq t_0)$. But as proved Desoer and Thomasian (1963), the latter definition and Definition 4.7.1 are equivalent. The results similar to Theorem 4.7.2 can be found in (Rugh, 1996, Chapter 8). About other relevant results see Notes from Chapter 3.

References for Chapter 4

Bellman, R.E. (1953). *Stability Theory of Differential Equations*, McGraw-Hill, New York.

Bellman R., Bentsmann, J., and Meerkov, S.M. (1985). Stability of fast periodic systems, *IEEE Transactions on Automatic Control, 30, No 3*, 289-291.

Daleckii, Yu. L. and Krein, M. G. (1974). *Stability of Solutions of Differential Equations in Banach Space*. Amer. Math. Soc., Providence, R. I.

Desoer, C.A. and Thomasian, A.J. (1963). A note on zero-state stability of linear systems. *Proceedings of the First Allerton Conference on Circuit and System Theory*, University of Illinois, Urbana, Illinois.

Dollard, J.D. and Friedman, Ch. N. (1979). *Product Integration with Applications to Differential Equations*. Encyclopedia of mathematics and its applications; v.10., London, Addison-Wesley Publ. Company.

Gantmaher, F. R. (1967). *Theory of Matrices*. Nauka, Moscow. In Russian

Gil', M. I. (1995). *Norm Estimations for Operator-valued Functions and Applications*. Marcel Dekker, Inc, New York.

Izobov, N. A. (1974). Linear systems of ordinary differential equations, *Itogi Nauki i Tekhniki. Mat. Analis, 12*, 71-146. In Russian.

Levin A. Yu. (1969). Non-oscillations of solutions of the equation $x^{(n)}(t) + a_1(t)x^{(n-1)}(t) + \ldots + p_n(t)x(t)$, *Russian Mathematical Surveys, 24(2)*, 43-96.

Lozinskii, S.M. (1958). Error estimates for numerical solutions of differential equations, *Izvestia VUZ. Matematika, N 5*, 52-90. In Russian.

Rugh, W.J. (1996). *Linear System Theory*. Prentice Hall, Upper Saddle River, New Jersey.

Winter, A. (1946). Asymptotic integration constant, *American Journal of Mathematics, 68*, 125-132.

Chapter 5

NONLINEAR FINITE DIMENSIONAL SYSTEMS WITH AUTONOMOUS LINEAR PARTS

In the present chapter nonlinear finite dimensional systems with separated autonomous linear parts are considered.

Section 5.1 is devoted to Aizerman's conjecture.

In Section 5.2 we discuss the generalized Aizerman conjecture.

A bound for the region of attraction of the stationary solution and global stability conditions are presented in Section 5.3.

Section 5.4 deals with some standard facts on stability and instability in the first approximation.

Conditions for input-output stability are derived in Section 5.5.

The Aizerman conjecture in the input-output version is discussed in Section 5.6.

Section 5.7 is devoted to global feedback stabilization.

Throughout the present chapter $\|.\| = \|.\|_{C^n}$ *means the Euclidean norm.*

5.1 The Aizerman conjecture

Consider in a real Euclidean space $\mathbf{R}^n$ the equation

$$\dot{y} = Ay + b\, f(s,t) \ (s = cy, \ t \geq 0), \tag{1.1}$$

where A is a constant Hurwitz $n \times n$-matrix, b is a column, c is a row, f maps $\mathbf{R}^1 \times [0, \infty)$ into $\mathbf{R}^1$ with the property

$$|f(s,t)| \leq q|s| \text{ for all } s \in \mathbf{R}^1 \text{ and } t \geq 0. \tag{1.2}$$

We will say that the zero solution of system (1.1) is absolutely exponentially stable in the class of nonlinearities (1.2) if there are scalar constants $N, \epsilon > 0$ which do not depend on a concrete form of f (but which depend on q) such that

$$|cy(t)| \leq N \; exp(-\epsilon t) \; \|y(0)\|_{C^n} \; (t \geq 0)$$

for any solution $y(t)$ of (1.1).

Introduce the linear equation

$$\dot{y} = Ay + q_1 bcy \tag{1.3}$$

In 1949 M. A. Aizerman conjectured the following conjecture: under the condition $f(s,t) \equiv f(s)$, for the absolute stability of the zero solution of (1.1) in the class of nonlinearities satisfying the condition

$$0 \leq f(s)/s \leq q \text{ for all } s \in \mathbf{R}^1, s \neq 0 \text{ and } t \geq 0$$

it is necessary and sufficient that (1.3) be asymptotically stable for any $q_1 \in [0, q]$ (Aizerman, 1949). Clearly, class (1.2) is wider than the class considered by Aizerman. This hypothesis caused great interest among the specialists. Counterexamples were set up that demonstrated it was not, in general, true (see (Naredra and Taylor, 1973), (Reissig et al., 1974), (Willems, 1971), and references therein).

Therefore, the following problem arose: to find the class of systems that satisfy Aizerman's hypothesis. To formulate the relevant theorem, let us introduce the transfer function $W(\lambda)$ of the linear part of system (1.1):

$$W(\lambda) = c(\lambda I - A)^{-1} b = P^{-1}(\lambda) L(\lambda) \; (\lambda \in \mathbf{C}).$$

Here $P(\lambda)$ and $L(\lambda)$ are polynomials of degrees n, and $m < n$, respectively. It is assumed that $P(\lambda)$ and $L(\lambda)$ have no common roots. In addition, let

$$K(t) \equiv \frac{1}{2\pi} \int_{-\infty}^{\infty} exp[i\omega t] W(\omega) \, d\omega$$

be the impulse function.

Theorem 5.1.1 *Let the condition*

$$K(t) \geq 0 \textit{ for all } t \geq 0. \tag{1.4}$$

be fulfilled. Then for the absolute exponential stability of the zero solution of (1.1) in the class of nonlinearities (1.2) it is necessary and sufficient that the polynomial $P(\lambda) - qL(\lambda)$ be Hurwitzian.

That theorem immediately follows from Theorem 5.2.1 which is proved in the next section.

Clearly, Theorem 5.1.1 singles out one of the classes of linear parts of systems that satisfy the conjecture of M.A. Aizerman. Now Corollary 2.4.2 and Theorem 5.1.1 imply

Corollary 5.1.2 *Let all roots of $P(\lambda)$ belong to a real segment $[a, b]$ and suppose that*

$$L^{(k)}(\lambda) \geq 0 \ (a \leq \lambda \leq b;\ k = 0, ..., deg L(\lambda)). \tag{1.5}$$

Then for the absolute exponential stability of the zero solution of (1.1) in the class of nonlinearities (1.2) it is necessary and sufficient that the polynomial $P(\lambda) - qL(\lambda)$ be Hurwitzian.

In particular, let $L(\lambda) \equiv 1$, and let all the roots of $P(\lambda)$ be real. Then for the absolute exponential stability of the zero solution of (1.1) in the class of nonlinearities (1.2) it is necessary and sufficient that the polynomial $P(\lambda) - q$ be Hurwitzian.

Now we combine Lemma 2.4.4 with Theorem 5.1.1. This yields.

Corollary 5.1.3 *Let $W(\lambda) = P_3^{-1}(\lambda)$, where $P_3(\lambda) = \lambda^3 + b_1\lambda^2 + b_2\lambda + b_3$ is a Hurwitz 3-rd order polynomial. Let, besides, a pair of its nonreal conjugate zeros (if they exist) lie to the left of the real root. Then for the absolute exponential stability of the zero solution of (1.1) in the class of nonlinearities (1.2) it is necessary and sufficient that the polynomial $P_3(\lambda) - q$ be Hurwitzian.*

Certainly one can combine Theorem 5.1.1 also with Theorem 2.4.1 and Corollary 2.4.5.

Example 5.1.4 *Consider the equation*

$$\frac{d^2x}{dt^2} + a_1\frac{dx}{dt} + a_2x = b_1\phi(x) + \frac{d\phi(x,t)}{dt} \ (a_1, a_2, b_1 = const > 0), \tag{1.6}$$

where $\phi(s, t)$ satisfies the condition

$$|\phi(s)| \leq q|s| \ (s \in \mathbf{R},\ t \geq 0). \tag{1.7}$$

Let the polynomial

$$P(\lambda) = \lambda^2 + a_1\lambda + a_2$$

have real roots $\lambda_1 \leq \lambda_2 < 0$. Then under the condition

$$b_1 + \lambda_1 \geq 0 \tag{1.8}$$

we have

$$L(\lambda) = \lambda + b_1 \geq 0,\ \frac{d}{d\lambda}L(\lambda) = 1 \ (\lambda_1 \leq \lambda \leq \lambda_2).$$

Thus, if the polynomial

$$P(\lambda) - qL(\lambda) = \lambda^2 + a_1\lambda + a_2 - q(b_1 + \lambda) = \lambda^2 + (a_1 - q)\lambda + a_2 - qb_1$$

is Hurwitzian, then by Corollary 5.1.2, the zero solution of equation (1.6) is absolutely exponentially stable in the class of nonlinearities (1.7). Therefore, conditions (1.8), $a_1 - q > 0$ and $a_2 - qb_1 > 0$ assure the absolute stability.

Example 5.1.5 *Consider the equation*

$$\frac{d^3x}{dt^3} + a_1\frac{d^2x}{dt^2} + a_2\frac{dx}{dt} + a_3x = b_1\phi(x) + \frac{d\phi(x)}{dt}$$

$$(a_1, a_2, a_3, b_1 = const > 0), \tag{1.9}$$

where $\phi(x)$ satisfies condition (1.7). Let the polynomial

$$P(\lambda) = \lambda^3 + a_1\lambda^2 + a_2\lambda + a_3$$

have real roots $\lambda_1 \leq \lambda_2 \leq \lambda_3 < 0$. Then under condition (1.8), relations

$$L(\lambda) = \lambda + b_1 \geq 0, \ \frac{d}{d\lambda}L(\lambda) = 1 \ \ (\lambda_1 \leq \lambda \leq \lambda_3)$$

hold. That is, if the polynomial

$$P(\lambda) - qL(\lambda) = \lambda^3 + a_1\lambda^2 + a_2\lambda + a_3 - q(b_1 + \lambda) =$$

$$\lambda^3 + a_1\lambda^2 + (a_2 - q)\lambda + a_3 - qb_1$$

is Hurwitzian, then by Corollary 5.1.2, the zero solution of equation (1.9) is absolutely exponentially stable in the class of nonlinearities (1.7).

5.2 The generalized Aizerman conjecture

5.2.1 Statement of the result

Now we consider the system

$$\dot{y} = Ay + BF(s,t), \ (t \geq 0, s = Cy) \tag{2.1}$$

where A is a linear operator in the real Euclidean space $\mathbf{R}^n$,

$$B : \mathbf{R}^m \to \mathbf{R}^n \text{ and } C : \mathbf{R}^n \to \mathbf{R}^m$$

are linear operators also, and $m \leq n$. Besides, A is a Hurwitz matrix, and $F(s,t)$ maps $\mathbf{R}^m \times [0,\infty)$ into $\mathbf{R}^m$ with the property

$$|F(x,t)| \leq Q|x| \ (x \in \mathbf{R}^m, \ t \geq 0). \tag{2.2}$$

Here Q is a non-negative $m \times m$-matrix ; $|F(x,t)|$ and $|x|$ are vectors whose coordinates are the moduli of the vectors $F(x,t)$ and x, respectively.

Let $W(\lambda)$ be the transfer matrix of the linear part of the system from the input F to the output y:

$$W(\lambda) = C(\lambda I - A)^{-1}B.$$

It is simple to check that $W(\lambda) = P^{-1}(\lambda)L(\lambda)$, where $L(\lambda)$ and $P(\lambda)$ are matrices whose elements are polynomials in λ. So each element $W_{jk}(\lambda)$ of $W(\lambda)$ $(j,k = 1,...,m)$ is a rational quotient function and the degree of the numerator of this function is less than the degree of the denominator. It is assumed that $W(\lambda)$ is non-degenerate, i.e. the poles of this matrix are not canceled by the zeros. In particular, (2.1) can possess the form

$$P(D)y = L(D)F(y,t) \quad (t \geq 0;\ D = d/dt).$$

Let

$$K(t) = \frac{1}{2\pi}\int_{-i\infty}^{i\infty} exp[i\omega t]W(i\omega)d\omega \tag{2.3}$$

be the impulse matrix of the linear part of the system.

We will say that the zero solution of the system (2.1) is *absolutely exponentially stable in the class of nonlinearities (2.2)* if there are scalar constants $N, \epsilon > 0$ independent of the form of F, such that

$$||Cy(t)||_{R^m} \leq N exp(-\epsilon t)||y(0)||_{R^n} \text{ for all } t \geq 0$$

and any solution $y(t)$ of system (2.1).

Theorem 5.2.1 *Let $K(t) \geq 0$ for all $t \geq 0$. Then for the absolute exponential stability of the zero solution of equation (2.1) in the class of nonlinearities (2.2) it is necessary and sufficient that the polynomial $det(P(\lambda) - L(\lambda)Q)$ be a Hurwitz one.*

The proof is presented in Subsection 5.2.3. Clearly, Theorem 5.2.1 generalizes Theorem 5.1.1.

Furthermore Corollary 2.4.2 and Theorem 5.2.1 imply

Corollary 5.2.2 *Let each element $W_{jk}(\lambda)$ of the transfer function $W(\lambda)$ $(j,k = 1,...,m)$ be equal to $W_{jk}(\lambda) = p_{jk}^{-1}(\lambda)m_{jk}(\lambda)$, where $p_{jk}(\lambda)$ and $m_{jk}(\lambda)$ are polynomials in λ. Let all roots of $p_{jk}(\lambda)$ belong to a real segment $[a,b]$ and suppose that*

$$\frac{d^\nu m_{jk}(\lambda)}{d\lambda^\nu} \geq 0 \ (\lambda \in [a,b],\ t \geq 0;\ \nu = 0,1,...,deg\ m_{jk}(\lambda)). \tag{2.4}$$

Then for the absolute exponential stability of the zero solution of (2.1) in the class of nonlinearities (2.2) it is necessary and sufficient that the polynomial $det(P(\lambda) - L(\lambda)Q)$ be a Hurwitz one.

In particular, let $L(\lambda) = I$, and let all the eigenvalues of A be real. Then for the absolute exponential stability of the zero solution of (2.1) in the class of nonlinearities (2.2) it is necessary and sufficient that the polynomial $det(P(\lambda) - Q)$ be a Hurwitz one.

Certainly one can combine Theorem 5.2.1 also with Theorem 2.4.1, Lemma 2.4.4 and Corollary 2.4.5.

5.2.2 Example

Consider the system

$$P_{11}(D)x_1 + P_{12}(D)x_2 = \phi_1(x_1, x_2, t),$$

$$P_{22}(D)x_2 = \phi_2(x_1, x_2, t), \tag{2.5}$$

where $P_{jk}(\lambda)$ are polynomials. The scalar-valued functions ϕ_1 and ϕ_2 satisfy the conditions

$$|\phi_j(z_1, z_2, t)| \le q_{j1}|z_1| + q_{j2}|z_2| \ (j = 1, 2) \tag{2.6}$$

for all $z_1, z_2 \in \mathbf{R}^1$ and $t \ge 0$. Assume that all the zeros of $P_{11}(\lambda)$ and $P_{22}(\lambda)$ belong to a real segment $[a, b]$ $(a < b \le 0)$. In the considered case

$$L(\lambda) = I;\ Q = (q_{jk})_{j,k=1}^2;\ W(\lambda) = (W_{jk}(\lambda))_{j,k=1}^2,$$

where

$$W_{jj}(\lambda) = P_{jj}^{-1}(\lambda)\ (j = 1, 2),\ W_{12}(\lambda) = -P_{11}^{-1}(\lambda)P_{12}(\lambda)P_{22}^{-1}(\lambda),\ W_{21}(\lambda) = 0.$$

In addition,

$$P(\lambda) = \begin{pmatrix} P_{11}(\lambda) & P_{12}(\lambda) \\ 0 & P_{22}(\lambda) \end{pmatrix}.$$

If

$$P_{12}^{(k)}(\lambda) \le 0\ (k = 0, 1, ..., deg P_{12}(\lambda)) \tag{2.7}$$

for all $\lambda \in [a, b]$, and $t \ge 0$, then due to Corollary 5.2.2 for the absolute exponential stability of the zero solution of system (2.5) in the class of nonlinearities (2.6) it is necessary and sufficient that $det(P(\lambda) - Q)$ be a Hurwitz polynomial.

Note that condition (2.7) is certainly satisfied if

$$P_{12}(\lambda) = c_1\lambda + c_2 \text{ and } c_1 b + c_2 \le 0.$$

5.2.3 Proof of Theorem 5.2.1

First, we shall prove the sufficiency of the conditions. Note that $K(t) = C\exp[At]B$. Furthermore, equation (2.1) is equivalent to following one

$$y(t) = exp[At]y(0) + \int_0^t exp[A(t-\tau)]BF(s(\tau), \tau)d\tau,\ (s(t) = Cy(t)).$$

Multiplying this equation by C, we have

$$s(t) = z(t) + \int_0^t K(t-\tau)F(s(\tau), \tau)d\tau$$

with $z(t) = Cexp[At]y(0)$. Taking into account condition (2.2), we arrive at the inequality

$$|s(t)| \le |z(t)| + \int_0^t K(t-\tau)Q|s(\tau)|d\tau$$

From Lemma 1.7.1 it follows that

$$|s(t)| \leq \eta(t). \tag{2.8}$$

where η is the solution of the equation

$$\eta(t) = |z(t)| + \int_0^t K(t-\tau)Q\eta(\tau)d\tau.$$

We will solve this equation by the Laplace transformation. Since the transform of the convolution is equal to the product of the transforms, after straightforward calculations and taking into account (2.3), we get the equation

$$\overline{\eta}(\lambda) = g(\lambda) + P^{-1}(\lambda)L(\lambda)Q\overline{\eta}(\lambda),$$

where $\overline{\eta}(\lambda)$ and $g(\lambda)$ are the Laplace transforms of $\eta(t)$ and of $|z(t)|$, respectively, λ is the dual variable. Thus we can write down

$$\eta(t) = \frac{1}{2\pi i}\int_{-i\infty}^{i\infty} \exp[\lambda t](P(\lambda) - L(\lambda)Q)^{-1}P(\lambda)g(\lambda)d\lambda.$$

It is easy to see that

$$|z(t)| \leq exp[\alpha(A)t]G(t),$$

where $G(t)$ is some polynomial non-negative matrix. Recall that $\alpha(A)$ is the real part of the extreme right-hand eigenvalue of A. Thus, thanks the residue theorem

$$\eta(t) \leq \exp(mt)G_1(t) \text{ for all } t \geq 0,$$

where $G_1(t)$ is a polynomial matrix, and $m = max\{\alpha(A), \beta\}$. Here β is the real part of the extreme right-hand zero of the polynomial $det(P(\lambda) - L(\lambda)Q)$. This and (2.8) prove the sufficiency of conditions.

To prove the necessity of the conditions put $F(s) = Qs$. Then (2.1) take the form

$$y = Ay + BQs, \ (t \geq 0, s = Cy).$$

Hence,

$$\lambda\overline{y}(\lambda) - y(0) = A\overline{y} + BQC\overline{y}(\lambda),$$

where $\overline{y}(\lambda)$ is the Laplace transform of $y(t)$. Consequently,

$$C\overline{y}(\lambda) = -C(A - I\lambda)^{-1}(y(0) + BQC\overline{y}(\lambda)) = h(\lambda) + W(\lambda)BQC\overline{y}(\lambda)$$

where $h(\lambda) = -C(A - I\lambda)^{-1}y(0)$. Therefore,

$$C\overline{y}(\lambda) = h(\lambda) + P^{-1}(\lambda)L(\lambda)QC\overline{y}(\lambda)$$

and

$$C\overline{y}(\lambda) = (P(\lambda) - L(\lambda)Q)^{-1}P(\lambda)h(\lambda).$$

From this the result immediately follows. □

5.3 Region of attraction and global stability

5.3.1 Statement of the results

Let us consider in $\mathbf{C}^n$ the equation

$$\dot{x} = Ax + F(x,t),\ t \geq 0, \tag{3.1}$$

where A is a constant $n \times n$-matrix, and F maps $\Omega(r) \times [0,\infty)$ into $\mathbf{C}^n$ with the property

$$\|F(h,t)\| \leq \nu\|h\| \text{ for all } h \in \Omega(r) \text{ and } t \geq 0. \tag{3.2}$$

Here $\nu = const > 0$, and $\Omega(r) = \{h \in \mathbf{C}^n : \|h\| \leq r\}$. Introduce the algebraic equation

$$z^n = \nu \sum_{j=0}^{n-1} \frac{g^j(A)}{\sqrt{j!}} z^{n-j-1}, \tag{3.3}$$

and denote by $z(\nu, A)$ the extreme right-hand (unique positive and simple) zero of that equation. Recall that

$$\Gamma(A) = \sum_{j=0}^{n-1} \frac{g^j(A)}{|\alpha(A)|^{j+1}\sqrt{j!}},$$

and

$$\chi(A) = \max_{t \geq 0} exp[\alpha(A)t] \sum_{j=0}^{n-1} \frac{g^j(A)t^j}{\sqrt{j!}}.$$

Theorem 5.3.1 *Under condition (3.2), let the matrix $A + z(\nu, A)I$ be a Hurwitz one. Then the zero solution of equation (3.1) is asymptotically stable. Moreover, the inequality*

$$\nu\Gamma(A) < 1 \tag{3.4}$$

is valid, and any vector x_0 satisfying the condition

$$(1 - \nu\Gamma(A))^{-1}\, \chi(A)\|x_0\| < r, \tag{3.5}$$

belongs to a region of attraction of the zero solution. Additionally, the solution $x(t)$ of (3.1) with $x(0) = x_0$ subordinates the estimate

$$\|x(t)\| \leq \frac{\chi(A)\|x_0\|}{1 - \nu\Gamma(A)}\ (t \geq 0). \tag{3.6}$$

In particular, let (3.2) hold with $\Omega(r) = \mathbf{C}^n$ (i.e. $r = \infty$), then the zero solution of (3.1) is globally stable if $A + z(\nu, A)I$ is a Hurwitz matrix.

Certainly, by Lemma 1.11.1 or Corollary 1.11.2 one can estimate the root $z(\nu, A)$.

5.3.2 Proofs

Lemma 5.3.2 *Under condition (3.2), let the inequality (3.4) hold. Then the zero solution of equation (3.1) is asymptotically stable. Moreover, any vector x_0 satisfying inequality (3.5) belongs to the region of attraction of the zero solution. Additionally, the solution $x(t)$ of (3.1) with $x(0) = x_0$ subordinates estimate (3.6).*

Proof: Rewrite (3.1) in the form

$$x(t) = e^{At}x(0) + \int_0^t e^{A(t-s)}F(x,s)ds.$$

Set for a fixed t_0

$$c = \sup_{t \le t_0} \|x(t)\|.$$

Take into account that solutions continuously depend on initial vectors. Then (3.5) gives $c \le r$ if t_0 is sufficiently small, and by (3.2)

$$c \le \|e^{A(t-s)}\|\|x(0)\| + \nu c \int_0^t \|e^{A(t-s)}\|ds \ (t \le t_0).$$

By Corollary 2.1.7 we get

$$c \le \chi(A)\|x(0)\| + \Gamma(A)\nu c,$$

Now (3.4) yields the relation

$$c \le \chi(A)(1 - \nu\Gamma(A))^{-1}\|x(0)\| \ (t \in [0, t_0]).$$

Due to (3.5) this bound can be extended from $[0, t_0]$ to the whole positive half-line. This bound yields the stability in the sense of Lyapunov. The asymptotic stability can be proved by a small perturbation of equation (3.1). □

Lemma 5.3.3 *Let $z(\nu, A)$ be the extreme right-hand zero of the algebraic equation (3.3). Then the condition*

$$\alpha(A) + z(\nu, A) < 0$$

implies inequality (3.4).

Proof: The hypothesis of the lemma entails the relation

$$|\alpha(A)| > z(\nu, A). \tag{3.7}$$

Dividing (3.3) by z^{n+1}, and taking into account that $z(\nu, A)$ is its root, we arrive at the equality

$$1 = \nu \sum_{j=0}^{n-1} \frac{g^j(A)}{z^{j+1}(\nu, A)\sqrt{j!}}.$$

But according to (3.7)

$$\nu\Gamma(A) = \nu \sum_{j=0}^{n-1} \frac{g^j(A)}{|\alpha(A)|^{j+1}\sqrt{j!}} <$$

$$\nu \sum_{j=0}^{n-1} \frac{g^j(A)}{z^{j+1}(\nu, A)\sqrt{j!}} = 1.$$

This is the stated result. □

The assertion of Theorem 5.3.1 immediately follows from Lemma 5.3.2 and Lemma 5.3.3

5.4 Stability and instability in the first approximation

5.4.1 Statement of the result

Again consider in $\mathbf{C}^n$ the equation

$$\dot{x} = Ax + F(x,t) \ (t \geq 0), \tag{4.1}$$

where A is a constant $n \times n$-matrix, and F maps $\mathbf{C}^n \times [0, \infty)$ into $\mathbf{C}^n$. We will say that (4.1) *is a quasilinear equation* if

$$\lim_{h \to 0} \frac{||F(h,t)||}{||h||} = 0 \tag{4.2}$$

uniformly in t.

Theorem 5.4.1 *(Lyapunov). Let (4.1) be a quasilinear equation. Then if A is a Hurwitz matrix, the zero solution of (4.1) is asymptotically stable. Conversely, if A has an eigenvalue lying in the interior of the right half-plane, then the zero solution of (4.1) is unstable.*

5.4.2 Proofs

Assume that

$$||F(h,t)|| \leq \nu||h|| \text{ for all } h \in \mathbf{C}^n \text{ and } t \geq 0. \tag{4.3}$$

Here $\nu \geq 0$ is a constant. In addition, suppose

$$\sigma(A) = \sigma_+ \cup \sigma_- \tag{4.4}$$

where σ_- lies in the interior of the left half-plane, and σ_+ lies in the interior of the right half-plane. Put

$$\alpha_\pm = max Re\sigma_\pm \text{ and } \beta_\pm = min Re\sigma_\pm.$$

Lemma 5.4.2 *Suppose (4.3) is fulfilled, and the spectral sets σ_+ and σ_- are not empty. Then there are $\nu_0 > 0$ and initial vectors $x_\pm(0)$ such that under $\nu \le \nu_0$ the corresponding solutions $x_\pm(t)$ of equation (4.1) are subject to estimates*

$$||x_\pm(t)|| \le c||x_\pm(0)||e^{\tilde\alpha_\pm t} \ (t \ge 0,\ c = const),$$

with $\tilde\alpha_+ \ge \alpha_+ > 0$, and $\alpha_- \le \tilde\alpha_- < 0$.

Proof: Let P_+ and P_- be the spectral projectors corresponding to σ_+ and σ_- . That is,

$$P_\pm = \frac{1}{2\pi i}\int_{\Gamma_\pm}(\lambda I - A)^{-1}d\lambda,$$

where the contour Γ_- (Γ_+) surrounds the set σ_- (σ_+), and does not surround σ_+ (σ_-). Multiplying (4.1) by P_+ and P_-, we get

$$dx_+/dt = A_+x_+ + F_+(x,t),$$

$$dx_-/dt = A_-x_- + F_-(x,t),\ (t \ge 0) \tag{4.5}$$

with the notation

$$x_\pm = P_\pm x,\ A_\pm = P_\pm A,\ F_\pm(x,t) = P_\pm F(x,t).$$

According to (4.3)

$$||F_\pm(x,t)|| \le q(||x_+|| + ||x_-||).$$

Here

$$q = \nu \max\{||P_+||, ||P_-||\}.$$

The system (4.5) can be written as

$$x_\pm(t) = e^{tA_\pm}x_\pm(0) + \int_0^t e^{(t-s)A_\pm}F_\pm(x(s),s)ds.$$

Therefore

$$||x_+(t)|| \le ||e^{tA_+}||||x_+(0)|| + q\int_0^t ||e^{(t-s)A_+}||(||x_+(s)|| + ||x_-(s)||)ds,$$

$$||x_-(t)|| \le ||e^{tA_-}||||x_-(0)|| + q\int_0^t ||e^{(t-s)A_-}||(||x_+(s)|| + ||x_-(s)||)ds.$$

But according to Corollary 2.1.6 for any $\tilde\alpha_\pm > \alpha_\pm$ there is a constant M such that

$$||e^{tA_\pm}|| \le Me^{t\tilde\alpha_\pm} \ (t \ge 0).$$

Take $\tilde\alpha_\pm$ in such a way that $\tilde\alpha_+ > 0$ and $\tilde\alpha_- < 0$. We get the system of inequalities

$$||x_+(t)|| \le Me^{t\tilde\alpha_+}||x_+(0)|| + Mq\int_0^t e^{(t-s)\tilde\alpha_+}(||x_+(s)|| + ||x_-(s)||)ds,$$

$$\|x_-(t)\| \leq Me^{t\tilde{\alpha}_-}\|x_-(0)\| + Mq\int_0^t e^{(t-s)\tilde{\alpha}_-}(\|x_+(s)\| + \|x_-(s)\|)ds.$$

By Lemma 1.7.1

$$e^{-t\tilde{\alpha}_-}\|x_-(t)\| \leq \eta_-(t), \text{ and } e^{-t\tilde{\alpha}_+}\|x_+(t)\| \leq \eta_+(t), \tag{4.6}$$

where $\eta_+(t), \eta_-(t)$ are solutions of the system

$$\eta_+(t) = M\|x_+(0)\| + Mq\int_0^t (\eta_+(s) + \eta_-(s))ds,$$

$$\eta_-(t) = M\|x_-(0)\| + Mq\int_0^t (\eta_+(s) + \eta_-(s))ds.$$

That system is equivalent to the following one:

$$\dot{\eta}_+(t) = Mq(\eta_+(t) + \eta_-(t)),$$

$$\dot{\eta}_-(t) = Mq(\eta_-(t) + \eta_+(t)). \tag{4.7}$$

The simple calculations show that the eigenvalues of the matrix B corresponding to system (4.7) are

$$\lambda_{1,2}(B) = O(q) \ (q \to 0).$$

Let d_+ and d_- be the normed eigen-vectors corresponding to $\lambda_1(B)$ and $\lambda_2(B)$, respectively. Then the solution $\eta_+(t)$ of system (4.7) with the initial condition $\eta_+(0) = d_+$ is

$$\eta_+(t) = d_+e^{\lambda_1(B)t} = d_+e^{0(q)t},$$

and the solution $\eta_-(t)$ with $\eta_-(0) = d_-$ equals

$$\eta_-(t) = d_-e^{\lambda_2(B)t} = d_-e^{0(q)t}.$$

Now the result is due to relation (4.6). □
Now we are going to establish lower solution estimates.

Lemma 5.4.3 *Suppose (4.3) is fulfilled, and the spectral sets σ_+ and σ_- are not empty. Then there are $\nu_0 > 0$ and initial vectors $x_\pm(0)$ such that under the condition $\nu \leq \nu_0$ the corresponding solutions $x_\pm(t)$ of equation (4.1) are subject to estimates*

$$\|x_\pm(t)\| \geq c_1\|x_\pm(0)\|e^{\tilde{\beta}_\pm t} \ (t \geq 0, \ c_1 = const). \tag{4.8}$$

with $\tilde{\beta}_+ > 0$, and $\tilde{\beta}_- < 0$.

Proof: Fox a fixed $T > 0$, put in (4.1)
$\tau = T - t$, $x(T-\tau) = y(\tau)$, and $-F(y, T-\tau) = F_1(y, \tau)$. We obtain the equation

$$dy(\tau)/d\tau = -Ay(\tau) + F_1(y(\tau), \tau), \ (0 \le \tau \le T).$$

The eigenvalues of $-A$ are $-\lambda_1(A), ..., -\lambda_n(A)$. Hence according to (4.4)

$$\sigma(-A) = (-\sigma_+) \cup (-\sigma_-)$$

and $\sup Re(-\sigma_\pm) = -\beta_\pm$. Moreover, due to (4.3)

$$||F_1(h, \tau)|| \le \nu||h|| \text{ for all } h \in \mathbf{C}^n \text{ and } \tau \le T.$$

Lemma 5.4.2 implies that there is ν_0 such that for a $\nu \le \nu_0$, there are corresponding solution $y_\pm(\tau)$ satisfying the inequality

$$||y_\pm(\tau)|| \le const||y_\pm(0)||e^{(-\beta_\pm+\epsilon)\tau}$$

with any $\epsilon > 0$. But $y_\pm(0) = x_\pm(T)$ and $y_\pm(T) = x_\pm(0)$. Therefore for $T = t$, we obtain the desired inequality. Taking t instead of T, we arrive at the result. □

Proof of Theorem 5.4.1: Due to (4.2) we can chose r in such a way that (4.2) holds with a sufficiently small ν. Now the sufficiency of conditions is due to Theorem 4.3.1.

Let us prove the necessity. Let (4.4) hold, and σ_+ is not empty. Take

$$F_0(h, t) = F(h, t) \text{ if } h \in \Omega(r) \text{ and } ||F_0(h, t)|| \le \nu||h|| \text{ for all } h \in \mathbf{C}^n.$$

This extension is correct thanks to the Urysohn lemma (see e.g. (Gemignani, 1972, p.108)).

First, assume that A has no the eigenvalues on the imaginary axis. Then thanks to Lemma 5.4.3 there is a solution of the equation

$$\dot{x} = Ax + F_0(x, t)$$

with an arbitrary small initial condition, which goes out $\Omega(r)$. But in $\Omega(r)$ solutions of the latter equation and equation (4.1) coincide. That is, there is a solution of equation (4.1) with an arbitrary small initial condition which goes out $\Omega(r)$. This proves the instability. If A has the eigenvalues on the imaginary axis, then we can reduce the considered case to the previous one by a small perturbation. □

5.5 Input-output stability

5.5.1 Statement of the result

Let us consider the system

$$\dot{x} = Ax + F(x, u, t), \tag{5.1}$$

$$y(t) = Cx(t) + Du(t) \quad (t \geq 0), \tag{5.2}$$

where $x : [0,\infty) \to \mathbf{C}^n$ is the state, $u : [0,\infty) \to \mathbf{C}^m$ $(m \leq n)$ is the input, $y : [0,\infty) \to \mathbf{C}^l$ $(l \leq n)$ is the output. Besides, A, C, D are constant matrices, and F maps $\mathbf{C}^n \times \mathbf{C}^m \times [0,\infty)$ into $\mathbf{C}^n$.

Definition 5.5.1 *We will say that system (5.1), (5.2) is bounded-input, bounded-output stable (BIBO-stable) if for any $\delta > 0$ there is a constant $\epsilon > 0$, such that the conditions*

$$x(0) = 0 \tag{5.3}$$

and $\|u(t)\|_{\mathbf{C}^m} \leq \delta$ $(t \geq 0)$, imply the inequality $\|y(t)\|_{\mathbf{C}^l} \leq \epsilon$ $(t \geq 0)$. System (5.1), (5.2) is BIBO-stable with finite gain, if there exist positive constants γ_0, b, such that for any measurable bounded input, the relation

$$\sup_{t\geq 0} \|y(t)\|_{\mathbf{C}^l} \leq \gamma_0 \sup_{t\geq 0} \|u(t)\|_{\mathbf{C}^m} + b$$

is valid provided condition (5.3) holds. Furthermore, system (5.1), (5.2) is BIBO-stable with finite gain and zero bias (wb) if there exists a positive constant γ_0, such that for any measurable bounded input, the relation

$$\sup_{t\geq 0} \|y(t)\|_{\mathbf{C}^m} \leq \gamma_0 \sup_{t\geq 0} \|u(t)\|_{\mathbf{C}^m}$$

is valid provided condition (5.3) holds.

According to Note 4.4 in the linear case this definition is in agreement with Definition 4.7.1.

We will assume that A is a Hurwitz matrix and that there are constants $\nu, \mu > 0$, such that

$$\|F(h,w,t)\|_{C^n} \leq \nu\|h\|_{C^n} + \mu\|w\|_{C^m} \quad (h \in \mathbf{C}^n;\ w \in \mathbf{C}^m,\ t \geq 0). \tag{5.4}$$

Theorem 5.5.2 *Let the conditions (5.4) and*

$$\nu\Gamma(A) \equiv \nu \sum_{j=0}^{n-1} \frac{g^j(A)}{\sqrt{j!}|\alpha(A)|^{j+1}} < 1$$

hold. Then system (5.1), (5.2) is BIBO-stable wb.

This theorem is a particular case of Theorem 6.3.1 below. So we omit the proof.

Let $z(A,\nu)$ be the extreme right-hand zero of the algebraic equation (3.3).

Theorem 5.5.3 *Let condition (5.4) be fulfilled. If, in addition, the matrix $A + z(A,\nu)I$ is Hurwitzian, then system (5.1), (5.2) is BIBO-stable wb.*

This theorem is a particular case of Corollary 6.4.4 below. Moreover, due to Lemma 1.11.1 the condition

$$w(\nu, A) \equiv \nu \sum_{j=0}^{n-1} \frac{g^j(A)}{\sqrt{j!}} < 1 \tag{5.5}$$

implies the inequality $z(A, \nu) \leq w^{1/n}(\nu, A)$. Now Theorem 5.5.3 yields

Corollary 5.5.4 *Let conditions (5.4) and (5.5) hold. If, in addition, the matrix $A + w^{1/n}(\nu, A)I$ is a Hurwitz one, then system (5.1), (5.2) is BIBO-stable wb.*

5.5.2 Example

Consider the single-input, single-output system

$$\ddot{y} + 2p\dot{y} + qy = f(t, y, \dot{y}, u(t)), \tag{5.6}$$

where p, q are positive constants. Besides, it is assumed that the scalar-valued function $f : [0, \infty) \times \mathbf{R}^3 \to \mathbf{R}$ satisfies the condition

$$f^2(t, y, z, w) \leq \nu^2(y^2 + z^2) + \mu^2 w^2 \tag{5.7}$$

for all $t \geq 0$; $y, z, w \in \mathbf{R}$. In the considered case according to Example 2.1.4, $g(A) \leq 1 + q$. In addition, $\alpha(A) = Re\,\{-p + (p^2 - q)^{1/2}\}$. Assume that

$$\nu\ (|\alpha(A)|^{-1} + (1+q)|\alpha(A)|^{-2}) < 1.$$

Then, due to Theorem 5.5.2, system (5.6) under condition (5.7) is BIBO - stable wb.

5.6 The input-output version of Aizerman's conjecture

5.6.1 Statement of the result

Consider the system

$$\dot{x}(t) = Ax(t) + b\ f(s(t), u(t), t)\ (s(t) = cx(t),\ t \geq 0), \tag{6.1}$$

$$y(t) = Mx(t) + Nu(t), \tag{6.2}$$

where $x(t) : [0, \infty) \to \mathbf{R}^n$ is the state, $y(t) : [0, \infty) \to \mathbf{R}^l$ is the output, $u(t) : [0, \infty) \to \mathbf{R}^m$ is the input; A, M, N are constant matrices, b is a column, c is a row, f maps $\mathbf{R}^1 \times \mathbf{R}^m \times [0, \infty)$ into $\mathbf{R}^1$ with the following property: there are constants $q_S, q_I \geq 0$ such that

$$|f(s, h, t)| \leq q_S|s| + q_I\|h\|_{R^m} \text{ for all } s \in \mathbf{R}^1, h \in \mathbf{R}^m \text{ and } t \geq 0. \tag{6.3}$$

In addition, A is a Hurwitz matrix.

The input-output version of Aizerman's conjecture is that system (6.1), (6.2) under (6.3) is BIBO-stable if (1.3) is asymptotically stable for any $q_1 \in [0, q_S]$. In this section it is proved that the considered system with positive impulse functions satisfies the Aizerman conjecture in the input-output version. Let $W(\lambda) = c(\lambda I - A)^{-1}b = P^{-1}(\lambda)L(\lambda)$ be the transfer function. Here $P(\lambda), L(\lambda)$ are polynomials. Recall that the impulse function $K(t)$ is defined in Section 5.1.

Theorem 5.6.1 *Let conditions (6.3) be fulfilled. In addition, let the impulse function of the linear part of (6.1) be positive. Then system (6.1), (6.2) is BIBO-stable wb provided the polynomial $P(\lambda) - q_S L(\lambda)$ is Hurwitzian.*

The proof is presented in Subsection 5.6.3. Note that it is simple to check that the Hurwitzness of the polynomial $P(\lambda) - q_S L(\lambda)$ is equivalent to the asymptotic stability of the linear system

$$\dot{x}(t) = Ax(t) + q_S bc\, x(t). \tag{6.4}$$

Corollary 2.4.2 and Theorem 5.6.1 imply

Corollary 5.6.2 *Let all the roots of $P(\lambda)$ belong to a real segment $[a, b]$ and suppose that*

$$L^{(k)}(\lambda) \geq 0 \ (a \leq \lambda \leq b;\ k = 0, ..., deg L(\lambda)\ )$$

Then system (6.1), (6.2) is BIBO-stable wb provided the polynomial $P(\lambda) - q_S L(\lambda)$ is Hurwitzian.

In particular, let $L(\lambda) \equiv 1$, and let all the roots of $P(\lambda)$ be real. Then system (6.1), (6.2) is BIBO-stable wb provided the polynomial $P(\lambda) - q_S$ is Hurwitzian.

Now we combine Lemma 2.4.4 with Theorem 5.6.1. This yields.

Corollary 5.6.3 *Let $W(\lambda) = P_3^{-1}(\lambda)$ where $P_3(\lambda) = \lambda^3 + b_1\lambda^2 + b_2\lambda + b_3$ is a Hurwitz 3-rd order polynomial. Let, besides, a pair of its nonreal conjugate zeros (if they exist) lie to the left of the real root. Then system (6.1), (6.2) is BIBO-stable wb provided the polynomial $P(\lambda) - q_S$ is Hurwitzian.*

Certainly one can combine Theorem 5.6.1 also with Theorem 2.4.1 and Corollary 2.4.5.

5.6.2 Examples

Example 5.6.4 *Consider the following single-input, single output system :*

$$\frac{d^2x}{dt^2} + a_1\frac{dx}{dt} + a_2x = b_1\phi(x, u) + \frac{d\phi(x, u)}{dt} \quad (a_1, a_2, b_1 = const > 0),$$

$$y(t) = Mx(t) + Nu(t), \tag{6.5}$$

where $\phi(s, u)$ satisfies the condition

$$|\phi(s, z)| \leq q_S|s| + q_I|z| \ (s, z \in \mathbf{R}). \tag{6.6}$$

Let the polynomial $P(\lambda) = \lambda^2 + a_1\lambda + a_2$ have real roots $\lambda_1 < \lambda_2 < 0$. Then under the condition

$$b_1 + \lambda_1 \geq 0 \tag{6.7}$$

we have $L(\lambda) = \lambda + b_1 \geq 0 \ (\lambda_1 \leq \lambda \leq \lambda_2)$, $\frac{d}{d\lambda}L(\lambda) = 1$. Thus, if the polynomial

$$P(\lambda) - qL(\lambda) = \lambda^2 + a_1\lambda + a_2 - q(b_1 + \lambda) = \lambda^2 + (a_1 - q)\lambda + a_2 - qb_1$$

is Hurwitzian, then by Corollary 5.6.2, system (6.5) under condition (6.6) is BIBO-stable wb provided the polynomial $P(\lambda) - q_S L(\lambda)$ is Hurwitzian. Therefore, the conditions (6.7), $a_1 - q > 0$ and $a_2 - qb_1 > 0$ assure the input-output stability w.b.

Example 5.6.5 *Consider the equation*

$$\frac{d^3x}{dt^3} + a_1\frac{d^2x}{dt^2} + a_2\frac{dx}{dt} + a_3x = b_1\phi(x) + \frac{d\phi(x, u)}{dt}, \tag{6.8}$$

where $a_1, a_2, a_3, b_1 = const > 0$, $\phi(x, z)$ satisfies the condition (6.6). Let the polynomial

$$P(\lambda) = \lambda^3 + a_1\lambda^2 + a_2\lambda + a_3$$

have real roots $\lambda_1 \leq \lambda_2 \leq \lambda_3 < 0$. Then under condition (6.7), relations

$$L(\lambda) = \lambda + b_1 \geq 0 \ (\lambda_1 \leq \lambda \leq \lambda_3), \ \frac{d}{d\lambda}L(\lambda) = 1$$

hold. That is, if the polynomial

$$P(\lambda) - qL(\lambda) = \lambda^3 + a_1\lambda^2 + a_2\lambda + a_3 - q(b_1 + \lambda) =$$

$$\lambda^3 + a_1\lambda^2 + (a_2 - q)\lambda + a_3 - qb_1$$

is Hurwitzian, then by Corollary 5.1.2, system (6.8) under conditions (6.6) and (6.7) is BIBO-stable wb.

5.6.3 Proof of Theorem 5.6.1

Equation (6.1) with $x(0) = 0$ is equivalent to following one:

$$x(t) = \int_0^t exp[A(t - \tau)]bf(s(\tau), u(\tau), \tau)d\tau.$$

Multiplying this equation by c, we have

$$s(t) = \int_0^t K(t - \tau)f(s(\tau), u(\tau), \tau)d\tau,$$

since $c\exp[At]b = K(t)$. Due to (6.3), we arrive at the inequality

$$|s(t)| \le \int_0^t K(t-\tau)(q_S|s(\tau)| + q_I\|u(\tau)\|_{C^m})d\tau.$$

Or

$$|s(t)| \le \int_0^t K(t-\tau)q_S|s(\tau)|d\tau + h(t)$$

with the notation

$$h(t) = q_I \int_0^t K(t-\tau)\|u(\tau)\|_{C^m}d\tau.$$

Since A is a Hurwitz matrix, there are constants $\alpha_0, c_1 > 0$, such that

$$0 \le c\exp[At]b = K(t) \le c_1 e^{-\alpha_0 t} \ (t \ge 0).$$

So

$$h(t) \le q_I c_1 \|u\|_{C([0,\infty),\mathbf{C}^n)} \int_0^t e^{-\alpha_0(t-\tau)}d\tau \le b(u), \tag{6.9}$$

where

$$b(u) = q_I c_1 \alpha_0^{-1} \|u\|_{C([0,\infty),\mathbf{C}^n)}.$$

Consequently,

$$|s(t)| \le \int_0^t K(t-\tau)q_S|s(\tau)|d\tau + b(u).$$

From Lemma 1.7.1 it follows that

$$|s(t)| \le \eta(t), \tag{6.10}$$

where η is the solution of the equation

$$\eta(t) = b(u) + \int_0^t K(t-\tau)q_S\eta(\tau)d\tau.$$

We will solve this equation by the Laplace transformation. Since the transform of the convolution is equal to the product of the transforms, after straightforward calculations and taking into account that $P^{-1}(\lambda)L(\lambda)$ is the Laplace transform of $K(t)$, we get the equation

$$\overline{\eta}(\lambda) = \frac{b(u)}{\lambda} + P^{-1}(\lambda)L(\lambda)q_S\overline{\eta}(\lambda),$$

where $\overline{\eta}(\lambda)$ is the Laplace transform of $\eta(t)$. Thus,

$$\eta(t) = \frac{1}{2\pi i}\int_{-i\infty}^{i\infty} \exp[\lambda t](P(\lambda) - q_S L(\lambda))^{-1}P(\lambda)\lambda^{-1}d\lambda\, b(u).$$

Since $P(\lambda) - q_S L(\lambda)$ is a Hurwitz polynomial, the function under the integral has no the poles in the right open half-plane, and has one simple pole on the imaginary axis. So, thanks to the residue theorem

$$\eta(t) \leq const\ b(u)\ (t \geq 0).$$

Now relations (6.9) and (6.10) yield the required result. □

5.7 Global feedback stabilization

5.7.1 Statement of the result

Let us consider the system

$$dx/dt = Ax + F(x,t) + Bu(t), \tag{7.1}$$

where $x : [0, \infty) \to \mathbf{C}^n$ is the state, $u : [0, \infty) \to \mathbf{C}^m$ $(m \leq n)$ is the input. In addition, A, B are constant matrices and F continuously maps $\mathbf{C}^n \times [0, \infty)$ into $\mathbf{C}^n$. Let us take the feedback law

$$u(t) = Lx(t), \tag{7.2}$$

where L is a constant $m \times n$-matrix. Then (7.1) takes the form

$$dx/dt = Ax + F(x,t) + BLx. \tag{7.3}$$

Definition 5.7.1 *System (7.1) is said to be globally exponentially stabilizable (at $x = 0$) by means of the feedback law (7.2), if there is a matrix L such that the zero solution of (7.3) is globally exponentially stable.*

It is assumed that there is a constant $\nu > 0$ such that the condition

$$\|F(h,t)\| \leq \nu\|h\|\ (h \in \mathbf{C}^n;\ t \geq 0) \tag{7.4}$$

holds. In addition, the pair (A, B) is stabilizable. That is, there is a constant matrix L such that $W_L = A + BL$ is a Hurwitz matrix: $\alpha(W_L) < 0$.

Theorem 5.7.2 *Let (A, B) be a stabilizable pair. Let there be a matrix L satisfying the condition*

$$\nu \sum_{j=0}^{n-1} \frac{g^j(A+BL)}{\sqrt{j!}\,|\alpha(A+BL)|^{j+1}} < 1. \tag{7.5}$$

Then system (7.1) under (7.4) is globally exponentially stabilizable by means of the feedback law (7.2).

This theorem easily follows from Theorem 5.3.1.

5.7.2 Example

Consider the system

$$\dot{y}_1 - a_{11}y_1 - a_{12}y_2 = f_1(t, y_1, y_2) + b_1 u(t),$$

$$\dot{y}_2 - a_{22}y_2 = f_2(t, y_1, y_2) + b_2 u(t), \tag{7.6}$$

where $a_{11} > 0, a_{22} > 0, a_{12}, b_1 > 0, b_2 > 0$ are constants. Besides, the functions $f_1, f_2 : R_+ \times \mathbf{R}^2 \to \mathbf{R}$ satisfy the condition

$$f_1^2(t, y, z) + f_2^2(t, y, z) \leq \nu^2(y^2 + z^2) \tag{7.7}$$

for all $t \geq 0$; $y, z \in \mathbf{R}$. Take $L = -(l_1, l_2)$ with

$$l_1 = \tau b_1,\ l_2 = \tau b_2,$$

where τ is a positive constant. Then

$$W_L = A + BL = \begin{pmatrix} w_{11} & w_{12} \\ w_{21} & w_{22} \end{pmatrix}$$

where

$$w_{jk} = b_j l_k + a_{jk} = -\tau b_j b_k + a_{jk}\ (a_{21} = 0;\ j, k = 1, 2).$$

Thanks to Proposition 2.1.1 we easily get $g(W_L) \leq |a_{12}|$. In addition,

$$\alpha(W_L) = Re\{(w_{11} + w_{22})/2 + [(w_{11} - w_{22})^2/4 + w_{12}w_{21}]^{1/2}\}.$$

Simple calculations show that for any $\alpha_0 < 0$ one can take $\tau > 0$ in such a way that $\alpha(W_L) = \alpha_0$. Thus, for any $\nu > 0$ there is $\alpha(W_L)$ such that

$$\nu \sum_{j=0}^{n-1} \frac{|a_{12}|^j}{\sqrt{j!}|\alpha(A + BL)|^{j+1}} < 1.$$

Therefore, due to Theorem 5.7.2, system (7.6) under condition (7.7) is globally exponentially stabilizable for any $\nu > 0$.

5.8 Notes

Note 5.1 As it was mentioned, Aizerman's conjecture was that if system (1.3) is asymptotically stable for each $q_1 \in [0, q]$, then (1.1) with $f(s,t) = f(s)$ is absolutely stable in the class of nonlinearities similar to (1.2). Kalman (1957) proposed to prove that if $f(s,t) = f(s)$ is differentiable, with

$$0 \leq f'(s) \leq q\ (s \in \mathbf{R}),$$

and the system (1.3) is asymptotically stable for each $q_1 \in [0, q]$, then (1.1) is absolutely stable. This conjecture was disproved as was that of Aizerman by the generation of counterexamples cf. (Naredra and Taylor, 1973), (Willems, 1971).

In 1966 N. Truchan showed that the Aizerman's hypothesis is satisfied by systems having linear parts in the form of single loop circuits with up to five stable aperiodic links connected in tandem. In 1981 the author showed that any system satisfies the Aizerman's hypothesis if its impulse function is a non-negative one (Gil', 1983a, 1983 b). That result includes Truchan's one.

Note 5.2 The generalized Aizerman hypothesis was considered in (Gil', 1985) and (Gil', 1994). The results presented in Section 5.2 supplement the classical absolute stability criteria cf. (Xiao-Xin, 1993), (Harris and Valenca, 1983), (Vidyasgar, 1993), etc.

Note 5.3 The material of Section 5.3 is taken from (Gil', 1982). Theorem 5.3.1 allows us to apply the Hurwitz criterion to nonlinear systems. Besides, that theorem supplements the global stability criterion from the paper (Krasnosel'skii and Pokrovskii, 1977).

About some other analytical methods of estimating the domain of attraction for differential equations see for instance (Levin, 1994), (Khalil, 1992), (Siljajak, 1978).

Note 5.4 The contents of Sections 5.5 and 5.6 are new. Theorem 5.5.1 supplements the well-known input-output stability criteria (see e.g. (Vidyasgar, 1993)).

Note 5.5 Theorem 5.7.1 is adapted from (Gil' and Ailon, 1997). Stabilizability of nonlinear systems have been discussed by many authors (see (Sontag, 1990), (Nijmeijer and Schaft, 1990), (Slotine and Li, 1991), (Tsinias, 1991), (Vidyasagar, 1993), and references given therein). The basic method for the stabilizability is the Lyapunov functions one.

References for Chapter 5

Aizerman, M.A. (1949). On a problem concerning global stability of dynamical systems, *Uspekhi Matematicheskikh Nauk. 4(4)*, 187-188. In Russian.

Daleckii, Yu. L. and M. G. Krein. (1974). *Stability of Solutions of Differential Equations in Banach Space*. Amer. Math. Soc., Providence, R. I.

Gemignani, M. G. (1972). *Elementary Topology*. Addison-Wesley Publishing Company, Massachusetts.

Gil', M. I. (1982). Bounds for the stability domain of quasilinear systems, *Sibirskij Mat. J. 23*, 219-221. In Russian.

Gil', M.I. (1983 a). On a class of one-contour systems which are absolutely stable in the Hurwitz angle, *Automation and Remote Control, No. 10*, 70-75.

Gil', M.I. (1983 b). On one class of absolutely stable systems, *Soviet Physics Doklady, 269(6)*, 1324-1327.

Gil', M.I. (1985). On one class of absolutely stable multivariable systems, *Soviet Physics Doklady, 280(4)*, 811-815.

Gil', M.I. (1994). Class of absolutely stable multivariable systems, *International Journal of Systems Sciences, 25(3)*, 613-617.

Gil', M.I. and Ailon, A. (1997). On global feedback stabilization of nonlinear nonautonomous systems, *International Journal of Control 68, N4*, 935-941.

Harris, C.J. and Valenca, J. (1983). *The Stability of Input-Output Dynamical Systems*. Academic Press, London-New York.

Kalman, R.E. (1957). Physical and mathematical mechanisms of instability of nonlinear automatic control systems, *Trans. ASME, 79*, 553-566.

Khalil, H.K. (1992). *Nonlinear Systems*. MakMillan, New York.

Krasnosel'skii, M. A. and Pokrovskii, A. (1977), The absent bounded solution principle, *Soviet Math. Doklady, 233*, 293-296.

Levin, A. (1994). Analytical method of estimating the domain of attraction for polynomial differential equations, *IEEE Transactions on Automatic Control, 39, No 12*, 2471-2476.

Naredra, K. S. and J. H. Taylor. (1973). *Frequency Domain Criteria for Absolute Stability*, Academic Press, New York and London.

Nijmeijer, H. and A. van der Schaft (1990). *Nonlinear Dynamical Control Systems*. Springer-Verlag, New York.

Reissig, R., Sansone, G. and R. Conti. (1974). *Nonlinear Differential Equations of High Order*. Noordhoff International Publishing, Leiden.

Siljak, D.D. (1978). *Large Scale Dynamic Systems. Stability and Structure*, North-Holland, New York.

Slotine, J.-J. and W. Li. (1991). *Applied Nonlinear Control*. Prentice-Hall. Englewood Cliffs, New Jersey.

Sontag, E.G. (1990). *Mathematical Control Theory: Deterministic Finite Dimensional Systems*. Springer-Verlag, New York.

Tsinias, J. (1991). A theorem on global stabilization of nonlinear systems by linear feedback, *Systems & Control Letters 17*, pp. 357-362.

Vidyasagar, M. (1993). *Nonlinear Systems Analysis*, second edition. Prentice-Hall. Englewood Cliffs, New Jersey.

Voronov. A. (1979). *Stability, Controllability, Observability*, Nauka, Moscow. In Russian.

Willems, J. C. (1971). *The Analysis of Feedback Systems*, M. I. T. Press, Cambridge, M. A.

Xiao-Xin, Liao. (1993). *Absolute Stability of Nonlinear Control Systems*. Kluwer, China.

Chapter 6

NONLINEAR FINITE DIMENSIONAL SYSTEMS WITH TIME-VARIANT LINEAR PARTS

The present chapter is devoted to systems with linear time-variant leading parts.

In Section 6.1 the asymptotic stability of systems with general nonautonomous leading parts is discussed.

Section 6.2 deals with systems whose linear parts satisfy the Lipschitz conditions. Conditions for asymptotic stability are established.

Explicit conditions for the boundedness of solutions are presented In Sections 6.3 and 6.4.

Input-output stability of systems whose linear parts satisfy Lipschitz's conditions is discussed in Section 6.5.

Conditions for the global feedback stabilization of systems with nonautonomous leading parts are investigated in Section 6.6.

Throughout the present paper $\|.\|$ *means the Euclidean norm.*

6.1 Asymptotic stability of systems with general linear parts

Let us consider in $\mathbf{C}^n$ the equation

$$\dot{x} = A(t)x + F(x,t) \ (t \geq 0), \tag{1.1}$$

where $A(t)$ is a piecewise continuous $n \times n$-matrix, and F maps $\Omega(r) \times [0, \infty)$ into $\mathbf{C}^n$ with the following property: there exists a non-negative continuous function $\nu(t)$ bounded on $[0, \infty)$, such that

$$||F(h,t)|| \leq \nu(t)||h|| \text{ for all } h \in \Omega(r) \text{ and } t \geq 0. \tag{1.2}$$

Recall that $\Omega(r) = \{h \in \mathbf{C}^n : ||h|| \leq r\}$, and for any $n \times n$-matrix A

$$p(t, A) \equiv exp[\alpha(A)t] \sum_{k=0}^{n-1} \frac{g^k(A)t^k}{(k!)^{3/2}}$$

with $\alpha(A) = \max_k Re\lambda_k(A)$. Put

$$q(t,s) \equiv ||A(t) - A(s)|| \ (t, s \geq 0),$$

and

$$\chi \equiv \sup_{t \geq 0} p(t, A(t)).$$

Theorem 6.1.1 *Let the conditions (1.2), $\chi < \infty$, and*

$$\zeta(A(.), F) \equiv \sup_{t \geq 0} \int_0^t p(t-s, A(t))[q(t,s) + \nu(s)]ds < 1 \tag{1.3}$$

be fulfilled. Then the zero solution of equation (1.1) is asymptotically stable. In addition, any initial vector x_0 satisfying the inequality

$$\frac{\chi ||x_0||}{1 - \zeta(A(.), F)} < r, \tag{1.4}$$

belongs to the region of attraction of the zero solution. Moreover, under (1.4) the estimate

$$||x(t)|| \leq \frac{\chi ||x_0||}{1 - \zeta(A(.), F)} \ (t \geq 0)$$

is true for a solution $x(t)$ with the initial vector x_0.

In particular, let inequality (1.2) hold with $\Omega(r) = \mathbf{C}^n$, i.e. $r = \infty$. Then under condition (1.3), the zero solution of (1.1) is globally asymptotically stable.

Proof: As follows from Corollary 2.1.7, the inequality

$$||exp[A(\tau)t]|| \leq p(t, A(\tau)) \ (t, \tau \geq 0) \tag{1.5}$$

is valid. We rewrite equation (1.1) in the form

$$dx/dt - A(\tau)x = [A(t) - A(\tau)]x + F(x,t),$$

regarding an arbitrary $\tau \geq 0$ as fixed. This equation is equivalent to the following one:

$$x(t) = exp[A(\tau)t]x(0) +$$

$$\int_0^t exp[(A(\tau)(t-s)][(A(s)-A(\tau))x(s)+F(x(s),s)]ds. \qquad (1.6)$$

Since the solutions continuously depend on the initial vector, the inequality

$$||x(t)|| \leq r \ (0 \leq t \leq t_0)$$

is true for a sufficiently small t_0. Due to (1.2) and (1.5), the latter inequality implies the relation

$$||x(t)|| \leq p(t, A(\tau))||x(0)|| + \int_0^t p(t-s, A(\tau))[q(\tau, s) + \nu(s)]||x(s)||ds \ (t \leq t_0).$$

Taking $\tau = t$, we get

$$||x(t)|| \leq p(t, A(t))||x(0)|| + \int_0^t p(t-s, A(t))[q(t, s) + \nu(s)]||x(s)||ds \ (t \leq t_0).$$

This relation according to the definitions of $\zeta(A(.), F)$ and χ yields

$$\sup_{s\leq t_0} ||x(s)|| < \chi||x(0)|| + \sup_{s\leq t_0} ||x(s)||\zeta(A(.), F).$$

Consequently, due to (1.3)

$$\sup_{s\leq t_0} ||x(s)|| \leq \chi||x(0)||(1-\zeta(A(.), F))^{-1}$$

for a sufficiently small t_0. But condition (1.4) ensures this bound for all $t \geq 0$. That bound provides the Lyapunov stability. The asymptotic stability can be proved by a small perturbation of system (1.1) as claimed. □

6.2 Systems with the Lipschitz property

6.2.1 Statement of the result

Assume that the following conditions are satisfied:

$$||A(t) - A(s)|| \leq q_0|t-s| \ (t, s \geq 0), \qquad (2.1)$$

and

$$||F(h, t)|| \leq \nu_0||h|| \ (h \in \Omega(r), t \geq 0). \qquad (2.2)$$

Here q_0 and ν_0 are non-negative constants. In addition,

$$v = \sup_{t\geq 0} g(A(t)) < \infty \text{ and } \rho \equiv -\sup_{t\geq 0} \alpha(A(t)) > 0. \qquad (2.3)$$

Theorem 6.2.1 *Let the conditions (2.1-2.3) and*

$$\zeta_0(A(.),\nu_0) \equiv \sum_{j=0}^{n-1} \frac{v^j}{\sqrt{j!}} \left(\frac{q_0(j+1)}{\rho^{j+2}} + \frac{\nu_0}{\rho^{j+1}}\right) < 1 \tag{2.4}$$

hold. Then the zero solution of equation (1.1) is asymptotically stable. In addition, any initial vector x_0 satisfying the inequality

$$\frac{\chi\|x_0\|}{1-\zeta_0(A(.),\nu_0)} < r, \tag{2.5}$$

belongs to the region of attraction of the zero solution. Moreover, the corresponding solution is subject to the estimate

$$\|x(t)\| \le \frac{\chi\|x_0\|}{1-\zeta_0(A(.),\nu_0)} \ (t \ge 0). \tag{2.6}$$

The proofs are presented in Subsection 6.2.3 below. Furthermore, denote

$$b_j = \frac{jq_0 v^{j-1}}{\sqrt{(j-1)!}} + \frac{\nu_0 v^j}{\sqrt{j!}} \text{ for } j = 1, \ldots, n-1,$$

$$\text{and } b_0 = \nu_0,\ b_n = q_0\frac{nv^{n-1}}{\sqrt{(n-1)!}}. \tag{2.7}$$

Let $z(A(.),\nu_0)$ be the extreme right-hand (unique positive and simple) zero of the algebraic equation

$$z^{n+1} = \sum_{j=0}^{n} b_j z^{n-j}. \tag{2.8}$$

Theorem 6.2.2 *Let conditions (2.1-2.3) hold. If, in addition, the matrix $A(t) + z(A(.),\nu_0)I$ is Hurwitzian for all $t \ge 0$, then the zero solution of equation (1.1) is asymptotically stable, and inequality (2.4) is valid. Moreover, any vector x_0 satisfying (2.5), belongs to the region of attraction of the zero solution and the estimate (2.6) is true for the corresponding solution.*

Corollary 6.2.3 *Let the conditions (2.1-2.3) hold with $\Omega(r) = \mathbf{C}^n$ (i.e. $r = \infty$), and let $A(t) + z(A(.),\nu_0)I$ be a Hurwitz matrix for all $t \ge 0$. Then the zero solution of (1.1) is globally asymptotically stable.*

6.2.2 Example

Consider the scalar equation

$$\ddot{y} + p(t)\dot{y} + w(t)y = f(t, y, \dot{y}), \tag{2.9}$$

where $p(t)$ and $w(t)$ are non-negative scalar-valued functions with the property

$$|p(t) - p(s)| + |w(t) - w(s)| \leq q_0 |t - s| \text{ for all } t, s \geq 0. \qquad (2.10)$$

It is assumed that f is a real function defined on $[0, \infty) \times \mathbf{R}^2$ and satisfying the condition

$$f^2(t, y, z) \leq \nu_0^2\,(y^2 + z^2)\ (t \geq 0;\ z, y \in \mathbf{R},\ y^2 + z^2 \leq r^2;\ r \leq \infty). \qquad (2.11)$$

Let $A(t)$ be the matrix of the linear part of equation (2.9), and

$$\alpha(A(t)) \leq -\rho < 0 \text{ for all } t \geq 0.$$

Let us suppose that $sup_{t \geq 0} w(t) < \infty$. Then according to Example 2.1.4

$$g(A(t)) \leq v \equiv 1 + \sup_{t \geq 0}\ w(t). \qquad (2.12)$$

In addition, assume that

$$\zeta_0(A(.), \nu_0) = q_0 \rho^{-2} + \nu_0 \rho^{-1} + v(2 q_0 \rho^{-3} + \nu_0 \rho^{-2}) < 1. \qquad (2.13)$$

Then by Theorem 6.2.1 the zero solution of (2.9) is asymptotically stable. Each vector $x(0) = (y(0), y'(0))$ satisfying (2.5), belongs to the region of attraction of the zero solution. Note that for the considered equation

$$\chi \leq \max_{t \geq 0} exp[-t\rho]\ (1 + vt).$$

6.2.3 Proofs of Theorems 6.2.1 and 6.2.2

Proof of Theorem 6.2.1: Using (1.5) and (2.3), we get

$$\int_0^t p(t-s, A(\tau))(t-s)ds \leq \int_0^t e^{-\rho(t-s)} \sum_{k=0}^{n-1} \frac{v^k (t-s)^{k+1}}{(k!)^{3/2}} ds \leq$$

$$\int_0^\infty e^{-\rho z} \sum_{k=0}^{n-1} \frac{v^k z^{k+1}}{(k!)^{3/2}} dz = \theta_1(A)\ (t, \tau \geq 0),$$

where

$$\theta_1(A) = \sum_{k=0}^{n-1} \frac{(k+1) v^k}{\sqrt{k!}\,\rho^{k+2}}.$$

Moreover,

$$\int_0^t p(t-s, A(\tau))ds \leq \int_0^t e^{-\rho(t-s)} \sum_{k=0}^{n-1} \frac{v^k (t-s)^k}{(k!)^{3/2}} ds \leq$$

$$\int_0^\infty e^{-\rho z} \sum_{k=0}^{n-1} \frac{v^k z^k}{(k!)^{3/2}} dz = \theta_0(A),$$

where

$$\theta_0(A) = \sum_{k=0}^{n-1} \frac{v^k}{\rho^{k+1}\sqrt{k!}} \ (t, \tau \geq 0).$$

Consequently, according to (1.3) and (2.4)

$$\zeta(A(.), F) \leq q_0\theta_1(A) + \nu_0\theta_0(A) = \zeta_0(A(.), \nu_0) < 1. \tag{2.14}$$

Now the result is due to Theorem 6.1.1. □

Lemma 6.2.4 *Let $z(A(.), \nu_0)$ be the extreme right-hand zero of the equation (2.8). Then the inequality*

$$\alpha(A(t)) + z(A(.), \nu_0) < 0 \text{ for all } t \geq 0$$

implies the inequality $\zeta_0(A(.), \nu_0) < 1$.

Proof: The hypothesis of the lemma entails the inequality

$$\rho \equiv -\sup \alpha(A(t)) > z(A(.), \nu_0).$$

Dividing (2.8) by z^{n+1}, and taking into account that $z(A(.), \nu_0)$ is its root, we arrive at the inequality

$$1 = \sum_{j=0}^{n} b_j z^{-j-1}(A(.), \nu_0) < \sum_{j=0}^{n} b_j \rho^{-j-1}.$$

But according to (2.7),

$$\zeta_0(A(.), \nu_0) = \sum_{j=0}^{n} b_j \rho^{-j-1} < 1.$$

As claimed. □

The assertion of Theorem 6.2.2 immediately follows from Theorem 6.2.1 and Lemma 6.2.4.

6.3 Boundedness of solutions of systems with general linear parts

6.3.1 Statement of the result

Recall that quantities $q(t, s), \Omega(r), p(t, A), \zeta(A(.), F)$, and χ are defined in Section 6.1. Again consider in $\mathbf{C}^n$ the equation

$$\dot{x} = A(t)x + F(x, t) \ (t \geq 0), \tag{3.1}$$

where $A(t)$ is a piecewise continuous $n \times n$-matrix, and F continuously maps $\Omega(r) \times [0, \infty)$ into $\mathbf{C}^n$ with the property

$$||F(h,t)|| \leq \nu(t)||h|| + l(t) \text{ for all } h \in \Omega(r) \text{ and } t \geq 0. \tag{3.2}$$

Here $\nu(t), l(t)$ are positive continuous scalar-valued functions bounded on $[0, \infty)$. Put

$$L(A(.), F) \equiv \sup_{t \geq 0} \int_0^t p(t-s, A(t)) l(s) ds.$$

Theorem 6.3.1 *Let the conditions $\chi < \infty$, (3.2) and (1.3) be fulfilled. Also let for a $x_0 \in \mathbf{C}^n$, the inequality*

$$\frac{\chi||x_0|| + L(A(.), F)}{1 - \zeta(A(.), F)} < r \tag{3.3}$$

hold. Then a solution $x(t)$ of equation (3.1) with the initial condition $x(0) = x_0$ is bounded. Moreover, the estimate

$$||x(t)|| \leq \frac{\chi||x(0)|| + L(A(.), F)}{1 - \zeta(A(.), F)} \text{ for all } t \geq 0 \tag{3.4}$$

is true.

In particular, let (3.2) hold with $\Omega(r) = \mathbf{C}^n$, i.e. $r = \infty$. Then under (1.3) condition (3.3) is automatically satisfied, and any solution of equation (3.1) is bounded.

Corollary 6.3.2 *Let the conditions (3.2) and (1.3) hold. In addition, let*

$$(1 - \zeta(A(.), F))^{-1} \, L(A(.), F) < r.$$

Then equation (3.1) possesses bounded solutions.

In fact, taking sufficiently small initial vectors with the property (3.3) and employing Theorem 6.3.1, we arrive at the required assertion.

6.3.2 Proof of Theorem 6.3.1

According to condition (3.3), for a sufficiently small $t_0 > 0$ we have $||x(t)|| \leq r$ for $t \in [0, t_0]$. Due to (1.5), (1.6) and (3.2) we have

$$||x(t)|| \leq p(t, A(\tau))||x(0)|| + \int_0^t p(t-s, A(\tau))[q(\tau, s) + \nu(s)]||x(s)||ds$$

$$+ \int_0^t p(t-s, A(\tau)) l(s) ds, \; t \leq t_0.$$

For $\tau = t$ and according to the definition of $L(A(.), F)$ and χ we derive

$$\sup_{t \leq t_0} ||x(t)|| < \chi ||x(0)|| + \sup_{s \leq t_0} ||x(s)|| \zeta(A(.), F) + L(A(.), F).$$

Now condition (1.3) yields the inequality

$$\sup_{t \leq t_0} ||x(t)|| \leq \frac{\chi ||x(0)|| + L(A(.), F)}{1 - \zeta(A(.), F)}$$

for a sufficiently small t_0. But condition (3.3) ensures this bound for all $t \geq 0$. This proves the result. □

6.4 Boundedness of solutions of systems with the Lipschitz property

Again, consider equation (3.1), assuming that F continuously maps $\Omega(r) \times [0, \infty)$ into $\mathbf{C}^n$. Assume also that the conditions (2.1), (2.3), and

$$||F(h, t)|| \leq \nu_0 ||h_0|| + l_0 \text{ for all } h \in \Omega(r), t \geq 0. \tag{4.1}$$

be fulfilled with constants $\nu_0, l_0 > 0$. Denote

$$L_0 \equiv l_0 \sum_{j=0}^{n-1} \frac{v^j}{\sqrt{j!}\rho^{j+1}}.$$

Recall that ζ_0 is defined by (2.4).

Theorem 6.4.1 *Let conditions (2.1), (2.3), (2.4) and (4.1) be fulfilled. In addition, for a $x_0 \in \mathbf{C}^n$, let the inequality*

$$(1 - \zeta_0(A(.), \nu_0))^{-1}(\chi ||x_0|| + L_0) < r \tag{4.2}$$

hold. Then a solution $x(t)$ of equation (3.1) with the initial condition $x(0) = x_0$ is bounded. Moreover, the estimate

$$||x(t)|| \leq (1 - \zeta_0(A(.), \nu_0))^{-1}(\chi ||x_0|| + L_0) \text{ for all } t \geq 0 \tag{4.3}$$

is true.

Proof: Take into account relation (2.14). Further, it is simple to check that under consideration, $L(A(.), F) \leq L_0$. Now the result is due to Theorem 6.3.1. □

Corollary 6.4.2 *Under conditions (2.1), (2.3), (2.4), (4.1), let the inequality*

$$(1 - \zeta_0(A(.), \nu_0))^{-1} L_0 < r, \tag{4.4}$$

hold. Then (3.1) possesses bounded solutions.

Recall that $z(A(.), \nu_0)$ denotes the extreme right-hand (unique positive and simple) zero of the algebraic equation (2.8).

Theorem 6.4.3 *Under conditions (2.1), (2.3), and (4.1), let the matrix $A(t) + z(A(.), \nu_0)I$ be Hurwitzian for all $t \geq 0$. Then the inequality (2.4) is valid, and under (4.2) a solution $x(t)$ of (3.1) with $x(0) = x_0$ is bounded. Moreover, estimate (4.3) is true.*

The proof of that theorem is exactly the same as the proof of Theorem 6.2.2.

Corollary 6.4.4 *Under conditions (2.1), (2.3), let inequality (4.1) hold with $\Omega(r) = \mathbf{C}^n$ (i.e. $r = \infty$), and let $A(t) + z(A(.), \nu_0)I$ be a Hurwitz matrix for all $t \geq 0$. Then (2.4) holds, and any solution of (3.1) is bounded, and estimate (4.3) is true.*

Furthermore, let $A(t) \equiv A$ be a constant matrix. Then we can write down

$$\zeta_0(A(.), \nu_0) = \nu_0 \sum_{j=0}^{n-1} \frac{g^j(A)}{|\alpha(A)|^{j+1}\sqrt{j!}} = \nu_0 \Gamma(A),$$

$$L_0 = l_0 \sum_{j=0}^{n-1} \frac{g^j(A)}{|\alpha(A)|^{j+1}\sqrt{j!}} = l_0 \Gamma(A),$$

and

$$\chi = \max_{t \geq 0} exp[\alpha(A)t] \sum_{j=0}^{n-1} \frac{g^j(A)t^j}{\sqrt{j!}} = \chi(A).$$

Moreover, (2.8) takes the form

$$z^{n+1} = \nu_0 \sum_{j=0}^{n-1} \frac{g^j(A)z^{n-j-1}}{\sqrt{j!}},$$

and $z(A(.), \nu_0)$ coincides with the extreme right-hand zero $z_0(A, \nu_0)$ of the last equation. Exploiting Theorem 6.4.3 and omitting simple calculations, we obtain the following result.

Corollary 6.4.5 *Under condition (4.1) let the matrix $A + z_0(A, \nu_0)I$ be a Hurwitz one. Then $\nu_0 \Gamma(A) < 1$. If, in addition,*

$$L_0 < r(1 - \nu_0 \Gamma(A)),$$

then equation (3.1) with $A(t) \equiv A$ possesses bounded solutions.

Example 6.4.6 *Consider the equation*

$$\ddot{y} + p(t)\dot{y} + w(t)y = f(t, y, \dot{y})$$

where $p(t)$ and $w(t)$ are the same as in Subsection 6.2.2 . Moreover, it is assumed that the scalar-valued function f satisfies the condition

$$f(t, y, z) \leq \nu_0\sqrt{y^2 + z^2} + l_0 \ (y^2 + z^2 \leq r^2;\ t \geq 0).$$

Let v be defined by (2.12), and conditions (2.13) and (4.4) hold. Then, due to Corollary 6.4.2, the considered equation possesses bounded solutions. Note that in the considered example

$$L_0 = l_0(\frac{1}{\rho} + \frac{v}{\rho^2}).$$

6.5 Input-output stability

6.5.1 Statement of the results

Let us consider the system

$$dx/dt = A(t)x + F(x, u(t), t), \tag{5.1}$$

$$y(t) = C(t)x(t) + D(t)u(t) \ \ (t \geq 0), \tag{5.2}$$

where $x : [0, \infty) \to \mathbf{C}^n$ is the state, $u : [0, \infty) \to \mathbf{C}^m$ $(m \leq n)$ is the input, $y : [0, \infty) \to \mathbf{C}^l$ $(l \leq n)$ is the output. Moreover, $A(t), C(t), D(t)$ are variable matrices, and F continuously maps $\mathbf{C}^n \times \mathbf{C}^m \times [0, \infty)$ into $\mathbf{C}^n$.

The definition of the input-output stability for system (5.1), (5.2) is the same as in Section 5.5. It is assumed that there are constants $\nu_0, \nu_I > 0$, such that the condition

$$||F(h, w, t)||_{C^n} \leq \nu_0||h||_{C^n} + \nu_I||w||_{C^m} \ (h \in \mathbf{C}^n;\ w \in \mathbf{C}^m;\ t \geq 0) \tag{5.3}$$

holds.

Theorem 6.5.1 *Let conditions (5.3), (2.1), (2.3) and (2.4) hold. In addition, let matrices $C(t), D(t)$ be uniformly bounded on $[0, \infty)$. Then system (5.1), (5.2) is BIBO - stable wb.*

The assertion of this theorem easily follows from Theorem 6.4.1 above.

Recall that numbers b_k $(k = 0, ..., n)$ are defined by (2.7), and $z(A(.), \nu_0)$ is the extreme right-hand root of the algebraic equation (2.8). Corollary 6.4.4 implies

Corollary 6.5.2 *Let conditions (2.1), (2.3) and (5.3) be fulfilled. In addition, let matrices $C(t), D(t)$ be uniformly bounded on $[0,\infty)$, and the matrix $A(t) + z_0(A(.),\nu_0)I$ is Hurwitzian for all $t \geq 0$, then system (5.1), (5.2) is BIBO-stable wb.*

Further, due to Lemma 1.11.1, the relation

$$w(q,\nu_0,v) \equiv \sum_{k=0}^{n} b_k < 1 \tag{5.4}$$

implies the inequality

$$z(A(.),\nu_0) \leq w^{1/(n+1)}(q,\nu_0,v).$$

Now Theorem 6.5.1 yields

Corollary 6.5.3 *Let conditions (2.1), (2.3), (5.3) and (5.4) hold. If, in addition the matrix $A(t)+w^{1/(n+1)}(q,\nu,v)I$ is Hurwitzian for all $t \geq 0$, then system (5.1), (5.2) is BIBO-stable wb.*

Example 6.5.4 *Consider the single input-single output system*

$$\ddot{x} + p(t)\dot{x} + w(t)x = f(t,x,\dot{x},u(t)),$$
$$y(t) = cx(t) + du(t) \ (c,d = const), \tag{5.5}$$

where $p(t), w(t)$ are positive scalar-valued bounded functions with the property (2.10). Moreover, it is assumed that the scalar-valued function

$$f : [0,\infty) \times \mathbf{R}^3 \to \mathbf{R}$$

is real and satisfies the condition

$$f^2(t,y,z,w) \leq \nu_0^2(y^2+z^2) + \nu_I^2 w^2$$

for all $t \geq 0$; $y,z,w \in \mathbf{R}$. Let v be defined by (2.12), and condition (2.13) holds. Then due to Theorem 6.5.1, system (5.5) is BIBO-stable wb.

6.6 Global feedback stabilization

6.6.1 Statement of the result

In the present section, we derive explicit sufficient conditions for the global exponential feedback stabilizability of a class of nonautonomous nonlinear systems.

Let us consider in $\mathbf{C}^n$ the system

$$dx/dt = A(t)x + F(x,t) + B(t)u(t), \tag{6.1}$$

where $x : [0, \infty) \to \mathbf{C}^n$ is the state, $u : [0, \infty) \to \mathbf{C}^m$ $(m \le n)$ is the input. Moreover, $A(t), B(t)$ are variable matrices having bounded derivatives, and F continuously maps $\mathbf{C}^n \times [0, \infty)$ into $\mathbf{C}^n$. Let us take the feedback law

$$u(t) = L(t)x(t), \tag{6.2}$$

where $L(t)$ is a variable $m \times n$-matrix. Then (6.1) takes the form

$$dx/dt = A(t)x + F(x,t) + B(t)L(t)x(t). \tag{6.3}$$

Definition 6.6.1 *System (6.1) is said to be globally exponentially stabilizable (at $x = 0$) by means of the feedback law (6.2) if there is a matrix $L(t)$ such that the zero solution of (6.3) is globally exponentially stable.*

Clearly, this definition agrees with Definition 5.7.1. It is assumed that there is a constant $\nu > 0$ such that the condition

$$\|F(h,t)\| \le \nu \|h\| \; (h \in \mathbf{C}^n;\ t \ge 0) \tag{6.4}$$

holds.

Let the pair $(A(t), B(t))$ be stabilizable for each $t \ge 0$. That is, there is a matrix $L(t)$ such that

$$W_L(t) = A(t) + B(t)L(t)$$

is a Hurwitz matrix for every fixed $t \ge 0$. Namely,

$$\rho_L \equiv -\sup_{t \ge 0} \alpha(W_L(t)) > 0. \tag{6.5}$$

Additionally, it is assumed that under (6.5), $A(t), B(t), L(t)$ are differentiable, and

$$q_L \equiv \sup_{t \ge 0} \|\dot{W}_L(t)\| < \infty, \text{ and } v_L \equiv \sup_{t \ge 0} g(W_L(t)) < \infty. \tag{6.6}$$

Denote

$$G(W_L, F) \equiv \sum_{j=0}^{n-1} \frac{v_L^j}{\sqrt{j!}} \left[\frac{q_L(j+1)}{\rho_L^{j+2}} + \frac{\nu}{\rho_L^{j+1}} \right]. \tag{6.7}$$

Theorem 6.6.2 *Let $(A(t), B(t))$ be a stabilizable pair for each $t \ge 0$. Let there be a matrix $L(t)$ satisfying the conditions (6.5), (6.6) and*

$$G(W_L, F) < 1. \tag{6.8}$$

Then system (6.1) under (6.4) is globally exponentially stabilizable by means of the feedback law (6.2).

Put

$$\mu(A(t), B(t), F) = \min_L G(W_L, F),$$

where the minimum is taken over all $m \times n$-matrices $L(t)$ satisfying (6.6) and (6.5).

Corollary 6.6.3 *Let $(A(t), B(t))$ be a stabilizable pair for each $t \geq 0$. Let the conditions (6.4), and*

$$\mu(A(t), B(t), F) < 1$$

hold. Then system (6.1) is globally exponentially stabilizable by means of the feedback law (6.2).

6.6.2 Example

Consider the single input-single output system

$$\ddot{y} - a_1(t)\dot{y} - a_2(t)y = f(t, y, \dot{y}) + bu(t), \tag{6.9}$$

where $a_1(t), a_2(t)$ are positive scalar-valued bounded functions with the property

$$q_0 \equiv \sup_{t \geq 0} |\dot{a}_1(t)| + |\dot{a}_2(t)| < \infty. \tag{6.10}$$

Moreover, it is assumed that the function $f : [0, \infty) \times \mathbf{R}^2 \to \mathbf{R}$ satisfies the condition

$$f^2(t, y, z)^2 \leq \nu^2(y^2 + z^2) \tag{6.11}$$

for all $t \geq 0$; $y, z \in \mathbf{R}$. Rewrite equation (6.9) in the form (6.1) with

$$A(t) = \begin{pmatrix} a_1(t) & a_2(t) \\ 1 & 0 \end{pmatrix}, \text{ and } B(t) \equiv B = \begin{pmatrix} b \\ 0 \end{pmatrix}.$$

In the present case $(A(t), B)$ is a controllable pair. Take

$$L(t) \equiv L = -(l_1, l_2) \text{ with } l_k > \max_{t \geq 0} a_k(t)/b_k \ (k = 1, 2).$$

Then

$$W_L(t) = \begin{pmatrix} -p(t) & -w(t) \\ 1 & 0 \end{pmatrix},$$

where

$$p(t) = bl_1 - a_1(t) > 0, \ w(t) = bl_2 - a_2(t) > 0.$$

Due to Example 2.1.4 we have

$$v_L \leq v(l_1, l_2) \equiv 1 + \sup_{t \geq 0} w(t).$$

Put

$$\rho_L = \rho(l_1, l_2) \equiv -\sup_{t \geq 0} \alpha(W_L(t)) = \inf_{t \geq 0} Re\{\frac{p(t)}{2} - \sqrt{\frac{p^2(t)}{4} - w(t)}\}.$$

Since, in the present case, $B(t)$ and $L(t)$ are constant, due to (6.10), we have $q_L = q_0$. Thus,

$$G(W_L, F) \leq G(l_1, l_2) \equiv q_0\rho^{-2}(l_1, l_2) + \nu_0\rho^{-1}(l_1, l_2) +$$

$$v(l_1, l_2)(2q_0\rho^{-3}(l_1, l_2) + \nu_0\rho^{-2}(l_1, l_2)).$$

If q_0 and ν_0 are sufficiently small, in such a way that for some l_1, l_2 we have $G(l_1, l_2) < 1$, then due to Theorem 6.6.1, system (6.9), under conditions (6.10) and (6.11), is globally exponentially stabilizable.

6.6.3 Proofs

First, consider in $\mathbf{C}^n$ the equation

$$dx/dt = K(t)x + F(x, t), \tag{6.12}$$

where $K(t)$ is a variable Hurwitz $n \times n$-matrix with the properties

$$\|K(t) - K(s)\| \leq q\ |t - s|\ (q \equiv const;\ t, s \geq 0), \tag{6.13}$$

and

$$v \equiv \sup_{t \geq 0} g(K(t)) < \infty, \text{ and } \rho \equiv -\sup_{t \geq 0} \alpha(K(t)) > 0. \tag{6.14}$$

Lemma 6.6.4 *Let conditions (6.4), (6.12) and (6.13) be fulfilled. In addition, let*

$$G(K(.), F) \equiv \sum_{j=0}^{n-1} \frac{v^j}{\sqrt{j!}}\left[\frac{q(j+1)}{\rho^{j+2}} + \frac{\nu}{\rho^{j+1}}\right] < 1. \tag{6.15}$$

Then the zero solution of (6.12) is globally exponentially stable.

This result is due to Theorem 6.2.1 above

Proof of Theorem 6.6.1: Rewrite equation (6.1) in the form (6.12) with $W_L(t) = A(t) + B(t)L(t)$ instead of $K(t)$. According to (6.6), conditions (6.13) and (6.14) hold. Moreover, (6.8) provides the existence of a matrix $L(t)$ such that condition (6.15) is fulfilled. This result is now due to the previous lemma. □

6.7 Notes

The materials of Sections 6.1 and 6.2 are adapted from the papers (Gil', 1984, 1985). In these papers, the results from Section 6.1 are applied to the predator-prey system.

The contents of Sections 6.3 and 6.4 are based on the papers (Gil', 1987) and (Gil' and Shargorodsky, 1986a). In papers (Gil' and Shargorodsky, 1986a, 1986b) also considered applications of Theorem 6.4.1 to limit cycles and oscillations of autogenerators.

The material of Section 6.6 is taken from (Gil' and Ailon, 1997).

Some other related results can be found in the literature cited in Chapter 5.

References for Chapter 6

Gil', M. I. (1984). On a stability domain of the predator-prey system, *Journal of General Biology, 2*, 396-401.

Gil', M. I. (1985). Two-sides estimates for solutions of nonlinear equations of a class, *Differential Equations, 21*, 891-894.

Gil', M. I. (1987). Conditions for dissipativity and existence of limit cycles, *Differential Equations, 23*, 712-714.

Gil', M.I. and Ailon, A. (1997). On global feedback stabilization of nonlinear nonautonomous systems, *International Journal of Control 68, N4:*, 935-941.

Gil', M. I. and Shargorodsky, L.L. (1986a). On one criterion of existence of limit cycles, *Soviet Mathematics, 30* , 12-14.

Gil', M. I. and Shargorodsky, L.L. (1986b). On oscillations of auto generators, *Radiotechnics and Electronics, 9*, 1818-1822. In Russian.

Chapter 7

ESSENTIALLY NONLINEAR FINITE DIMENSIONAL SYSTEMS

This chapter is devoted to systems without linear leading parts.

In Section 7.1, the freezing method for linear systems is extended to nonlinear systems having the Lipschitz property.

In Section 7.2 we consider nonlinear systems with differentiable right parts.

In Section 7.3 we derive solution estimates which generalize the Lozinskii and Wazewski inequalities

Nonlinear systems with linear majorants are discussed in Section 7.4.

Section 7.5 is devoted to nonlinear triangular systems.

Perturbations of general nonlinear systems are investigated in Section 7.6.

In Section 7.7 we investigate the asymptotic stability of systems which are "close" to triangular ones.

Nonlinear scalar equations with real variable characteristic roots are examined in Section 7.8.

The input-output stability of essentially nonlinear systems is investigated in Section 7.9. In the present chapter $||.|| = ||.||_{C^n}$ means the Euclidean norm.

7.1 The freezing method for nonlinear systems with the Lipschitz property

7.1.1 Statement of the result

Let us consider in $\mathbf{C}^n$ the equation

$$\dot{x}(t) = B(x(t),t)x(t) \quad (t \geq 0), \tag{1.1}$$

where

$$B(h,t) = (b_{jk}(h,t))_{j,k=1}^n$$

is an $n \times n$-matrix for every $h \in \mathbf{C}^n$ and $t \geq 0$.

As above, $\Omega(r)$ is a ball in $\mathbf{C}^n$ with some radius $r \leq \infty$ and with the center at zero. It is assumed that there are constants $\mu(r), \nu(r)$, such that the relations

$$\|B(h_1,t)-B(h,t)\| \leq \nu(r)\|h_1-h\| \text{ and } \|B(h_1,t)-B(h,s)\| \leq \mu(r)|t-s| \tag{1.2}$$

hold for all $h, h_1 \in \Omega(r); t, s \geq 0$. Further, let us suppose that

$$\rho(r) \equiv - \sup_{h\in\Omega(r),t\geq 0} \alpha(B(h,t)) > 0 \tag{1.3}$$

and

$$v(r) \equiv \sup_{h\in\Omega(r),t\geq 0} g(B(h,t)) < \infty. \tag{1.4}$$

Recall that the quantity $g(A), \alpha(A)$ for a matrix A is introduced in Section 2.1. Clearly,

$$p(t, B(h,\tau)) \equiv e^{\alpha(B(h,\tau))t} \sum_{k=0}^{n-1} \frac{g^k(B(h,\tau))t^k}{(k!)^{3/2}} \leq e^{-\rho(r)t} \sum_{k=0}^{n-1} \frac{v^k(r)t^k}{(k!)^{3/2}},$$

and

$$\chi(r) \equiv \sup_{h\in\Omega(r);\ \tau,t\geq 0} p(t, B(h,\tau)) \leq \sup_{t\geq 0} e^{-\rho(r)t} \sum_{k=0}^{n-1} \frac{v^k(r)t^k}{(k!)^{3/2}}.$$

Finally, put

$$M(r) = \sup_{h\in\Omega(r),t\geq 0} \|B(h,t)h\| \text{ and } q_0(r) = \nu(r)M(r) + \mu(r).$$

Theorem 7.1.1 *For some positive $r \leq \infty$, let the conditions (1.2) - (1.4), and*

$$\Gamma(r) \equiv q_0(r) \sum_{j=0}^{n-1} \frac{(j+1)v^j(r)}{\sqrt{j!}\rho^{j+2}(r)} < 1 \tag{1.5}$$

hold. Then the zero solution of equation (1.1) is exponentially stable. Moreover, any initial vector satisfying the condition

$$\|x(0)\|\frac{\chi(r)}{1-\Gamma(r)} < r, \tag{1.6}$$

belongs to the region of attraction of the zero solution, and the corresponding solution $x(t)$ of (1.1) is subject to the estimate

$$\|x(t)\| \le \|x(0)\| \frac{\chi(r)}{1-\Gamma(r)} \text{ for all } t \ge 0. \tag{1.7}$$

Note that if $B(h,t)$ is normal for any $h \in \Omega(r)$, $t \ge 0$, then $v(r) = 0$, and (1.5) takes the form $q_0(r) < \rho(r)$.

Corollary 7.1.2 *Let conditions (1.2)-(1.5) hold with $\Omega(r) = \mathbf{C}^n$. Then the zero solution of equation (1.1) is globally exponentially stable. Moreover, any solution $x(t)$ of (1.1) is subject to estimate (1.7).*

Denote by $z_0(r)$ the extreme right-hand root of the algebraic equation

$$z^{n+1} = q_0(r)\sum_{j=0}^{n-1} \frac{v^j(r)(j+1)}{\sqrt{j!}} z^{n-j-1}. \tag{1.8}$$

Theorem 7.1.3 *For some positive $r \le \infty$, let conditions (1.2) and (1.4) hold. If, in addition, the matrix $B(h,t) + Iz_0(r)$ is Hurwitzian for all $h \in \Omega(r)$ and $t \ge 0$, then the zero solution of (1.1) is exponentially stable. Besides, the relations (1.3) and (1.5) hold. Moreover, any initial vector satisfying the condition (1.6), belongs to the region of attraction of the zero solution and the corresponding solution of (1.1) is subject to estimate (1.7).*

Corollary 7.1.4 *Let conditions (1.2) and (1.4) hold with $\Omega(r) = \mathbf{C}^n$. If, in addition, the matrix $B(h,t) + Iz_0(r)$ is Hurwitzian for all $h \in \mathbf{C}^n$ and $t \ge 0$, then the zero solution of equation (1.1) is globally exponentially stable. Moreover, any solution $x(t)$ of (1.1) is subject to estimate (1.7).*

Recall that

$$w_n = \sum_{k=0}^{n-1} \frac{k+1}{\sqrt{k!}}.$$

Setting $z = v(r)y$ in (1.8) and applying Lemma 1.11.1, we can assert that the following estimates are true:

$$z_0(r) \le v^{1-2/(n+1)}(r)[q_0(r)w_n]^{1/(n+1)} \text{ if } q_0(r)w_n \le v^2(r).$$

This inequality and Theorem 7.1.3 yield

Corollary 7.1.5 *For some positive $r \leq \infty$, let the conditions (1.2), (1.4) and $q_0(r)w_n \leq v(r)$ hold. If, in addition, the matrix*

$$B(h, t) + Iv^{1-2/(n+1)}(r)[q_0(r)w_n]^{1/(n+1)}$$

is a Hurwitz one for all $h \in \Omega(r)$ and $t \geq 0$, then the zero solution of (1.1) is exponentially stable. In addition, the relations (1.3) and (1.5) hold. Moreover, any initial vector satisfying condition (1.6) belongs to the region of attraction of the zero solution and the corresponding solution $x(t)$ of (1.1) is subject to estimate (1.7).

7.1.2 Proof of Theorem 7.1.1

Let us introduce the linear equation

$$dy_h/dt = B(h(t), t)y_h, \tag{1.9}$$

where $h(t) : [0, \infty) \to \Omega(r)$ is a differentiable function. If $h(t) = x(t)$ is a solution of (1.1), then (1.9) and (1.1) coincide.

Further, the continuous dependence of solutions on initial data implies that under the condition $||x(0)|| < r$ there is t_0, such that

$$x(t) \in \Omega(r) \text{ for } 0 \leq t \leq t_0 \tag{1.10}$$

for any solution of (1.1). Now we use Theorem 3.2.1 to estimate a solution of (1.9). If we put $A(t) = B(h(t), t)$ with some function $h(t)$, then (1.9) takes the form $\dot{x} = A(t)x$. Due to (1.2) we have

$$||B(h(t), t) - B(h(s), s)|| \leq ||B(h(t), t) - B(h(s), t)||+$$

$$||B(h(t), s) - B(h(s), s)|| \leq \nu(r)||h(t) - h(s)|| + \mu(r)|t - s|.$$

Hence,

$$||B(x(t), t) - B(x(s), s)|| \leq \nu(r)||x(t) - x(s)|| + \mu(r)|t - s| \leq$$

$$(\nu(r) \sup_{t \leq t_0} ||\dot{x}(t)|| + \mu(r))|t - s| \ (s, t \leq t_0).$$

But according to (1.1) $||\dot{x}(t)|| \leq M(r)$ $(t \leq t_0)$. Thus,

$$||B(x(t), t) - B(x(s), s)|| \leq q_0(r)|t - s| \ (s, t \leq t_0).$$

Applying Theorem 3.2.1 with $A(t) = B(x(t), t)$ to equation (1.1) we have the bound

$$||x(t)|| \leq \chi(r)(1 - \Gamma(r))^{-1}||x(0)|| \text{ for } t \leq t_0.$$

But condition (1.6) allows us to extend this bound to all $t \geq 0$. So estimate (1.7) is proved. It yields the Lyapunov stability. To prove the exponential stability, it is sufficient to substitute the expression $x_1(t) = x(t)e^{\epsilon t}$ with a sufficiently small $\epsilon > 0$ into equation (1.1) and to repeat the reasoning of the present proof. □

Proof of Theorem 7.1.3: Repeating the reasoning of the proof of Theorem 3.2.2, one can show that the Hurwitzness of the matrix $B(h,t) + z(r)I$ implies the inequality (1.5). Now Theorem 7.1.1 yields the required result. □

7.2 Systems with differentiable right parts

7.2.1 Statement of the result

Consider equation (1.1), assuming that $B(\ldots)$ is continuously differentiable with respect to all its variables. For any $h = (h_k) \in \Omega(r)$ and $t \geq 0$ we construct the matrix $T(h,t) = (T_{jk}(h,t))$ with the elements

$$T_{jk}(h,t) = \frac{\partial}{\partial t} b_{jk}(h,t) + \sum_{m=1}^{n} F_m(h,t) \frac{\partial}{\partial h_m} b_{jk}(h,t) \; (j,k,m = 1,\ldots,n).$$

Here

$$F_m(h,t) = b_{m1}(h,t)h_1 + \ldots + b_{mn}(h,t)h_n.$$

It is assumed that

$$q(r) \equiv \sup\{\|T(h,t)\| : h \in \Omega(r), t \geq 0\} < \infty. \tag{2.1}$$

Theorem 7.2.1 *For some positive $r \leq \infty$, let the conditions (1.3), (1.4), (2.1) and*

$$\zeta(r) \equiv q(r) \sum_{j=0}^{n-1} \frac{(j+1)v^j(r)}{\sqrt{j!}\rho^{j+2}(r)} < 1 \tag{2.2}$$

hold. Then the zero solution of equation (1.1) is exponentially stable. Moreover, any initial vector satisfying the condition

$$\|x(0)\| \frac{\chi(r)}{1-\zeta(r)} < r, \tag{2.3}$$

belongs to a region of attraction of the zero solution and the corresponding solution $x(t)$ of (1.1) is subject to the estimate

$$\|x(t)\| \leq \|x(0)\| \frac{\chi(r)}{1-\zeta(r)} \text{ for all } t \geq 0. \tag{2.4}$$

Note that if $B(h,t)$ is normal for any $h \in \Omega(r)$, $t \geq 0$, then $v(r) = 0$, and (2.2) takes the form $q(r) < \rho(r)$.

Corollary 7.2.2 *Let conditions (1.3)-(1.4), (2.1) and (2.2) hold with $\Omega(r) = \mathbf{C}^n$. Then the zero solution of equation (1.1) is globally exponentially stable. Moreover, any solution $x(t)$ of (1.1) is subject to estimate (2.4).*

Denote by $z(r)$ the extreme right-hand root of the algebraic equation

$$z^{n+1} = q(r) \sum_{j=0}^{n-1} \frac{v^j(r)(j+1)}{\sqrt{j!}} z^{n-j-1}. \tag{2.5}$$

Theorem 7.2.3 *For some positive $r \leq \infty$, let conditions (1.2), (1.4) and (2.1) hold. Then, if the matrix $B(h,t) + Iz(r)$ is Hurwitzian for all $h \in \Omega(r)$ and $t \geq 0$, the zero solution of (1.1) is exponentially stable. Moreover, relations (1.3) and (2.2) hold. In addition, any initial vector satisfying condition (2.3), belongs to the region of attraction of the zero solution and the corresponding solution of (1.1) is subject to the estimate (2.4).*

Corollary 7.2.4 *Let conditions (1.2), (1.4) and (2.1) hold with $\Omega(r) = \mathbf{C}^n$. If, in addition, the matrix $B(h,t) + Iz(r)$ is Hurwitzian for all $h \in \mathbf{C}^n$ and $t \geq 0$, then the zero solution of equation (1.1) is globally exponentially stable. Moreover, (2.2) is valid and any solution $x(t)$ of (1.1) is subject to estimate (2.4).*

7.2.2 Example

Let us consider the scalar equation

$$\ddot{y} + f(y, \dot{y}, t)\dot{y} + w(y, \dot{y}, t)y = 0, \tag{2.6}$$

assuming that the scalar-valued functions $f = f(y, z, t)$ and $w = w(y, z, t)$ are positive and continuously differentiable with respect to all their arguments under the conditions

$$t \geq 0,\ y^2 + z^2 \leq r^2$$

with a fixed positive number $r \leq \infty$. Take

$$b_{11} = -f,\ b_{12} = -w,\ b_{21} = 1 \text{ and } b_{22} = 0.$$

$(b_{jk} = b_{jk}(y, z, t))$. Due to Example 2.1.4, one can write

$$v(r) = 1 + \sup_{y^2+z^2 \leq r^2;\ t \geq 0} w(y, z, t).$$

Clearly, it is assumed that the supremum is finite. In the considered case

$$F_1 = F_1(y, z, t) = b_{11}z + b_{12}y = -fz - wy,\quad F_2 = F_2(y, z, t) = b_{21}z + b_{22}y = z,$$

and the matrix $T(x,t) = (T_{jk}(x,t))_{j,k=1}^2$ with $x = (y, z)$, has the entries

$$T_{21} = T_{22} = 0,\ T_{11} = -f_t' - f_y'z + (fy + wz)f_z'$$

and

$$T_{12} = -w'_t - w'_y z + (fz + wy)w'_z .$$

So we can take

$$q(r) = \sup\{|T_{11}(x,t)| + |T_{12}(x,t)| : \|x\|^2 = y^2 + z^2 \le r^2, t \ge 0\}.$$

Furthermore,

$$\rho(r) = \inf_{y^2+z^2 \le r^2, t \ge 0} Re\{f(y,z,t)/2 - \sqrt{f(y,z,t)/4 - w(y,z,t)}\}.$$

If the inequality

$$\zeta(r) = q(r)[\rho^{-1}(r) + v(r)\rho^{-2}(r)] < 1$$

holds, then due to Theorem 7.2.1, the zero solution of equation (2.6) is exponentially stable, and any vector $x(0)$ satisfying (2.3) with

$$\chi(r) = \max_{t \ge 0} exp[-\rho(r)t]\ (1 + v(r)t),$$

belongs to the region of attraction of the zero solution. Moreover, the estimate (2.4) is true.

7.2.3 Proofs

Lemma 7.2.5 *If for some $t_0 > 0$, $x(t) : [0, t_0] \to \Omega(r)$ is a solution of (1.1), then*

$$\|\frac{dB(x(t),t)}{dt}\| \le q(r)\ (0 \le t \le t_0).$$

Proof: We have

$$\frac{dB(x(t),t)}{dt} = \frac{\partial B(x(t),t)}{\partial t} + \sum_{m=1}^{n} \dot{x}_m B_m(x(t),t),$$

where

$$B_m(h,t) = \frac{\partial B(h,t)}{\partial h_m}\ (h = (h_m) \in \mathbf{C}^n),$$

and $x_m(t)$ are coordinates of $x(t)$. But according to (1.1)

$$\dot{x}_m(t) = \sum_{j=1}^{n} b_{mj}(x,t)x_j(t) \equiv F_m(x(t),t).$$

Thus,

$$\frac{dB(x(t),t)}{dt} = T(x(t),t).$$

This gives the stated result. □

Proof of Theorem 7.2.1: Take into account that (1.9) and (1.1) coincide when $h(t) = x(t)$. Further, the continuous dependence of solutions on initial data implies the relation (1.10).

Applying Theorem 3.2.1 to equation (1.9) and omitting simple calculation we have, by virtue of Lemma 7.2.5,

$$||x(t)|| \leq \chi(r)(1 - \zeta(r))^{-1}||x(0)|| \text{ for } t \leq t_0.$$

But condition (2.3) allows us to extend this bound to all $t \geq 0$. So estimate (2.4) is proved. It yields the Lyapunov stability. The exponential stability can be proved by small perturbations (see the proof of Theorem 7.1.1). □

Proof of Theorem 7.2.3: Repeating the reasoning of the proof of Theorem 3.2.2, one can show that the Hurwitzness of the matrix $B(e,t) + z(r)I$ ($h \in B(r)$ implies the inequality $\zeta(r) < 1$. Now Theorem 7.2.1 yields the required result. □

7.3 The generalized Lozinskii and Wazewski inequalities

Throughout the present section $||.||_n$ is an arbitrary norm in $\mathbf{C}^n$. Again, let $B(h,t)$ be an $n \times n$-matrix for all $h \in \mathbf{C}^n$ and $t \geq 0$.

Theorem 7.3.1 *For any $h \in \Omega(r)$ and all sufficiently small $\delta > 0$, let the relation*

$$||I + \delta B(h,t)||_n \leq 1 + a(t)\delta \ (t \geq 0) \tag{3.1}$$

hold, where $a(t)$ is a Riemann-integrable function independent of h and having the property

$$\int_0^t a(s)ds \leq 0 \ (t \geq 0). \tag{3.2}$$

Then any solution $x(t)$ of (1.1) satisfies the estimate

$$||x(t)||_n \leq ||x(s)||_n \, exp[\int_s^t a(\tau)d\tau] \ (t \geq s \geq 0). \tag{3.3}$$

provided $||x(s)||_n < r$.

Proof: The continuous dependence of solutions on initial data implies that under the condition $||x(0)|| < r$ there is t_0 such that relation (1.10) holds. Again consider equation (1.9) with an arbitrary function. $h(t) : [0, \infty) \to \Omega(r)$. Using Theorem 4.2.1 with $A(t) = B(h(t), t)$, we have the required estimate (3.3) for $t \leq t_0$. But condition (3.2) permits us to extend it to the whole positive half-line. □

Let

$$\sup_{h\in\Omega(r)} \lim_{\delta\downarrow 0} \frac{\|I+B(h,t)\delta\|_n - 1}{\delta} = a(t) < \infty,\ t \geq 0.$$

Then, obviously,

$$\|I+B(h,t)\delta\|_n \leq 1 + a(t)\delta + o(\delta)\ (h \in \Omega(r),\ \delta \downarrow 0).$$

Now by virtue of Theorem 7.3.1 we easily get the inequality (3.3). Thus under condition (3.2) Theorem 7.3.1 is a generalization of the Lozinskii inequality (see Section 4.2).

Denote by $(.,.)$ the scalar product in $\mathbf{C}^n$, and define a new scalar product

$$(h,v)_A = (Ah,v)\ (h.v \in \mathbf{C}^n)$$

with a positive definite matrix A. Put $\|.\|_A = \sqrt{(.,.)_A}$, and

$$\Omega_A(r) = \{h \in \mathbf{C}^n : \|h\|_A \leq r\}.$$

Theorem 7.3.2 *For any $t \geq 0$ and $h \in \Omega_A(r)$, let*

$$Re\ (B(h,t)h,h)_A \leq \Lambda(t)(h,h)_A, \tag{3.4}$$

where $\Lambda(t)$ is the Riemann-integrable function independent of h. Then under the condition

$$\int_0^t \Lambda(s)ds \leq 0\ (t \geq 0)$$

any solution $x(t)$ of equation (3.1) satisfies the estimate

$$\|x(t)\|_A \leq \|x(s)\|_A exp[\int_s^t \Lambda(\tau)\,d\tau]\ (t > s \geq 0)$$

provided $\|x(s)\|_n < r$.

Proof: Doing the scalar product in (3.1) by u and considering that

$$\frac{d}{dt}(x(t),x(t))_A = 2\ Re(\dot{x}(t),x(t))_A,$$

we get

$$\frac{d}{dt}(x(t),x(t))_A = 2\ Re(B(x(t),t)x(t),x(t))_A. \tag{3.5}$$

For a sufficiently small t_0 we have relation (1.10) and, therefore,

$$\frac{d}{dt}(x(t),x(t))_A \leq 2\Lambda(t)(x(t),x(t))_A,\ 0 \leq t \leq t_0.$$

Solving this inequality, we arrive at the inequality

$$(x(t),x(t))_A \leq exp[2\int_0^t \Lambda(s)ds](x(0),x(0))_A,\ 0 \leq t \leq t_0.$$

This proves the theorem. The details are left to the reader. □

Clearly, the latter theorem is a generalization of the Wazewski inequality (see Section 4.2).

Theorem 7.3.3 *For any $t \geq 0$ and $h \in \mathbf{C}^n$, let the inequality*

$$Re(B(h,t)h,h)_A \geq \beta(t)(h,h)_A$$

hold, where $\beta(t)$ is a Riemann-integrable function independent of h. Then any solution $x(t)$ of (1.1) satisfies the estimate

$$||x(t)||_A \geq ||x(0)||_A exp[\int_0^t \beta(\tau)d\tau] \ (t \geq 0).$$

Proof: By (3.5) the inequality

$$\frac{d}{dt}(x(t),x(t))_A \geq 2\beta(t)(x(t),x(t))_A$$

holds. Solving this inequality, we easily obtain the result. □

7.4 Nonlinear systems with linear majorants

Consider in $\mathbf{C}^n$ equation (1.1), where $B(h,t)$ is an $n \times n$-matrix for each $h \in \mathbf{C}^n$ and all $t \geq 0$, again.

Definition 7.4.1 *For all sufficiently small positive δ and $h \in \Omega(r)$ $(r \leq \infty)$, let there be a variable matrix $M(t)$ such that the relation*

$$|v + \delta B(h,t)v| \leq (I + M(t)\delta)|v| \ (v \in \mathbf{C}^n, \ t \geq 0) \tag{4.1}$$

is valid. Then we will say that system (1.1) has in set $\Omega(r)$ a linear majorant $M(t)$.

Inequality (4.1) means that

$$|b_{jk}(h,t)| \leq m_{jk}(t) \text{ for } j \neq k,$$

and

$$|1 + \delta b_{kk}(h,t)| \leq 1 + \delta m_{kk}(t) \ (h \in \Omega(r), t \geq 0),$$

and $M(t) = (m_{jk}(t))_{j,k=1}^n$. Let us introduce the equation

$$\dot{z}(t) = M(t)z(t) \ (t \geq 0) \tag{4.2}$$

and assume that this equation is stable. This implies that the Cauchy operator $V(t)$ of equation (4.2) is bounded:

$$l = \sup_{t \geq 0} ||V(t)|| < \infty. \tag{4.3}$$

Lemma 7.4.2 *Let system (1.1) have a linear majorant $M(t)$ in the ball $\Omega(r)$. Then, under condition (4.3), any solution $x(t)$ of (1.1) is subject to the inequality*

$$|x(t)| \leq V(t)|x(0)| \ (t \geq 0), \tag{4.4}$$

provided that

$$||x(0)|| < rl^{-1}, \tag{4.5}$$

and, consequently, the zero solution of equation (1.1) is stable in the Lyapunov sense.

Proof: Clearly, $l > 1$. So $||x(0)|| < r$, and condition (1.10) holds. Set $A(t) = B(x(t), t)$, where $x(t)$ is the solution of equation (1.1). Now the inequality (4.4) for $0 \leq t \leq t_0$ is due to Lemma 4.3.1. Since $x(0)$ is in the interior of $\Omega(rl^{-1})$, the solution remains in $\Omega(r)$. This proves the result. □

Corollary 7.4.3 *Let system (1.1) have a linear constant majorant M in the set $\Omega(r)$. In addition, let M be a Hurwitz matrix, such that*

$$l \equiv \sup_{t \geq 0} ||e^{Mt}|| < \infty.$$

Then any solution $x(t)$ of (1.1) with an initial vector $x(0)$ satisfying (4.5) is subject to the inequality

$$|x(t)| \leq exp\,[Mt]|x(0)| \ (t \geq 0),$$

and, therefore, the zero solution of equation (1.1) is exponentially stable.

Note that Corollary 2.1.7 yields the inequality

$$l \leq \max_{t \geq 0} e^{\alpha(M)t} \sum_{k=0}^{n-1} \frac{g^k(M)t^k}{(k!)^{3/2}}.$$

7.5 Nonlinear triangular systems

Let us consider the triangular system

$$\dot{u}_j(t) = \sum_{k=1}^{j} a_{jk}(u, t)u_k(t) \ (u = (u_k(t)),\ t \geq 0;\ j = 1, ..., n) \tag{5.1}$$

with continuous functions

$$a_{jk}(.,.) : \mathbf{R}^n \times [0, \infty) \to \mathbf{R} \ (1 \leq k \leq j \leq n).$$

Assume that for a positive $r \leq \infty$ the following relations are valid:

$$\alpha(t,r) \equiv \max_{w\in\Omega(r)} \max_{j=1,\ldots,n} a_{jj}(w,t) < 0, \tag{5.2}$$

and

$$g(t,r) \equiv \max_{w\in\Omega(r)} \sqrt{\sum_{1\leq j<k\leq n} a_{jk}^2(w,t)} < \infty. \tag{5.3}$$

Moreover, with the notation

$$\psi_r(t,\tau) = e^{\int_\tau^t \alpha(s,r)ds} \sum_{k=0}^{n-1} (k!)^{-1}[\int_\tau^t g(t_1,r)dt_1]^k \ (0 \leq \tau \leq t),$$

the condition

$$\tilde{\chi}(r) \equiv \sup_{t\geq\tau\geq 0} \psi_r(t,\tau) < \infty \tag{5.4}$$

holds.

Theorem 7.5.1 *Let conditions (5.2-5.4) be fulfilled. Then the zero solution of equation (5.1) is stable. If an initial vector u_0 satisfies the inequality*

$$\tilde{\chi}(r)\|u_0\| < r, \tag{5.5}$$

then a solution $u(t)$ of (5.1) with the initial condition $u(\tau) = u_0$ satisfies the inequality

$$\|u(t)\| \leq \psi_r(t,\tau)\|u(\tau)\| \ (t \geq \tau \geq 0). \tag{5.6}$$

Proof: Consider the triangular linear system

$$\dot{v}_j(t) = \sum_{k=1}^{j} a_{jk}(h(t),t)v_k(t), \ (t \geq 0;\ j = 1,\ldots,n) \tag{5.7}$$

with a continuous function $h : [0,\infty) \to \Omega(r)$.

By Theorem 3.4.1, the inequality

$$\|v(t)\| \leq \psi_r(t,\tau)\|v(\tau)\| \ (t \geq \tau \geq 0) \tag{5.8}$$

is true for any solution $v(t)$ of system (5.7).

Clearly, $\tilde{\chi}(r) \geq 1$. So condition (5.5) implies the inequality $\|u(\tau)\| < r$. Then for a sufficiently small $t_0 - \tau$, $\|u(t)\| \leq r$ $(0 \leq \tau \leq t \leq t_0)$. But equation (5.7) coincides with (5.1) if $h(t) = u(t)$, where $u(t)$ is the solution of (5.1). So by (5.8), $\|u(t)\| \leq \tilde{\chi}(r)\|u_0\| \leq r$ $(\tau \leq t \leq t_0)$. According to (5.6) this inequality can be extended to $[0,\infty)$. This proves the result. □

Corollary 7.5.2 *Let the conditions (5.2)-(5.4), and*

$$g(t,r) \leq g_0 < \infty \ (g_0 = const,\ t \geq 0) \tag{5.9}$$

be fulfilled. Then the zero solution of system (5.1) is stable provided the conditions

$$a_{jj}(w,t) \leq \alpha_0 < 0 \ (w \in \Omega(r);\ t \geq 0;\ j = 1, ..., n)$$

are valid.

7.6 Perturbations of nonlinear equations

Throughout the present section, $||.||_n$ denotes an arbitrary norm in $\mathbf{C}^n$. Consider in $\mathbf{C}^n$ the equation

$$\dot{x}(t) = B(x(t),t)x(t) + C(x(t),t)x(t) \ \ (t \geq 0), \tag{6.1}$$

where $B(z,t)$ and $C(z,t)$ are $n \times n$-matrices for any $z \in \mathbf{C}^n$ and $t \geq 0$.

Let $U_h(t,s)$ $(t,s \geq 0)$ be the evolution operator of the linear equation

$$\dot{y}(t) = B(h(t),t)y(t) \ \ (t \geq 0). \tag{6.2}$$

with a differentiable function

$$h : [0,\infty) \to \Omega(r) \ (r \leq \infty).$$

Let us assume that there are positive continuous functions $\phi(r,t,s)$ and $m(t,r)$ independent of h, such that

$$||U_h(t,s)||_n \leq \phi(r,t,s) \ \text{ and } ||C(w,t)||_n \leq m(t,r) \ (w \in \Omega(r);\ t,s \geq 0). \tag{6.3}$$

Additionally, suppose that

$$\varpi(r) \equiv \sup_{t \geq 0} \phi(r,\ t,0) < \infty \text{ and } \eta(r) \equiv \sup_{t \geq 0} \int_0^t \phi(r,t,s)m(s,r)ds < 1. \tag{6.4}$$

Theorem 7.6.1 *Let conditions (6.3) and (6.4) hold. Then the zero solution of equation (6.1) is stable. If an initial vector $x(0)$ satisfies the inequality*

$$\varpi(r)||x(0)||_n < r(1-\eta(r)), \tag{6.5}$$

then the corresponding solution is subject to the estimate

$$||x(t)||_n \leq \frac{||x(0)||_n \varpi(r)}{1-\eta(r)} \ (t \geq 0).$$

Proof: By the property of the evolution operator, equation (6.2) is equivalent to the following one:

$$y(t) = U_h(t,0)y(0) + \int_0^t U_h(t,s)C(h(s),s)y(s)ds \quad (t \geq 0).$$

Taking $h(t) = x(t)$-the solution of (6.1), we get

$$x(t) = U_x(t,0)x(0) + \int_0^t U_x(t,s)C(x(s),s)x(s)ds \quad (t \geq 0)$$

Clearly, (6.5) implies $||x(0)||_n < r$. Then for sufficiently small $t_0 > 0$ we have $||x(t)||_n \leq r$ if $t \leq t_0$. So

$$||x(t)||_n \leq ||U_x(t,0)||_n ||x(0)||_n + \int_0^t ||U_x(t,s)||_n m(s,r)||x(s)||_n ds \leq$$

$$\varpi(r)||x(0)||_n + \int_0^t \phi(r,t,s)m(s,r)||x(s)||_n ds \quad (t \leq t_0).$$

Hence,

$$\sup_{0 \leq t \leq t_0} ||x(t)||_n \leq \varpi(r)||x(0)||_n + \sup_{0 \leq t \leq t_0} ||x(t)||_n \eta(r) \quad (t \leq t_0),$$

and therefore,

$$\sup_{0 \leq t \leq t_0} ||x(t)||_n \leq \varpi(r)||x(0)||_n \; (1 - \eta(r))^{-1} \leq r.$$

According to (6.5) this inequality can be extended to $[0,\infty)$. This proves the result. □

7.7 Nonlinear systems which are "close" to triangular ones

Consider system (1.1). It is assumed that $B(w,t) = (b_{jk}(w,t))_{j,k}^n$ is continuous in $w \in \mathbf{C}^n$ and piecewise continuous in t. Also consider the corresponding lower-triangular matrix $A(w,t) = (a_{jk}(w,t))_{j,k=1}^n$ with the entries

$$a_{jk}(w,t) = b_{jk}(w,t) \text{ if } j \geq k; \text{ and } a_{jk}(w,t) = 0 \; (w \in \mathbf{C}^n) \text{ if } j < k. \quad (7.1)$$

Assume that conditions (5.2-5.4) hold. According to (5.8), the evolution operator of the system (5.7) with a differentiable h defined on $[0,\infty)$ with values in $\Omega(r)$ satisfies the inequality

$$||U_h(t,s)|| \leq \psi_r(t,s) \quad (t,s \geq 0).$$

Put

$$m(t,r) = \sup_{h\in\Omega(r)} ||C(h,t)||,$$

where $C(h,t) = A(h,t) - B(h,t)$ is the upper triangular matrix with the entries

$$c_{jk}(h,t) = b_{jk}(h,t) \ (j < k) \text{ and } c_{jk}(h,t) = 0 \ (j \geq k).$$

In addition, suppose that

$$\tilde{\eta}(r) \equiv \sup_{t\geq 0} \int_0^t \psi_r(t,s)m(s,r)ds < 1. \tag{7.2}$$

Employing Lemma 7.6.1, we arrive at the following result.

Theorem 7.7.1 *Let conditions (5.2-5.4), and (7.2) hold. Then the zero solution of equation (1.1) is stable. In addition, if an initial vector u_0 satisfies the inequality*

$$\chi(r)||u_0|| < r(1 - \tilde{\eta}(r)), \tag{7.3}$$

then the corresponding solution $u(t)$ of system (1.1) is subject to the inequality

$$||u(t)|| \leq \frac{\chi(r)||u_0||}{1-\tilde{\eta}(r)} \leq r \ (t \geq 0).$$

Note that the following relations hold:

$$m(t,r) \leq \max_{j=1,\dots,n} \sum_{k=j+1}^{n} \sup_{h\in\Omega(r)} |b_{jk}(h,t)|.$$

7.8 Nonlinear scalar equations with real characteristic roots

Let us consider the scalar equation

$$y^{(n)} + p_1(y,\dots,y^{(n-1)},t)y^{(n-1)} + \dots + p_n(y,\dots,y^{(n-1)},t)y = 0 \ (t \geq 0) \tag{8.1}$$

whose coefficients $p_k(x_1,\dots,x_n,t) = p_k(x,t)$ are real scalar-valued functions continuous in $x = (x_k) \in \mathbf{R}^n$ and piecewise continuous in $t \geq 0$. Introduce the algebraic equation

$$\lambda^n + p_1(x,t)\lambda^{n-1} + \dots + p_n(x,t) = 0 \ (x \in \mathbf{R}^n;\ t \geq 0). \tag{8.2}$$

Theorem 7.8.1 *Suppose that, for all $x \in \mathbf{R}^n$ and a sufficiently large $t_0 > 0$, the roots $\lambda_1(x,t),\dots,\lambda_n(x,t)$ of (8.2) are real and satisfy the inequalities*

$$\nu_0 \leq \lambda_1(x,t) \leq \nu_1 \leq \lambda_2(x,t) \leq \dots \leq \nu_{n-1} \leq \lambda_n(x,t) \leq \nu_n < 0 \ (t \geq t_0), \tag{8.3}$$

where ν_i are some constants $\nu_0 < \nu_1 < \dots < \nu_n < 0$. Then equation (8.1) is stable.

This result immediately follows from Corollary 4.6.2.

7.9 Input-output stability of essentially nonlinear systems

In the present section, $\|.\|_n$ is an arbitrary norm in $\mathbf{C}^n$. Let us consider the system

$$\dot{x}(t) = B(x(t), t)x(t) + M(u(t), t) \tag{9.1}$$

$$y(t) = D(x(t), u(t)) \quad (t \geq 0). \tag{9.2}$$

where $x : [0, \infty) \to \mathbf{C}^n$ is the state, $u : [0, \infty) \to \mathbf{C}^m$ $(m \leq n)$ is the input, $y : [0, \infty) \to \mathbf{C}^l$ $(l \leq n)$ is the output; $B(w, t)$ is an $n \times n$-matrix dependent on $w \in \mathbf{C}^n, t \geq 0$; function M continuously maps $\mathbf{C}^m \times [0, \infty)$ into $\mathbf{C}^n$, and D continuously maps $\mathbf{C}^n \times \mathbf{C}^m$ into $\mathbf{C}^l$. The definitions of the input-output stability for system (9.1), (9.2) are the same as in Section 5.7.

Lemma 7.9.1 *Let the evolution operator $U_h(t, s)$ of the linear equation*

$$\dot{y} = B(h(t), t)y \quad (t \geq 0) \tag{9.3}$$

with an arbitrary differentiable function $h : [0, \infty) \to \mathbf{C}^n$, satisfy the inequality

$$\|U_h(t, s)\|_n \leq c_1 e^{-\nu(t-s)} \quad (t > s \geq 0), \tag{9.4}$$

where constants $c_1, \nu = const > 0$ are independent of h. In addition, let there be a continuous nondecreasing scalar-valued function ϕ, such that the inequality

$$\|M(w, t)\|_n \leq \phi(\|w\|_m) \quad (w \in \mathbf{C}^m;\ t \geq 0) \tag{9.5}$$

holds. Then system (9.1), (9.2) is BIBO-stable.

Proof: Consider the equation

$$\dot{z}(t) = B(h(t), t)z(t) + M(u(t), t) \tag{9.6}$$

with the zero initial vector. It can be written as

$$z(t) = \int_0^t U_h(t, s)M(z(s), u(s), s)ds.$$

Hence,

$$\|z(t)\|_n \leq \int_0^t c_1 e^{-\nu(t-s)} \phi(\|u(s)\|_m)ds \leq c_1 \nu^{-1} \sup_{s \geq 0} \phi(\|u(s)\|_m) \quad (t \geq 0).$$

Take into account that equation (9.3) coincides with (9.1) when $h(t) = x(t)$ is the solution of (9.1). Then the latter inequality gives.

$$\|x(t)\|_n \leq c_1 \nu^{-1} \sup_{s \geq 0} \phi(\|u(s)\|_m) \quad (t \geq 0).$$

Taking into account that ϕ is a continuous mapping, we get the result. □

The last lemma and Theorem 7.3.1 imply

Corollary 7.9.2 *Under condition (9.5), for all sufficiently small $\delta > 0$, let the condition*

$$\|I + B(w,t)\delta\|_n \leq 1 - \nu\delta \ (\nu = const > 0, \ w \in \mathbf{C}^n; \ t \geq 0)$$

be fulfilled. Then system (9.1), (9.2) is BIBO-stable.

Combining Lemma 7.9.1 with Corollary 4.2.5, we get

Corollary 7.9.3 *Let the condition*

$$((B(w,t) + B^*(w,t))w, w) \leq -\nu(w,w) \ (t \geq 0; \ w \in \mathbf{C}^n)$$

be fulfilled. Then system (9.1), (9.2) is BIBO-stable.

7.10 Notes

The material of Sections 7.1 and 7.2 are adapted from the papers (Gil', 1989a and 1989b). The contents of Sections 7.3-7.4 are based on the multiplicative representation of solutions from Section 4.1. About the theory of perturbations of nonlinear systems see for instance (Rajalaksmy and Sivasundaram, 1992), (Ladde et al., 1977), etc. The contents of Section 7.8 is adapted from the paper (Levin, 1969) (see also, (Gil', 1987)). About some other related results see Notes from Chapter 5.

References for Chapter 7

Gil', M. I. (1987). Single-loop systems with several nonlinearities satisfying the generalized Aizerman-Calman conjecture. *Soviet Physics Doklady, 32(2)*, 1315-1318.

Gil', M. I. (1989a). Stability of essentially nonlinear nonstationary systems, *Soviet Physics Doklady, 34*, 753-755.

Gil', M. I. (1989b). The freezing method for nonlinear equations. *Differential Eqs. 25*: 912-918.

Ladde, G.S., Lakshmikantham, V. and S. Leela (1977). A new technique in perturbation theory, *Rocky Mountain Journal Math.* , *6*, 133-140.

Levin, A. Yu. (1969). Non-oscillations of solutions of the equation $x^{(n)}(t) + a_1(t)x^{(n-1)}(t) + ... + p_n(t)x(t)$, *Russian Mathematical Surveys, 24(2)*, 43-96.

Rajalaksmy S. and Sivasundaram, S. (1992). Vector Lyapunov functions and the technique in perturbation theory, *Journal of Mathematical Analysis and Applications, 164*, 560-570.

Chapter 8

LINEAR AUTONOMOUS SYSTEMS WITH DELAY

In the present chapter we consider linear autonomous systems with delay.

In Section 8.1 we recall some basic notions of theory of Banach spaces and introduce the notions of characteristic matrices and characteristic functions of linear autonomous systems with delay.

Section 8.2 is devoted to the representation of solutions by virtue of the Laplace transformation.

L^2-norm estimates for solutions of nonhomogeneous time-invariant equations are presented in Section 8.3.

Estimates for the Green functions of retarded autonomous systems are derived in Sections 8.4 and 8.5.

Conditions for exponential stability are examined in Section 8.6.

Estimates for the C-norm of solutions of nonhomogeneous time-invariant equations are presented in Section 8.7.

Perturbations of matrix-valued functions and bounds for roots of characteristic functions of retarded systems are discussed in Sections 8.8 and 8.9.

Sections 8.10 and 8.11 are devoted to the stability of time-invariant retarded systems with small delays and with arbitrary ones, respectively.

8.1 Banach spaces and characteristic matrices

8.1.1 Banach spaces and bounded operators

A linear space X over **C** or **R** is called *a linear normed space* if for any $x \in X$ a non-negative number $||x||_X = ||x||$ is defined, called the norm of x, having the following properties:

1. $||x|| = 0$ iff $x = 0$,
2. $||\alpha x|| = |\alpha|||x||$,

3. $||x+y|| \leq ||x|| + ||y||$ for every $x, y \in X$, $\alpha \in \mathbf{C}$.

A sequence $\{h_n\}$ of elements of X converges *strongly* (in the norm) to $h \in H$ if

$$\lim_{n\to\infty} ||h_n - h|| = 0.$$

A sequence $\{h_n\}$ of elements of X is called the fundamental (Cauchy) one if

$$||h_n - h_m|| \to 0 \text{ as } m, n \to \infty.$$

If any fundamental sequence converges to an element of X, then X is called *a Banach space.*

Example 8.1.1 *The set $L^p(\omega, \mathbf{C}^n)$ $(1 \leq p < \infty)$ of all functions with values in $\mathbf{C}^n$ defined on a set $\omega \subseteq \mathbf{R}^m$ with the finite norm*

$$||h||_{L^p(\omega,\mathbf{C}^n)} = [\int_\omega ||h(x)||^p_{C^n} dx]^{1/p}$$

is a Banach space. As above, $||.||_{C^n}$ is the Euclidean norm.

For the sake of simplicity sometimes we will write in the sequel $||.||_{L[a,b]}$ instead of $||.||_{L^2([a,b],\mathbf{C}^n)}$.

The set $C(\omega, \mathbf{C}^n)$ of all continuous functions with values in $\mathbf{C}^n$ defined on a set $\omega \subseteq \mathbf{R}^m$ and equipped with the sup-norm

$$||h||_{C(\omega,\mathbf{C}^n)} = \sup_{x\in\omega} ||h(x)||_{C^n}$$

is a Banach space.

Furthermore, an operator A acting from a linear manifold $D \subseteq X$ into a Banach space Y is called a linear one if

$$A(\alpha x_1 + \beta x_2) = \alpha A x_1 + \beta A x_2$$

for any $x_1, x_2 \in D$ and $\alpha, \beta \in \mathbf{C}$. A linear manifold on which A is defined is called its domain. Usually we will denote the domain of A by $D(A)$. If $D(A) = X$, and there is a constant a, such that the inequality

$$||Ah||_Y \leq a||h||_X \text{ for all } h \in X$$

holds, then the operator is said to be bounded. The quantity

$$||A||_{X\to Y} \equiv \sup_{h\in X} \frac{||Ah||_Y}{||h||_X}$$

is called the norm of A. If $X = Y$ we will write $||A||_{X\to X} = ||A||_X$.

A linear operator *is said to be completely continuous (compact)* if it is defined on the entire space X and maps each bounded set in X into a compact one in Y.

Under the natural definitions of addition and multiplication by a scalar, and the norm, the set $B(X,Y)$ of all bounded linear operators acting from X

into Y becomes a Banach space. If $Y = X$ we will write $B(X,X) = B(X)$. A sequence $\{A_n\}$ of bounded linear operators from $B(X,Y)$ converges *in the uniform operator topology* (in the operator norm) to an operator A if

$$\lim_{n\to\infty} \|A_n - A\|_{X\to Y} = 0.$$

A sequence $\{A_n\}$ of bounded linear operators *converges strongly* to an operator A if the sequence of elements $\{A_n h\}$ strongly converges to Ah for every $h \in X$.

If ϕ is a linear operator from X into $\mathbf{C}$, then it is called a linear functional. It is bounded (continuous) if $\phi(x)$ is defined for any $x \in X$, and there is a constant a such that the inequality

$$\|\phi(h)\|_Y \le a\|h\|_X \text{ for all } h \in X$$

holds. The quantity

$$\|\phi\|_X \equiv \sup_{h\in X} \frac{|\phi(h)|}{\|h\|_X}$$

is called *the norm of the functional* ϕ. All linear bounded functionals on X form a Banach space with that norm. This space is called the space *dual* to X and is denoted by X^*.

In the sequel $I_X = I$ is the identity operator in $X : Ih = h$ for any $h \in X$.

The operator A^{-1} is the inverse one to $A \in B(X,Y)$ if $AA^{-1} = I_Y$ and $A^{-1}A = I_X$.

A point λ of the complex plane is said to be a regular point of an operator $A \in B(X)$ if the operator

$$R_\lambda(A) \equiv (A - I_X\lambda)^{-1} \in B(X)$$

(the resolvent) exists and is bounded. The set of all regular points of A is said to be *the resolvent set* of A. The complement $\sigma(A)$ of all regular points of A in the complex plane is the *spectrum* of A.

The quantity

$$r_s(A) = sup_{s\in\sigma(A)}|s|$$

is called the *spectral radius* of A. Gel'fand's formula

$$r_s(A) = \overline{lim}_{n\to\infty} \sqrt[n]{\|A^n\|}$$

is true. Furthermore, if there is a nontrivial solution $e \in X$ of the equation $Ae = \lambda(A)e$, where $\lambda(A)$ is a number, then this number is called *an eigenvalue* of the operator A, and e is *an eigenvector* of A corresponding to $\lambda(A)$.

Any eigenvalue is a point of the spectrum. An eigenvalue $\lambda(A)$ has the algebraic multiplicity $r \le \infty$ if

$$dim(\cup_{k=1}^{\infty} ker(A - I\lambda(A))^k) = r.$$

Here $kerB$ denotes the kernel of a mapping B.

Let $A \in B(X,Y)$. Consider a linear bounded functional f defined on Y. Then on X the linear bounded functional $g(x) = f(Ax)$ is defined. The operator realizing the relation $f \to g$ is called the operator A^* *dual (adjoint)* to A. By the definition

$$(A^*f)(x) = f(Ax) \ (x \in X).$$

The operator A^* is a bounded linear operator acting from Y^* to X^*.

8.1.2 Characteristic matrices

Let $R(\tau)$ be a continuous on the left $n \times n$-matrix defined on a real segment $[0,\eta]$, $\eta < \infty$, and having a bounded variation. Let us consider the system

$$\dot{x}(t) = \int_0^\eta dR(\tau)x(t-\tau) + f(t) \ (t \geq 0) \tag{1.1}$$

with a continuous vector-valued function f defined on $[0,\infty)$. Take the initial condition

$$x(t) = \Phi(t) \text{ for } -\eta \leq t \leq 0, \tag{1.2}$$

where $\Phi(t)$ is a given continuous vector-valued function defined on $[-\eta, 0]$.

A solution of (1.1) with the initial function Φ is a continuous vector-valued function $x(t)$ defined on $[-\eta,\infty)$, differentiable on $(0,\infty)$ which satisfies that equation for all $t > 0$, and (1.2) holds. Besides at $t = 0$, $\dot{x}(t)$ is the right derivative. Equation (1.1) has a unique solution for any initial function due to Lemma 9.1.1 below. We will give an explicit representation of solutions to equation (1.1).

Moreover, the matrix-valued function $K(p)$ defined by the expression

$$K(p) = \int_0^\eta exp(-ps)dR(s) - Ip \ (p \in \mathbf{C}) \tag{1.3}$$

is called *the characteristic matrix* of equation (1.1) and the determinant $det K(p)$ is called *the characteristic function* of equation (1.1).

Notice that the system

$$\dot{x}(t) = A_0 x(t) + A_1 x(t-h_1) + ... + A_m x(t-h_m) + \int_0^\eta B(s)x(t-s)ds \tag{1.4}$$

with $0 = h_0 < h_1 < ... < h_m \leq \eta$, constant matrices A_k and a variable matrix $B(s)$ can be written in the shape (1.1). Besides, the characteristic function has the form

$$K(p) = A_0 + A_1 e^{-ph_1} + ... + A_m e^{-ph_m} + \int_0^\eta B(s)e^{-ps}ds - Ip. \tag{1.5}$$

Equation (1.1) with $f \equiv 0$ *is said to be exponentially stable*, if there are constants $C, \nu > 0$ such that

$$\|x(t)\|_{C^n} \leq Ce^{-\nu t}\|\Phi\|_{C([-\eta,0],\mathbf{C}^n)} \ (t \geq 0)$$

for any solution x with a continuous initial function Φ.

Below we will show that if all the roots of $detK(p)$ are in the open left half-plane, then (1.1) is exponentially stable.

Let A be an $n \times n$ - matrix. We will use the following properties of $g(A)$ (see Section 2.1): $g^2(A) \leq N^2(A) - |TraceA^2|$. If $A = (a_{jk})_{j,k=1}^n$ is a triangular matrix:

$$a_{jk} = 0 \; (1 \leq k < j \leq n),$$

then

$$g^2(A) = \sum_{1 \leq j < k \leq n} |a_{jk}|^2. \tag{1.6}$$

If A is normal: $A^*A = AA^*$, then $g(A) = 0$. Moreover, due to Proposition 2.1.1, the relations

$$g(A) \leq \sqrt{1/2}N(A^* - A) \text{ and } g(Ae^{i\tau} + zI) = g(A) \; (\tau \in \mathbf{R}, \; z \in \mathbf{C}) \tag{1.7}$$

are true. Our results in the present chapter are based on the following estimate for the resolvent.

Lemma 8.1.2 *Let A be a linear operator in $\mathbf{C}^n$ (matrix). Then the inequality*

$$\|(A - \lambda I)^{-1}\|_{C^n} \leq \sum_{k=0}^{n-1} \frac{g^k(A)}{\sqrt{k!}\rho^{k+1}(A, \lambda)} \text{ for all regular } \lambda \tag{1.8}$$

is valid, where $\rho(A, \lambda)$ is the distance between the spectrum of A and a complex point λ.

For the proof see Section 17.3 below or (Gil', 1995, Corollary 1.2.4). Let us apply this lemma to the characteristic matrix $K(p)$. Denote

$$\Gamma(K(p)) = \sum_{k=0}^{n-1} \frac{g^k(B(p))}{\sqrt{k!}d^{k+1}(K(p))} \quad (p \in \mathbf{C}), \tag{1.9}$$

where $d(K(p))$ is the smallest modulus of eigenvalues of $K(p)$:

$$d(K(p)) = \min_{k=1,\dots,n} |\lambda_k(K(p))|.$$

It is simple to show that all the zeros of $detK(\lambda)$ lie in some half-plane $Rez \leq c_1$ with a constant c_1. It is well-known that either $detK(\lambda)$ is a polynomial, or it has infinitely many zeros $z_1, z_2, \dots$ with the property $Rez_m \to -\infty$ as $m \to \infty$. We will say that p_0 is *a regular point* of $K(.)$ if $detK(p_0) \neq 0$. A number λ is called *the characteristic value* of $K(.)$ if $detK(\lambda) = 0$.

Lemma 8.1.3 *For any regular point p of $K(.)$, the inequality*

$$\|[K(p)]^{-1}\|_{C^n} \leq \Gamma(K(p))$$

holds.

Proof: For any invertible $n \times n$-matrix A, Lemma 8.1.1 yields

$$\|A^{-1}\|_{C^n} \leq \sum_{k=0}^{n-1} \frac{g^k(A)}{\sqrt{k!}d^{k+1}(A)},$$

where $d(A)$ is the smallest modulus of eigenvalues of A:

$$d(A) = \min_{k=1,\ldots,n} |\lambda_k(A)|.$$

Thus,

$$\|[K(p)]^{-1}\|_{C^n} \leq \sum_{k=0}^{n-1} \frac{g^k(K(p))}{\sqrt{k!}d^{k+1}(K(p))} \quad (p \in \mathbf{C}).$$

But from (1.7) it follows that $g(K(p)) = g(B(p) - pI) = g(B(p))$ $(p \in \mathbf{C})$. Hence we get the required result. □

If $B(p)$ is a normal matrix, then $g(B(p)) = 0$, and

$$\|[K(p)]^{-1}\|_{C^n} \leq d^{-1}(K(p)).$$

For example that inequality holds, if $K(p) = Be^{-ph} - Ip$, where B is a Hermitian matrix.

Furthermore, equation (1.1) has a sufficient complete system of solutions of the form

$$x(t) = e^{\lambda t}C,$$

where C is a constant vector. A substitution of that equality to (1.1) leads to equation

$$K(\lambda)C = 0.$$

So λ is a characteristic value, and C is an eigenvalue of the characteristic matrix. If the multiplicity of λ is $m_0 > 1$, then (1.1) can possess also solutions of the form

$$x(t) = e^{\lambda t}(C_0 + tC_1 + \ldots + t^{m-1}C_{m-1}) \ (m \leq m_0).$$

Clearly, if $Re\lambda < 0$, then the solution tends to the zero as $t \to \infty$.

8.2 Representation of solutions

Let us recall the basic properties of the Laplace transformation. Let a vector-valued function $F(t)$ with a locally bounded variation defined on $[0, \infty)$ satisfy the condition

$$\int_0^\infty e^{-c_0 t}\|F(t)\|_{C^n} dt < \infty \tag{2.1}$$

with a constant c_0. Then the integral

$$\tilde{F}(p) = \int_0^\infty e^{-pt} F(t)dt$$

with $Rep > c_0$ converges and represents in the half-plane $Rep > c_0$ an analytical function. Besides, $\tilde{F}(p)$ is the Laplace transform of $F(t)$, and $F(t)$ is the Laplace original of $\tilde{F}(p)$. Also, for any real $c > c_0$, there exists the limit

$$\lim_{a\to\infty} \frac{1}{2\pi i} \int_{c-ai}^{c+ai} \tilde{F}(p)e^{pt}dp = \frac{F(t+0)+F(t-0)}{2}.$$

Lemma 8.2.1 *Let a function $f(t)$ have the property $f(t) = O(e^{tb_0})$ $(t \to \infty)$ with a real constant b_0. Then any solution $x(t)$ of equation (1.1) with an initial function $\Phi(t)$, admits the representation*

$$x(t) = -\frac{1}{2\pi i} \int_{c-i\infty}^{c+i\infty} e^{tp} K^{-1}(p) \, [\Phi(0) + \int_0^\eta dR(\tau) \int_{-\tau}^0 e^{-(s+\tau)p}\Phi(s)ds +$$

$$\tilde{f}(p)]dp \ (t \geq 0, \ c = const), \tag{2.2}$$

where $\tilde{f}(p)$ is the Laplace transform of $f(t)$.

Proof: Integrating equation (1.1), we get

$$x(t) = x(0) + \int_0^t \int_0^\eta dR(s)x(t_1 - s)dt_1 + I(t) \ (t \geq 0).$$

where

$$I(t) = \int_0^t f(t_1)dt_1.$$

Hence, with the notation

$$V(R) = \int_0^\eta \|dR(s)\|_{C^n} = \lim_{max|s_{k+1}-s_k|\to 0} \sum_{k=0}^m \|R(s_{k+1}) - R(s_k)\|_{C^n}$$

$$(0 = s_0 < s_1 < \ldots < s_m = \eta),$$

we have

$$\|x(t)\| \leq \|x(0)\| + \|I(t)\| + \int_0^\eta \|dR(s)\| \int_{-s}^{t-s} \|x(\tau)\|d\tau \leq$$

$$\|x(0)\| + \|I(t)\| + V(R) \int_{-\eta}^t \|x(\tau)\|d\tau =$$

$$\|x(0)\| + \|I(t)\| + V(R) \int_{-\eta}^0 \|\Phi(\tau)\|d\tau + V(R) \int_0^t \|x(\tau)\|d\tau \ (t \geq 0).$$

Here $\|.\| = \|.\|_{C^n}$. But if $b_0 \geq 0$, then

$$\|I(t)\| \leq \int_0^t e^{b_0 t_1} dt_1 \leq C_\epsilon e^{(b_0+\epsilon)t} \ (C_\epsilon = const;\ t \geq 0)$$

for any $\epsilon > 0$. If $b_0 < 0$, then $\|I(t)\| \leq 1/|b_0|$. Thanks to the Gronwall lemma the solution satisfies condition (2.1) with $c_0 = \epsilon + V(R) + \max\{0, b_0\}$. So we can apply the Laplace transformation to equation (1.1). It gives

$$\int_0^\infty e^{-pt}\dot{x}(t)dt = -x(0) + p\tilde{x}(p), (Rep > c_0),$$

where $\tilde{x}(p)$ is the Laplace transform of $x(t)$. Further,

$$\int_0^\infty e^{-pt} \int_0^\eta dR(\tau)x(t-\tau)dt = \int_0^\eta dR(\tau) \int_0^\infty e^{-pt} x(t-\tau)dt =$$

$$\int_0^\eta dR(\tau) \int_{-\tau}^\infty e^{-(s+\tau)p} x(s)ds =$$

$$\int_0^\eta dR(\tau) \int_{-\tau}^0 e^{-(s+\tau)p} \Phi(s)ds + \int_0^\eta e^{-p\tau} dR(\tau)\tilde{x}(p).$$

Due to (1.1) we thus get,

$$-x(0) + p\tilde{x}(p) = \int_0^\eta dR(\tau) \int_{-\tau}^0 e^{-(s+\tau)p}\Phi(s)ds + \int_0^\eta e^{-p\tau} dR(\tau)\tilde{x}(p) + \tilde{f}(p).$$

Therefore, by virtue of the characteristic matrix , one can write

$$\tilde{x}(p) = -K^{-1}(p)[\Phi(0) + \int_0^\eta dR(\tau) \int_{-\tau}^0 e^{-(s+\tau)p}\Phi(s)ds + \tilde{f}(p)].$$

Now the inverse Laplace transformation yields the required representation. □

Corollary 8.2.2 *Let all the roots of the function* $detK(p)$ *lie in a half-plane*

$$\{p \in \mathbf{C} : Rep < \alpha\}$$

with a real α. *Then any solution* $\phi(t)$ *of the homogeneous equation*

$$\dot{\phi}(t) = \int_0^\eta dR(\tau)\phi(t-\tau) \ (t \geq 0) \tag{2.3}$$

with an initial continuous function $\Phi(t)$ *admits the representation*

$$\phi(t) = -\frac{1}{2\pi i} \int_{c_1 - i\infty}^{c_1 + i\infty} e^{tp} K^{-1}(p)[\Phi(0) + \int_0^\eta dR(\tau) \int_{-\tau}^0 e^{-(s+\tau)p}\Phi(s)ds]dp$$

$$(t \geq 0) \tag{2.4}$$

with an arbitrary $c_1 = const > \alpha$.

In fact, the function under the integral in (2.2) tends to zero as $Im\ p \to \infty$ uniformly with respect to $Rep > \alpha$. Therefore due to the residue theorem, we can take in (2.2) arbitrary $c_1 > \alpha$ as claimed. □

8.3 L^2-norm estimates for solutions of nonhomogeneous equations

8.3.1 Statement of the result

Let $K(p)$ be the characteristic matrix of the equation (1.1), again. *In the sequel it is assumed that all the zeros of det $K(p)$ lie in the open left half-plane.* To formulate the result, denote

$$\Gamma_0(K) \equiv \sup_{\omega \in \mathbf{R}} \Gamma(K(\omega i)).$$

Recall that $\Gamma(K(p))$ was introduced in Section 8.1.

Theorem 8.3.1 *Any solution $x(t)$ of the equation (1.1) with $f \in L^2([0,\infty), \mathbf{C}^n)$, and the initial condition (1.2) satisfies the inequality*

$$||x||_{L^2([0,\infty),\mathbf{C}^n)} \le \Gamma_0(K)||f||_{L^2([0,\infty),\mathbf{C}^n)} + ||\phi||_{L^2([0,\infty),\mathbf{C}^n)},$$

where $\phi(t)$ is a solution of the homogeneous equation (2.3) with an initial function Φ.

Below we will check that ϕ belongs to $L^2([0,\infty), \mathbf{C}^n)$, really (see Lemma 8.6.2).

All the proofs are presented in the next subsection.

Let us give a few estimates for $\Gamma_0(K)$. Recall that

$$V(R) = \int_0^\eta ||dR(s)||_{C^n},$$

and put

$$v(B) \equiv \sup_{\omega \in \mathbf{R}} g(B(i\omega)), \text{ and } d_V(K) \equiv \inf_{-2V(R) \le \omega \le 2V(R)} d(K(i\omega)).$$

Lemma 8.3.2 *The relationship*

$$\inf_{-\infty \le \omega \le \infty} d(K(i\omega)) = d_V(K) \le V(R)$$

is valid for the function $K(i\omega)$ defined by (1.3).

The latter lemma yields

$$\Gamma_0(K) \le \overline{\Gamma}_0(K) \equiv \sum_{k=0}^{n-1} \frac{v^k(B)}{\sqrt{k!}d_V^{k+1}(K)}.$$

Thanks to the definition of $g(A)$, one can write

$$g(B(i\omega)) \le N(B(i\omega)) \le \sqrt{n}\|B(i\omega)\|_{C^n} \le \sqrt{n}\int_0^\eta \|dR(s)\|_{C^n} = \sqrt{n}V(R).$$

$$(\omega \in \mathbf{R}).$$

Let us give a more precise estimate for $g(B(i\omega))$ in the case

$$K(p) = \sum_{k=0}^{m} A_k exp(-ph_k) - pI, \tag{3.1}$$

where A_k are constant matrices; $h_0 = 0$ and $h_k > 0$ $(k = 1, ..., m)$ are constants. Due to (1.7), one can write

$$g(B(i\omega)) \le \sqrt{1/2}\sum_{k=0}^{m} N(A_k exp(-i\omega h_k) - A_k^* exp(i\omega h_k)) \ (\omega \in \mathbf{R}).$$

Hence,

$$v(B) \le \sqrt{1/2}N(A_0 - A_0^*) + \sqrt{2}\sum_{k=1}^{m} N(A_k).$$

Note that in the case (3.1)

$$V(R) = \sum_{k=0}^{m} \|A_k\|_{C^n}.$$

8.3.2 Proofs

Lemma 8.3.3 *Let all the zeros of det $K(p)$ lie in the open left half-plane. In addition, let the condition*

$$\tilde{b}_0 = \sup_{\omega \in \mathbf{R}} \|K^{-1}(i\omega)\|_{C^n} < \infty$$

hold. Then a solution $x(t)$ of equation (1.1) with $f \in L^2([0,\infty), \mathbf{C}^n)$, and the initial condition (1.2) satisfies the inequality

$$\|x\|_{L[0,\infty)} \le \tilde{b}_0\|f\|_{L[0,\infty)} + \|\phi\|_{L[0,\infty)}$$

where ϕ is a solution of (2.3) with the same initial function Φ.

Proof: Due to Parseval's equality $\tilde{f}(i\omega) \in L^2((-\infty,\infty),\mathbf{C}^n)$, and the equality

$$\frac{1}{2\pi}\int_{-\infty}^{\infty}\|[K(iw)]^{-1}\tilde{f}(iw)\|^2_{C^n}dw = \|Q\|^2_{L[0,\infty)}$$

is valid with

$$Q(t) = \frac{1}{2\pi}\int_{-\infty}^{\infty}[K(iw)]^{-1}\tilde{f}(iw)exp(iwt)dw.$$

Thanks to Lemma 8.2.1, we can write down that $y(t) = Q(t) + \phi(t)$. Once more exploiting Parseval's equality, we have

$$\|Q\|^2_{L[0,\infty)} \leq \tilde{b}_0^2\frac{1}{2\pi}\int_{-\infty}^{\infty}\|\tilde{f}(iw)\|^2_{C^n}dw = \tilde{b}_0^2\|f\|^2_{L[0,\infty)}.$$

This proves the result. □

Lemma 8.3.4 *Let $\Gamma_0(K) < \infty$. Then the inequality $\tilde{b}_0 \leq \Gamma_0(K)$ is true.*

This result follows from Lemma 8.1.2.

The assertion of Theorem 8.3.1 follows from Lemmata 8.3.3 and 8.3.4.

Proof of Lemma 8.3.2: For any eigenvalue $\lambda_j(K(i\omega))$ of $K(i\omega)$ with a fixed ω , one can write

$$\lambda_j(K(i\omega)) = (K(i\omega)h_j(i\omega), h_j(i\omega)),$$

where $h_j(\omega) \in \mathbf{C}^n$ is the corresponding eigenvector with

$$\|h_j(\omega)\|_{C^n} = 1$$

and $(.,.)$ is the scalar product in $\mathbf{C}^n$. Similarly,

$$\overline{\lambda}_j(K(i\omega)) = (K^*(i\omega)h_j(i\omega), h_j(i\omega)).$$

Thus

$$Re\lambda_j(K(i\omega)) = (K_R(i\omega)h_j(i\omega), h_j(i\omega)),$$
$$Im\lambda_j(K(i\omega) = (K_I(i\omega)h_j(i\omega), h_j(i\omega)), \tag{3.2}$$

where

$$K_R(i\omega) = (K^*(i\omega) + K(i\omega))/2,$$

and

$$K_I(i\omega) = (K(i\omega) - K^*(i\omega))/2i.$$

But

$$K_R(i\omega) = \frac{1}{2}(\int_0^{\eta} e^{-is\omega}dR^*(s) + \int_0^{\eta} e^{is\omega}(i\omega s)dR(s)).$$

That is,

$$|Re\lambda_j(K(i\omega))| \leq V(R). \tag{3.3}$$

Besides,

$$K_I(i\omega) = (\int_0^{\eta} e^{-is\omega} dR^*(s) - \int_0^{\eta} e^{is\omega}(i\omega s) dR(s))/2i - \omega I.$$

Taking into account (3.2), we can easily check that $Im\lambda_j(i\omega)$ has zeros. Thus,

$$\min_{-\infty \le \omega \le \infty} |\lambda_k(K(i\omega)| \le |Re\ \lambda_k(K(i\omega_0)|,$$

where ω_0 is a zero of $Im\lambda_k(K(i\omega))$. Now (3.3) implies

$$\min_{-\infty \le \omega \le \infty} |\lambda_k(K(i\omega))| \le V(R). \tag{3.4}$$

On the other hand,

$$\|K(i\omega)h_j(i\omega)\|_{C^n} = \|(\int_0^{\eta} exp(i\omega s) dR(s) - Ii\omega) h_j(i\omega)\|_{C^n}.$$

That is,

$$\|K(i\omega)h_j(i\omega)\|_{C^n} \ge |\omega| - \|\int_0^{\eta} exp(i\omega s) dR(s) h_j(i\omega)\|_{C^n}$$

$$\ge |\omega| - V(R).$$

if $|\omega| \ge V(R)$. Consequently,

$$|\lambda_j(K(i\omega))| = \|K(i\omega)h_j(i\omega)\|_{C^n} \ge V(R)$$

if $|\omega| \ge 2V(R)$. This and (3.4) mean that the minimum of $|\lambda_j(K(i\omega)|$ is attained on the segment $|\omega| \le 2V(R)$ as claimed. □

8.4 Norm estimates for the Green function

The Green function (the fundamental solution) of equation (1.1) is a matrix-valued function $G(t)$ satisfying the equation

$$\dot{G}(t) = \int_0^{\eta} dR(\tau) G(t-\tau) \ (t \ge 0) \tag{4.1}$$

and the initial condition

$$G(0+) = I, \ G(t) = 0 \ (t < 0). \tag{4.2}$$

By virtue of the Laplace transformation, one can write

$$G(t) = -\frac{1}{2\pi} \int_{-\infty}^{\infty} [K(i\omega)]^{-1} exp(i\omega t) d\omega. \tag{4.3}$$

Recall that all the zeros of $det K(p)$ are in the left half-plane. Thanks to the properties of the convolution, Lemma 8.2.1 yields the representation

$$y(t) = \int_0^t G(t-s)f(s)ds \tag{4.4}$$

for any solution $y(t)$ of equation (1.1) with the zero initial conditions. Denote

$$W_0(K) \equiv \sqrt{2\Gamma_0(K)[V(R)\Gamma_0(K)+1]}.$$

Theorem 8.4.1 *The Green function $G(t)$ of equation (1.1) satisfies the inequality $||G||_L \leq W_0(K)$.*

First, we will prove the following

Lemma 8.4.2 *Let $f \in L^2([0,\infty), \mathbf{C}^n)$, and let a solution $x(t)$ of equation (1.1) with an initial function Φ, belong to $L^2([0,\infty), \mathbf{C}^n)$. Then $\dot{x}(t) \in L^2([0,\infty), \mathbf{C}^n)$, and the inequality*

$$||\dot{x}(t)||_{L[0,\infty)} \leq V(R)\ (||x||_{L[0,\infty)} + ||\Phi||_{L[-\eta,0]}) + ||f||_{L[0,\infty)}$$

is valid.

Proof: Obviously,

$$|| \int_0^\eta dR(s)x(t-s)||^2_{L[0,\infty)} = \int_0^\infty || \int_0^\eta dR(s)x(t-s)||^2_{C^n} dt \leq$$

$$(\int_0^\eta ||dR(s)||_{C^n})^2 \sup_{0\leq s\leq\eta} \int_0^\infty ||x(t-s)||^2_{C^n} dt =$$

$$V^2(R)\ (||x||^2_{L[0,\infty)} + ||\Phi||^2_{L[-\eta,0]}).$$

Now according to (1.1) we arrive at the required result. □

Proof of Theorem 8.4.1: Set $\psi(p) = (p+b)^{-1}I$ $(p \in \mathbf{C})$ with a positive constant b. Hence, due to (4.3)

$$G(t) = -\frac{1}{2\pi}\int_{-\infty}^{\infty} [K(i\omega)]^{-1} exp(i\omega t)(i\omega + b)\psi(i\omega)d\omega.$$

By using that equality, one can write

$$G(t) = \dot{w}(t) + bw(t), \tag{4.5}$$

where

$$w(t) = -\frac{1}{2\pi}\int_{-\infty}^{\infty} [K(i\omega)]^{-1} exp(i\omega t)\psi(i\omega)d\omega.$$

Since Ie^{-bt} is the Laplace original for $v(p)$, it is not hard to check that $w(t)$ is the solution of the equation

$$\dot{w}(t) = \int_0^\eta dR(s)w(t-s) + Ie^{-bt}$$

with the zero initial condition $w(t) = 0$ for $t \leq 0$ and $w(+0) = 0$. Thanks to Theorem 8.3.1, the relations

$$||w||_{L[0,\infty)} \leq \Gamma_0(K)||e^{-bt}||_{L[0,\infty)} = \Gamma_0(K)(2b)^{-1/2}$$

hold. Furthermore, due to Lemma 8.4.2 and the latter inequality, one can write

$$||\dot{w}||_{L[0,\infty)} \leq V(R)||w||_{L[0,\infty)} + ||e^{-tb}||_{L[0,\infty)} \leq \frac{V(R)\Gamma_0(K)+1}{\sqrt{2b}}.$$

Now (4.5) gives

$$||G||_{L[0,\infty)} \leq ||\dot{w}||_{L[0,\infty)} + b||w||_{L[0,\infty)} \leq \frac{V(R)\Gamma_0(K)+1+b\Gamma_0(K)}{\sqrt{2b}}.$$

Taking $b = \Gamma_0^{-1}(K)(V(R)\Gamma_0(K)+1)$, we obtain the required inequality. □

Theorem 8.4.3 *The derivative of the Green function of equation (1.1) satisfies the inequality*

$$||\dot{G}||_{L[0,\infty)} \leq W_0(K)V(R).$$

Proof: Take into account that $G(t)$ is the solution of (4.1) with the initial condition $G(0) = I$, $G(t) = 0$ $(t < 0)$. Then

$$||\dot{G}||^2_{L[0,\infty)} = ||\int_0^\eta dR(s)G(t-s)||^2_{L[0,\infty)} = \int_0^\infty ||\int_0^\eta dR(s)G(t-s)||^2_{C^n}dt.$$

But $G(t-s) = 0$ $(t < s)$. Hence,

$$||\dot{G}||^2_{L[0,\infty)} \leq (\int_0^\eta ||dR(s)||_{C^n})^2 \sup_{-\eta\leq s\leq 0}\int_0^\infty ||G(t-s)||^2_{C^n}dt = V^2(R)\ ||G||^2_{L[0,\infty)}.$$

Now the previous lemma yields the result. □

To prove an estimate for the sup-norm of the Green function we will apply the following trivial

Lemma 8.4.4 *Let a continuous vector-valued function h and its derivative $\dot{h}$ belong to space $L^2([0,\infty), \mathbf{C}^n)$. Then*

$$||h(t)||^2_{C^n} \leq 2[\int_t^\infty ||h(s)||^2_{C^n}ds \int_t^\infty ||\dot{h}(s)||^2_{C^n}ds]^{1/2}.$$

Proof: Since $h(t)$ is a continuous function and $||h||_{L[0,\infty)} < \infty$, due to the mean value theorem it can be written

$$\int_t^{t+1} ||h(s)||^2_{C^n} ds = ||h(\theta)||^2_{C^n} \to 0 \text{ as } t \to \infty \ (t \leq \theta \leq t+1).$$

Consequently, $h(\infty) = 0$. Obviously,

$$||h(t)||^2_{C^n} = -\int_t^\infty \frac{d||h(s)||^2_{C^n}}{ds} ds = -2\int_t^\infty ||h(s)||_{C^n} \frac{d||h(s)||_{C^n}}{ds} ds.$$

Taking into account that

$$|\frac{d}{ds}||h(s)||_{C^n}| \leq ||\dot{h}(s)||_{C^n},$$

we get the result due to Schwarz's inequality. □

By the latter lemma,

$$\sup_{t \geq 0} ||G(t)||_{C^n} \leq \sqrt{2||G||_{L[0,\infty)}||\dot{G}||_{L[0,\infty)}}.$$

Now Theorems 8.4.1 and 8.4.3 imply

Corollary 8.4.5 *The Green function of equation (1.1) satisfies the inequality*

$$\sup_{t \geq 0} ||G(t)||_{C^n} \leq a_0(K), \text{ where } a_0(K) \equiv 2\sqrt{V(R)\Gamma_0(K)[V(R)\Gamma_0(K)+1]}.$$

8.5 The Lyapunov exponent of the Green function

Let all the zeros of $det K(p)$ lie in the half-plane $\mathbf{C}_{\alpha_0} = \{z \in \mathbf{C} : Re z < -\alpha_0\}$ with a positive α_0. Furthermore, for a non-negative number ν, put

$$\Gamma_\nu(K) \equiv \sup_{\omega \in \mathbf{R}} \Gamma(K(-\nu + \omega i)),$$

and

$$V_\nu(R) \equiv \nu + \int_0^\eta e^{\nu s} ||dR(s)||_{C^n},$$

and

$$a_\nu(K) \equiv 2\sqrt{V_\nu(R)\Gamma_\nu(K)[V_\nu(R)\Gamma_\nu(K)+1]}.$$

That is, $V_0(R) = V(R)$. Assume that

$$0 < \nu < \alpha_0, \tag{5.1}$$

and substitute the equality

$$G_\nu(t) \equiv e^{\nu t}G(t) \tag{5.2}$$

in equation (4.1). Then

$$\dot{G}_\nu(t) = \nu G_\nu(t) + \int_0^\eta e^{\nu s} dR(s) G_\nu(t-s) \ (t \geq 0). \tag{5.3}$$

Since $G_\nu(t) = G(t) = 0$ $(t < 0)$ and $G_\nu(0) = G(0) = I$, function $G_\nu(t)$ is the Green one of equation (5.3). The corresponding characteristic matrix is $K_\nu(p) = B_\nu(p) - pI$ with

$$B_\nu(p) = \nu I + \int_0^\eta e^{(\nu-p)s} dR(s) = B_\nu(p) = \int_0^\eta e^{-ps} dR_\nu(s)$$

with

$$R_\nu(t) = \nu\chi(t)I + \int_0^t e^{\nu s} dR(s),$$

where $\chi(0) = 0$, and $\chi(t) = 1$ if $0 < t \leq \eta$. We can write

$$\int_0^\eta \|dR_\nu(s)\|_{C^n} \leq \nu + \int_0^\eta e^{\nu s} \|dR_\nu(s)\|_{C^n} = V_\nu(R).$$

Clearly, $K_\nu(i\omega) = K(-\nu + i\omega)$. Applying Corollary 8.4.5 to equation (5.3), and taking into account the second relation from (1.7), we get the estimate $\|G_\nu(t)\|_{C^n} \leq a_\nu(K)$ $(t \geq 0)$. Hence (5.2) yields the following

Lemma 8.5.1 *Let all the zeros of* $det K(p)$ *lie in half-plane* $\mathbf{C}_{\alpha_0}$. *Then under condition (5.1), the Green function* $G(t)$ *of equation (1.1) is subject to the estimate* $\|G(t)\|_{C^n} \leq a_\nu(K)e^{-\nu t}$ $(t \geq 0)$.

The Lyapunov exponent of the Green function is the number

$$\kappa(G) = \overline{\lim}_{t\to\infty} t^{-1} ln\|G(t)\|_{C^n}.$$

Lemma 8.5.1 yields

Corollary 8.5.2 *Let all the zeros of* $det K(p)$ *lie in half-plane* $\mathbf{C}_{\alpha_0}$. *Then the inequality* $\kappa(G) \leq -\alpha_0$ *is valid.*

8.6 Stability conditions

Lemma 8.6.1 *Let* $\phi(t)$ *be a solution of the homogeneous equation (2.3) with the initial condition (1.2). Then the representation*

$$\phi(t) = G(t)\Phi(0) + \int_0^\eta dR(s) \int_{-s}^0 G(t-\tau-s)\Phi(\tau)d\tau \tag{6.1}$$

is true

Proof: Since $-K(p)$ is the Laplace transform of $G(t)$, the required result is due to Corollary 8.2.2 and the representation (4.3). □

Equality (6.1) and Lemma 8.5.1 imply the inequality

$$\|\phi(t)\|_{C^n} \le a_\nu(K)e^{-\nu t}\|\Phi(0)\|_{C^n} +$$

$$\int_0^\eta \|dR(s)\|_{C^n} \int_{-s}^0 a_\nu(K)e^{-\nu(t-\tau-s)}\|\Phi(\tau)\|_{C^n}d\tau.$$

Therefore,

$$\|\phi(t)\|_{C^n} \le C_\nu e^{-\nu t}\|\Phi(\tau)\|_{C[-\eta,0]} \ (t \ge 0)$$

with

$$C_\nu = a_\nu(K)(1 + \int_0^\eta \|dR(s)\|_{C^n} \int_{-s}^0 e^{\nu(\tau+s)}d\tau).$$

We thus can assert that equation (2.3) is exponentially stable provided all the zeros of $\det K(p)$ *lie in the open left half-plane.*

Let us derive solution estimates for equation (2.3) which will be used in the next chapters.

Lemma 8.6.2 *Let $\phi(t)$ be a solution of the homogeneous equation (2.3) with the initial condition (1.2), and let all the zeros of $\det K(p)$ lie in the open left half-plane. Then*

$$\|\phi\|_{L[0,\infty)} \le W_0(K)\|\Phi(0)\|_{C^n} + \Gamma_0(K)\|\Phi\|_{L[-\eta,0]}V(R).$$

Proof: Denote

$$J_s(t) \equiv \int_{-s}^0 G(t-\tau-s)\Phi(\tau)d\tau.$$

Clearly,

$$J_s(t) = \int_0^s G(t-z)\Phi(z-s)dz = \int_0^\infty G(t-z)\Phi_s(z)dz$$

with the notation $\Phi_s(z) = \Phi(z-s)$ for $0 \le z \le s$, and $\Phi_s(z) = 0$ for $z \ge s$. Since $G(t) = 0$ for $t < 0$, we get

$$J_s(t) = \int_0^t G(t-z)\Phi_s(z)dz.$$

That is, $J_s(t)$ is a solution of the equation

$$\dot{x}(t) = \int_0^\eta dR(\tau)x(t-\tau) + \Phi_s(t)$$

with the zero initial condition. Using Theorem 8.3.1, we get

$$\|J_s\|_{L[0,\infty)} \le \Gamma_0(K)\|\Phi_s(.)\|_{L[0,\infty)}.$$

But

$$\|\Phi_s(.)\|^2_{L[0,\infty)} = [\int_0^s \|\Phi(z-s)\|^2_{C^n} dz] \leq \|\Phi\|^2_{L[-\eta,0]} \ (0 \leq s \leq \eta).$$

Thus $\|J_s\|_{L[0,\infty)} \leq \Gamma_0(K)\|\Phi\|_{L[-\eta,0]}$ and thanks to (6.1) we have

$$\|\phi\|_{L[0,\infty)} \leq \|G\|_{L[0,\infty)}\|\Phi(0)\|_{C^n} + \|\int_0^\eta dR(s)J_s(t)\|_{L[0,\infty)} \leq$$

$$\|G\|_{L[0,\infty)}\|\Phi(0)\|_{C^n} + \int_0^\eta \|dR(s)\|_{C^n} \max_{0\leq s\leq \eta} \|J_s(.)\|_{L[0,\infty)} \leq$$

$$\|G\|_{L[0,\infty)}\|\Phi(0)\|_{C^n} + V(R)\Gamma_0(K)\|\Phi\|_{L[-\eta,0]}.$$

Now Theorem 8.4.1 yields the required result. □
Lemmas 8.6.2 and 8.4.2 imply

Lemma 8.6.3 *Let $\phi(t)$ be a solution of the homogeneous equation (2.3) with the initial condition (1.2), and let all the zeros of $\det K(p)$ lie in the open left half-plane. Then $\dot\phi$ is in $L^2([0,\infty), \mathbf{C}^n)$. Moreover, the inequality*

$$\|\dot\phi(t)\|_L \leq V(R)\ [W_0(K)\|\Phi(0)\|_{C^n} + (\Gamma_0(K)V(R)+1)\|\Phi\|_{L[-\eta,0]}]$$

is valid.

Furthermore. due to Lemma 8.6.1 and Corollary 8.4.5,

$$\|\phi(t)\|_{C^n} \leq a_0(K)(\|\Phi(0)\|_{C^n} + \int_0^\eta \|dR(s)\|_{C^n} \int_{-s}^0 \|\Phi(\tau)\|_{C^n} d\tau). \tag{6.2}$$

Hence, we easily get

Corollary 8.6.4 *Let all the zeros of the characteristic function of (2.3) lie in the open left half-plane. Then any solution $\phi(t)$ of (2.3) with the initial condition (1.2) is subject to the inequality*

$$\|\phi\|_{C([0,\infty),\mathbf{C}^n)} \leq a_0(K)\|\Phi\|_{C([-\eta,0],\mathbf{C}^n)}(1+V(R)\eta).$$

8.7 Estimates for the C-norm of solutions of nonhomogeneous equations

Theorem 8.3.1 and Lemma 8.6.2 yield

Corollary 8.7.1 *Let all the zeros of the characteristic function of equation (1.1) lie in the open left half-plane, and let $f \in L^2([0,\infty), \mathbf{C}^n)$. Then any solution $x(t)$ of that equation with the initial condition (1.2) satisfies the inequality*

$$\|x\|_{L[0,\infty)} \leq \Gamma_0(K)(V(R)\|\Phi\|_{L[-\eta,0]} + \|f\|_{L[0,\infty)}) + W_0(K)\|\Phi(0)\|_{C^n}.$$

Now, thanks to Lemma 8.4.2, we get

Corollary 8.7.2 *Let all the zeros of the characteristic function of equation (1.1) lie in the open left half-plane, and let $f \in L^2([0,\infty), \mathbf{C}^n)$. Then the derivative $\dot{x}(t)$ of any solution of that equation with the initial condition (1.2) satisfies the inequality*

$$||\dot{x}(t)||_{L[0,\infty)} \leq V(R)\,[W_0(K)||\Phi(0)||_{C^n} + (\Gamma_0(K)V(R)+1)||\Phi||_{L[-\eta,0]}] +$$
$$(1 + V(R)\Gamma_0(K))||f||_{L[0,\infty)}.$$

Clearly, a solution of (1.1) is equal to

$$x(t) = y(t) + \varphi(t),$$

where $y(t)$ is a solution of (1.1) with the zero initial condition, and ϕ is a solution of (2.3). Due to (4.4) and Theorem 8.4.1, the Schwarz inequality implies

$$||y||_{C([0,\infty),\mathbf{C}^n)} \leq W_0(K)||f||_{L[0,\infty)}.$$

Now Corollary 8.6.4 yields

Theorem 8.7.3 *Let all the zeros of the characteristic function of equation (1.1) lie in the open left half-plane. Then any solution $x(t)$ of (1.1) with $f \in L^2([0,\infty), \mathbf{C}^n)$ and the initial condition (1.2) is bounded on $[0,\infty)$. Moreover, the inequality*

$$||x||_{C([0,\infty),\mathbf{C}^n)} \leq W_0(K)||f||_{L[0,\infty)} + a_0(K)||\Phi||_{C([-\eta,0],\mathbf{C}^n)}(1 + V(R)\eta)$$

is valid.

8.8 Perturbations of matrix-valued functions

8.8.1 General matrix-valued functions

Everywhere in this subsection $A(\lambda)$ and $B(\lambda)$ $(\lambda \in \mathbf{C})$ are entire $n \times n$-matrix-valued functions. Put

$$q(\lambda) \equiv ||A(\lambda) - B(\lambda)||,$$

where $||.||$ denotes the Euclidean norm.

A point $\mu \in \mathbf{C}$ is called *the characteristic value* of $A(.)$ if $det A(\mu) = 0$, and a point $\lambda_0 \in \mathbf{C}$ is called *the regular one* of $A(.)$ if $det A(\lambda_0) \neq 0$.

Furthermore, denote by $d(A(\lambda))$ the smallest modulus of eigenvalues

$$\lambda_k(A(\lambda)) \; (k = 1, ..., n)$$

of the matrix $A(\lambda)$ with a fixed $\lambda \in \mathbf{C}$. That is,

$$d(A(\lambda)) = \min_{k=1,\ldots,n} |\lambda_k(A(\lambda))|.$$

Theorem 8.8.1 *Each characteristic value μ of $B(.)$ satisfies the inequality*

$$q(\mu)\sum_{k=0}^{n-1}\frac{g^k(A(\mu))}{d^{k+1}(A(\mu))\sqrt{k!}} \geq 1. \tag{8.1}$$

First we will prove the following simple lemma

Lemma 8.8.2 *Each characteristic value μ of $B(.)$ satisfies the inequality*

$$q(\mu)\|A^{-1}(\mu)\| \geq 1.$$

Proof: Suppose that for some characteristic value μ of $B(.)$, the inequality

$$q(\mu)\|A^{-1}(\mu)\| < 1 \tag{8.2}$$

holds. We can write

$$A^{-1}(\mu) - B^{-1}(\mu) = B^{-1}(\mu)(B(\mu) - A(\mu))A^{-1}(\mu).$$

Therefore,

$$\|B^{-1}(\mu)\| \leq \|A^{-1}(\mu)\| + q(\mu)\|B^{-1}(\mu)\|\|A^{-1}(\mu)\|.$$

Condition (8.2) yields the relation

$$\|B^{-1}(\mu)\| \leq \|A^{-1}(\mu)\|(1 - q(\mu))^{-1}.$$

Thus, (8.2) implies that μ is a regular point of $B(.)$. That contradiction proves the result. □

Proof of Theorem 8.8.1: Lemma 8.1.1 implies

$$\|A^{-1}(\lambda)\| \leq \sum_{k=0}^{n-1}\frac{g^k(A(\lambda))}{d^{k+1}(A(\lambda))\sqrt{k!}} \text{ for all regular } \lambda.$$

Now applying Lemma 8.8.2, we arrive at the inequality (8.1). As claimed. □

Theorem 8.8.3 *Let all characteristic values of $A(.)$ lie in a domain $D \subset \mathbf{C}$ with a boundary ∂D and let the condition*

$$q(\lambda)\sum_{k=0}^{n-1}\frac{g^k(A(\lambda))}{d^{k+1}(A(\lambda))\sqrt{k!}} < 1 \tag{8.3}$$

hold for any $\lambda \in \partial D$. Then all characteristic values of $B(.)$ are in D as well.

Proof: Consider the matrix-valued function

$$B_t(\lambda) = A(\lambda) + t(B(\lambda) - A(\lambda)) \ (0 \leq t \leq 1).$$

On the contrary suppose that $B(\lambda)$ have an eigenvalue in $\mathbf{C}/D$. Obviously zeros of $detB_t(\lambda)$ continuously depend on t. Consequently, there is $t_0 \in [0,1]$ such that $B_{t_0}(\lambda)$ has an characteristic value $\mu \in \partial D$. But

$$||A(t) - B_t(\lambda)|| = t||B(\lambda) - A(\lambda)|| \leq tq(\lambda).$$

Therefore, according to Theorem 8.8.1, we must write

$$t_0 q(\mu) \sum_{k=0}^{n-1} \frac{g^k(A(\mu))}{d^{k+1}(A(\mu))\sqrt{k!}} \geq 1.$$

This contradicts condition (8.3). □

8.8.2 Perturbations of triangular matrix-valued functions

Let $T(\lambda) = (T_{jk}(\lambda))_{j,k}^n$ be the upper-triangular part of an entire matrix-valued function $B(\lambda) = (b_{jk}(\lambda))_{j,k=1}^n$. That is,

$$T_{jk}(\lambda) = b_{jk}(\lambda) \text{ if } j \leq k \text{ and } T_{jk}(t) = 0 \text{ if } j > k.$$

Since diagonal entries of a triangular matrix are its eigenvalues, we get

$$d(T(\lambda)) = min_{k=1,\dots,n}|b_{kk}(\lambda)|.$$

According to (1.6),

$$g(T(\lambda)) = [\sum_{1 \leq j < k \leq n} |b_{jk}(\lambda)|^2]^{1/2}.$$

Furthermore, denote

$$q_B(\lambda) = ||B(\lambda) - T(\lambda)||.$$

Then Theorem 8.8.3 with $A(\mu) = T(\mu)$ yields

Corollary 8.8.4 *Let all roots of the diagonal functions $b_{jj}(\lambda)$ $(j = 1, ..., n)$ of $B(.)$ lie in an open domain $D \subset \mathbf{C}$ with a boundary ∂D, and let*

$$q_B(\lambda) \sum_{k=0}^{n-1} \frac{g^k(T(\lambda))}{|b_{jj}(\lambda)|^{k+1}\sqrt{k!}} < 1 \ (j = 1, ..., n)$$

for any $\lambda \in \partial D$. Then all characteristic values of $B(\lambda)$ are in $D \subset \mathbf{C}$ as well.

8.9 Bounds for roots of characteristic functions of retarded systems

The present section is devoted to applications of the results obtained in the previous section to a characteristic matrix of the form

$$K(\lambda) = \sum_{\nu=0}^{m} A_\nu e^{-\lambda h_\nu} - \lambda I \ (m < \infty, \ \lambda \in \mathbf{C}), \tag{9.1}$$

where A_ν ($\nu = 0, ..., m$) are real constant $n \times n$-matrices, and

$$0 = h_0 < h_1 < ... < h_m < \infty$$

are constant numbers.

Let $T_\nu = (T_{jk}^{(\nu)})_{j,k=1}^n$ be the upper-triangular part of the matrix

$$A_\nu = (a_{jk}^{(\nu)})_{j,k=1}^n \ (\nu = 0, ..., m).$$

I.e.,

$$T_{jk}^{(\nu)} = a_{jk}^{(\nu)} \text{ if } j \leq k \text{ and } T_{jk}^{(\nu)} = 0 \text{ if } j > k \ (\nu = 0, ..., m). \tag{9.2}$$

Put

$$T(\lambda) = \sum_{\nu=0}^{m} T_\nu e^{-\lambda h_\nu} - \lambda I.$$

Thus, $T(\lambda)$ is the upper-triangular part of $K(\lambda)$, and the diagonal entries of $T(\lambda)$ are

$$b_{kk}(\lambda) = \sum_{\nu=0}^{m} a_{kk}^{(\nu)} e^{-h_\nu \lambda} - \lambda. \tag{9.3}$$

Besides, according to (1.6)

$$g(T(i\omega)) \leq g_T \equiv \sum_{\nu=0}^{m} g(T_\nu),$$

where

$$g(T_\nu) = [\sum_{1 \leq j < k \leq n} |a_{jk}^{(\nu)}|^2]^{1/2}. \tag{9.4}$$

Obviously,

$$||K(i\omega) - T(i\omega)||_{C^n} \leq q_0 \equiv \sum_{\nu=0}^{m} ||W_\nu||_{C^n},$$

where

$$W_\nu = (W_{jk}^{(\nu)})_{j,k=1}^n$$

are the lower-triangular parts of matrices

$$A_\nu \ (\nu = 0, ..., m).$$

That is,

$$W_{jk}^{(\nu)} = a_{jk}^{(\nu)} \text{ if } j > k \text{ and } W_{jk}^{(\nu)} = 0 \text{ if } j \leq k.$$

Corollary 8.8.4 yields

Theorem 8.9.1 *Let all the roots of functions $b_{kk}(\lambda)$ $(k = 1, ..., n)$ defined by (9.3) lie in the open left half-plane, and let*

$$q_0 \sum_{k=0}^{n-1} \frac{g_T^k}{|b_{jj}(i\omega)|^{k+1}\sqrt{k!}} < 1 \; (j = 1, ..., n)$$

for every $\omega \in \mathbf{R}$. Then all the roots of function $det K(\lambda)$, defined according to (9.1), are also in the open left half-plane.

8.10 Systems with small delays

Let us consider the equation

$$\dot{x}(t) = \sum_{k=0}^{m} A_k x(t - h_k) \;\; (t \geq 0), \tag{10.1}$$

where A_k $(k = 0, ..., m)$ are constant matrices and

$$0 = h_0 < h_1 < ... < h_m = \eta \equiv const < \infty$$

are numbers. Furthermore, set

$$\tilde{A} = \sum_{k=0}^{m} A_k, \text{ and } V_2 = \sum_{k=1}^{m} h_k \|A_k\|_{C^n},$$

and

$$\Gamma_1(\tilde{A}) = (\omega^2(\tilde{A}) + \alpha^2(\tilde{A}))^{1/2} \sum_{k=0}^{n-1} \frac{g^k(\tilde{A})}{|\alpha(\tilde{A})|^{k+1}\sqrt{k!}},$$

where

$$\omega(\tilde{A}) = \max_k |Im\lambda_k(\tilde{A})|; \; \alpha(\tilde{A}) = \max_k Re\lambda_k(\tilde{A}).$$

Theorem 8.10.1 *Let $\tilde{A}$ be a Hurwitz matrix. Let the condition*

$$\Gamma_1(\tilde{A}) V_2 < 1 \tag{10.2}$$

hold. Then equation (10.1) is exponentially stable.

This theorem is a particular case of Theorem 10.7.2 below.

8.11 Stability with respect to arbitrary delay

In some problems it is interesting to have the conditions under which a retarded system is stable with arbitrary delay. Let us consider the system

$$\dot{x} = Ax + Bx(t-h)\ (t \geq 0;\ h = const > 0), \tag{11.1}$$

where A, B are constant $n \times n$-matrices. We remind that

$$\Gamma(A) = \sum_{k=0}^{n-1} \frac{g^k(A)}{\sqrt{k!}|\alpha(A)|^{k+1}}.$$

Theorem 8.11.1 *Let A be a Hurwitz matrix, and the relation $\|B\|_{C^n}\Gamma(A) < 1$ hold. Then equation (11.1) is exponentially stable.*

This theorem is a particular case of Theorem 10.9.1 below. Clearly, the stability conditions do not depend on the delay h.

8.12 Notes

Note 8.1 Numerous systems of automatic control are described by differential equations with delaying arguments. In particular, in the late thirties and early forties, Minorsky (1942) in his study of ship stabilization and automatic steering, pointed out very clearly the importance of considering delay in feedback mechanisms. The great interest in control theory during those and later years has certainly contributed significantly to the rapid development of the theory of differential equations with dependence on the past state. For a discussion of the physical applications of the differential difference equation see (Minorsky, 1962).

A class of difference-delay equations was encountered by Ergen (1954) in the theory of a circulating fuel nuclear reactor and has been extensively studied by Levin and Nohel (1964).

Taking into account the transmission time in the triode oscillator, Rubanik (1969) has employed the Van der Pol equation with the delayed argument. This list can be lengthened.

Note 8.2 The theory of the stability of linear autonomous retarded systems is extensive (see (Bellman and Cooke, 1963), (Pinney, 1958), (Kolmanowskii and Nosov, 1986), and references therein).

In the case of discrete delays, the stability analysis can be reduced to an analysis of a quasipolynomial of the form

$$D(z) = \sum_{i=0}^{m}\sum_{j=0}^{n} a_{ij} z^i exp(b_j z)$$

with positive b_j. In particular, a quasipolynomial with commensurable delays may be represented as

$$D(z) = \sum_{i=0}^{m} \sum_{j=0}^{n} a_{ij} z^i exp(jz).$$

The most general results on roots of that quasipolynomial are due to (Pontryagin, 1955). We state the results of Pontryagin (Theorems 8.12.1-8.12.4) without proofs. Put

$$P(z, w) = \sum_{i=0}^{m} \sum_{j=0}^{n} a_{ij} z^i w^j$$

Then $D(z) = P(z, e^z)$. A term $a_{i_0 j_0} z_0^i w_0^j$ is called the principal term of the polynomial $P(z, w)$, if $a_{i_0 j_0} \neq 0$, and if, for each other term $a_{ij} z^i w^j$ with $a_{ij} \neq 0$ we have either $j < j_0, i < i_0$, or $j = j_0, i < i_0$, or $j < j_0, i = i_0$.

Theorem 8.12.1 *If the polynomial $P(z, w)$ has no a principal term, then the equation $P(z, e^z) = 0$ has an infinite number of zeros with arbitrarily large positive real parts.*

The basic results for applications are the next three theorems.

Theorem 8.12.2 *Let $D(z) = P(z, e^z)$ and suppose $P(z, w)$ is a polynomial with a principal term $a_{rs} z^r w^s$. All of the zeros of $D(z)$ have negative real parts if and only if*

(i) The complex vector $D(i\omega)$ rotates in the positive direction with a positive velocity for ω ranging in $(-\infty, \infty)$.

(ii) For $\theta \in [-2k\pi, 2k\pi]$, $k \geq 0$ is an integer, there is an $\epsilon_k \to 0$ as $k \to \infty$ such that $D(i\omega)$ subtends an angle $4k\pi s + \pi r + \epsilon_k$.

Theorem 8.12.3 *Let $D(z) = P(z, e^z)$, where $P(z, w)$ is a polynomial with a principal term. Suppose $D(i\omega)$ $(\omega \in \mathbf{R})$ is separated into its real and imaginary parts, $D(i\omega) = F(\omega) + iG(\omega)$. If all the zeros of $D(z)$ have negative real parts, then all the zeros of $F(\omega)$ and $G(\omega)$ are real, simple, alternate and*

$$G'(\omega)F(\omega) - F'(\omega)G(\omega) > 0 \qquad (12.1)$$

for every $\omega \in \mathbf{R}$. Conversely, all the zeros of $P(z, e^z)$ have negative real parts provided that either of the following conditions are satisfied:

(i). All the zeros of $F(\omega)$ and $G(\omega)$ are real, simple, alternate and inequality (12.1) holds for at last one ω.

(ii). All the zeros of $F(\omega)$ are real and, for each zero inequality (12.1) is satisfied.

(iii). All the zeros of $G(\omega)$ are real and, for each zero inequality (12.1) is satisfied.

Now suppose that $f(z, u, v)$ is a polynomial in z, u, v with real coefficients, which has the form

$$f(z, u, v) = \sum_{m=0}^{r} \sum_{n=0}^{s} z^m \phi_m^{(n)}(u, v) \tag{12.2}$$

where $\phi_m^{(n)}(u, v)$ is a homogeneous polynomial of degree n in u, v. *The principal term* in the polynomial $f(z, u, v)$ is the term $z^r \phi_r^{(s)}(u, v)$, for which either $r > m$, $s > n$, or $r = m$, $s > n$, or $r > m$, $s = n$ for all other terms in (12.2).

Let $z^r \phi_r^{(s)}(u, v)$ denote the principal term of $f(z, u, v)$ in (12.2), let $\phi_*^{(s)}(u, v)$ denote the coefficient of z^r in $f(z, u, v)$,

$$\phi_*^{(s)}(u, v) = \sum_{n=0}^{s} \phi_r^{(n)}(u, v),$$

and let

$$\Phi_*^{(s)}(z) = \phi_*^{(s)}(\cos z, \sin z).$$

Theorem 8.12.4 *Let $f(z, u, v)$ be a polynomial with a principal term*

$$z^r \phi_r^{(s)}(u, v).$$

If ϵ is such that $\Phi_*^{(s)}(\epsilon + i\omega) \neq 0$, $\omega \in \mathbf{R}$, then for sufficiently large integers k, the function $F(z) = f(z, \cos z, \sin z)$ will have exactly $4ks + r$ zeros in the strip $-2k\pi + \epsilon \leq Re\ z \leq 2k\pi + \epsilon$. Consequently, the function $F(z)$ will have only real roots if and only if, for sufficiently large integers k, it has exactly $4ks + r$ zeros in the strip $-2k\pi + \epsilon \leq Re\ z \leq 2k\pi + \epsilon$.

The following result is due to Hayes (1950) (see also (Hale, 1977, Appendix)).

Theorem 8.12.5 *Let a, b be real. Then all roots of the equation*

$$(z + a)e^z + b = 0$$

have negative real parts if and only if

$$a > -1;$$

$$a + b > 0;$$

$$b < z_0 \sin z_0 - a \cos z_0$$

where z_0 is the root of the equation $z = -a \tan z$ $(0 < z < \pi)$ if $a \neq 0$, and $z_0 = \pi/2$ if $a = 0$.

The following theorem is proved in (Hale, 1977, Appendix).

Theorem 8.12.6 *All the roots the of equation*

$$(z^2 + az)e^z + 1 = 0$$

have negative real parts if and only if $a > (\sin z)/z$ *where* z *is a unique root of the equation* $z^2 = \cos z$, $0 < z < \pi$.

Note 8.3 To determinate the conditions under which quasipolynomials have no roots in the right half-plane, the method of D-subdivision is also widely applied. An excellent description of this method can be found in (Pinney, 1958), (Kolmanovskii and Nosov, 1986), etc. Frequency methods are also used in the theory of automatic control to investigate the stability of autonomous systems. For retarded systems frequency methods were developed by Kabakov (1946) and Tsypkin (1946). These methods are based on the argument principle from complex analysis. Consider the characteristic determinant

$$d(z) = det[zI - \int_{-\eta}^{0} e^{-sz} dR(s)]. \tag{12.3}$$

As above, $R(s)$ is an $n \times n$-matrix having a bounded variation.

Theorem 8.12.7 *(The Michailov criterion). For exponential stability of linear n-th order equation with the characteristic determinant* $d(i\omega)$, *it is necessary and sufficient that variation of* $arg\ d(i\omega)$ *is equal to* $n\pi/2$ *when* ω *varies from* 0 *to* ∞. *I.e.,*

$$arg\ d(i\omega)|_0^\infty = n\pi/2.$$

For the proof see (Kabakov, 1946) and (Tsypkin, 1946). The usage of the Michailov criterion for retarded systems is difficult because of the complexity of the corresponding curves. Let us mention the so called integral frequency criterion (Melkumjan, 1981) which is more effective. Let $d(i\omega)$ be defined by (12.3). Besides $U(\omega)$ and $V(\omega)$ are its real and imaginary parts, respectively: $d(i\omega) = U(\omega) + iV(\omega)$. Define the logarithmic derivative

$$L(\omega) = \frac{U(\omega)V'(\omega) - V(\omega)U'(\omega)}{U^2(\omega) + V^2(\omega)}$$

Theorem 8.12.8 *(The integral stability criterion). For exponential stability of an autonomous linear retarded system with the characteristic determinant* $d(i\omega)$, *it is necessary and sufficient that*

$$\int_0^\infty L(\omega) d\omega = \frac{n\pi}{2}.$$

A detailed description of frequency methods for retarded systems is presented in (Kolmanovski and Nosov, 1986).

Note 8.4 The representation of solutions of autonomous retarded systems by the Laplace transformation considered in Section 8.2 is well-known cf. (Myshkis, 1972). Sections 8.3-8.7 are based on the works (Gil', 1994), (Gil', 1997). The material in Sections 8.8 and 8.9 is adapted from (Gil', 1995, Chapter 4). Theorem 8.10.1 on the stability of systems with a small delay is taken from (Gil, 1997). Theorem 8.11.1 on the stability of systems with an arbitrary delay is taken from (Gil, 1998).

Various algorithms are devoted to computing delay margins for stability of linear delay systems (see (Chen, *et all*, 1995), (Chiasson, 1988), and references therein). Mainly, systems with one delay or with commensurable delays are considered. The issue of delay margins is closely related to the problem of stability for arbitrary delay or stability independent of delay. The following result is due to Boese (1991).

Theorem 8.12.9 *Let $P(z)$ and $Q(z)$ be entire functions, and*

$$|Q(iy)| < |P(iy)| \ (y \in \mathbf{R}).$$

If, in addition, $P(z)$ has no zeros in the closed right half-plane, then the function $P(z) + Q(z)e^{-zh}$ $(h > 0)$ has no zeros in the closed right half-plane.

This result particularly includes the Tsypkin criterion (Kolmanovskii and Nosov, 1992, p.65).

For some other relevant results see, for instance, (Stepán, 1989), (Cooke and van den Drieshe, 1986), (Gopasalmy, 1984), (MacDonald, 1989).

References for Chapter 8

Bellman, R. and K.L. Cooke (1963). *Differential-Difference Equations.* Academic Press, New York.

Boese, F. G. (1991). Delay-independent stability of a special sequence of neutral differential-difference equations, *J. Diff. Equations, 90*, 397-407.

Chen, J., Gu, G., and C.N. Nett (1995). A new method for computing delay margins for stability of linear delay systems, *System & Control Letters, 26*, 107-117.

Chiasson J.N. (1988). A method for computing the intervals of delay values for which difference-delay system is stable, *IEEE Trans. Autom. Control, AC-33*, 1176-1178.

Cooke, K.L. and P. van den Drieshe (1986). On zeros of some transcendental equations, *Funkcialaj Ekvacioj 29*, 77-90.

Ergen, W.K. (1954). Kinetic of the circulating fuel nuclear reactor, *J. Appl. Phys. 25*, 702-711.

Gil', M.I. (1994). On absolute stability of differential-delay systems, *IEEE, Trans. Automatic Control, 39, N 12*, 2481-2484.

Gil'. M. I. (1995). *Norm Estimations for Operator-valued Functions and Applications*. Marcel Dekker, Inc. New York.

Gil', M.I. (1997). On the freezing method for differential equations with delay, *Analysis, 17*, 25-33.

Gil', M.I. (1998). On exponential stability of nonlinear systems with delaying argument, *Nonlinear Analysis. Theory, Methods and Applications, 31, No 5/6*, 755-764 .

Gopasalmiy, K. (1984). Some delay does not matter, *Bull. Austral. Math. Soc., 30*, 91-98.

Gopasalmy, K. (1992). *Stability and Oscillations in Delay Differential Equations of Population Dynamics*. Kluwer Academic Publishers, Dordrecht.

Hale, J. (1977). *Theory of Functional Differential Equations*, Springer- Verlag, New York.

Hale, J. K. and Verduyn Lunel, S. M. (1993). *Introduction to Functional Differential Equations*. Springer-Verlag, New York.

Hayes, N. D. (1950). Roots of the transcendental equation associated with a certain differential difference equation, *J. London Math. Soc. 25*, 226-232

Kabakov, I. P. (1946). Concerning the control process for the steam pressure, *Ingener. Sbornik, 2*, 27-76. In Russian.

Kolmanovskii, V.B. and V.K. Nosov (1986). *Stability of Functional Differential Equations*. Ac. Press, New York.

Levin J.J. and J. Nohel (1964). On a nonlinear delay equations, *J. Math. An. Appl., 8*, 31-44.

MacDonald, N. (1989). *Biological Delay Systems: Linear Stability Theory*. Cambridge Univ. Press, Cambridge.

Melkumjan, D.O. (1981). *Analysis of Systems by the Method of the Logarithmic Derivative*. Energoatomizdat, Moscow. In Russian

Minorsky, N. (1942). Self-excited oscillations in dynamical systems possessing retarded actions, *J. Appl. Mech. 9*, 65-71.

Minorsky, N. (1962). *Nonlinear Oscillations.* D.Van Nostrand Company, Inc., Princeton.

Myshkis, A.D. (1972). *Linear Differential Equations with Delaying Argument.* Nauka, Moscow. In Russian.

Pinney, E. (1958). *Ordinary Difference-Differential Equations.* University Calif. Press. Berkeley.

Pontryagin, L. S. (1955). On zeros of some elementary transcendental functions, *Trans. of AMS, Series 2, v. 1*, 96-110.

Rubanik, V.P. (1969). *Oscillations of Quasilinear Systems with Retardation.* Nauka, Moscow. In Russian.

Stepán, G. (1989). *Retarded Dynamical Systems: Stability and Characteristic Functions.* Longman Scientific, Technical, Pitman Research Notes in Math. series, no. 210.

Tsypkin, Ja. Z. (1946). Stability of systems with delayed feedback, *Autom. and Remote Control 7(2)*, 107-129.

Chapter 9

LINEAR TIME-VARIANT SYSTEMS WITH DELAY

In the present chapter we consider linear nonautonomous systems with delay.

In section 9.1 we introduce the corresponding definitions of stability.

Section 9.2 is devoted to the stability of systems with bounded coefficients and separated autonomous parts.

The freezing method for systems with delay is developed in Sections 9.3 and 9.4.

Integrally small perturbations of nonautonomous systems are examined in Section 9.5.

In Section 9.6 we establish the generalized Wazewski and Lozinskii inequalities.

Linear time-variant retarded systems with small delays are investigated in Section 9.7.

9.1 Definitions

Consider the linear retarded system

$$\dot{x}(t) = \int_0^\eta d_s W(t,s)x(t-s) \ (0 < \eta < \infty;\ t \geq 0) \tag{1.1}$$

Here $W(t,s)$ is a matrix-valued function defined on $[0,\infty) \times [0,\eta]$ and having for all $t \geq 0$ a bounded variation in s. In addition, W is continuous in t and $W(t,0) \equiv 0$. For example, the following equation can be written in the form (1.1):

$$\dot{x}(t) = A_0(t)x(t) + A_1(t)x(t-\eta) + \int_0^\eta T(t,s)x(t-s)ds,$$

where A_0, A_1 are variable matrices, and $T(t,s)$ is a matrix-valued function defined on $[0,\infty) \times [-\eta, 0]$. Take the initial condition

$$x(t) = \Phi(t) \ (\Phi \in C([-\eta,0];\mathbf{C}^n); \ -\eta \le t \le 0). \tag{1.2}$$

A solution of the initial problem (1.1), (1.2) is defined as in Section 8.1.

Lemma 9.1.1 *Under the above assumptions, problem (1.1), (1.2) has a unique solution.*

This lemma is a particular case of the well-known Theorem 6.1.1 from the book (Hale, 1977).

We will say that equation (1.1) is stable if there is a constant N_0 such that

$$\|x(t)\|_{\mathbf{C}^n} \le N_0 \|\Phi\|_{C([-\eta,0],\mathbf{C}^n)} \ (t \ge 0)$$

for any solution $x(t)$ of problem (1.1), (1.2). Recall that $\|.\|_{\mathbf{C}^n}$ is the Euclidean norm.

We will say that equation (1.1) is uniformly stable if for any $t_0 \ge 0$ there is a constant N independent of t_0 such that

$$\|x(t)\|_{\mathbf{C}^n} \le N \sup_{t_0-\eta \le t \le t_0} \|x(t)\|_{\mathbf{C}^n} \ (t \ge t_0)$$

Equation (1.1) *is asymptotically stable* if it is stable and any its solution $x(t)$ has the property $x(t) \to 0$ as $t \to \infty$. Equation (1.1) *is uniformly asymptotically stable* if it is uniformly stable and $x(t) \to 0$ as $t \to \infty$.

The exponential stability for problem (1.1), (1.2) is defined as in Section 8.1.

9.2 Stability of retarded systems with bounded coefficients

Again, $R(\tau)$ is an $n \times n$-variable matrix defined on finite segment $[0,\eta]$, and having a bounded variation, and $W(t,s)$ is the same as in the previous section

Consider in $\mathbf{C}^n$ an equation with a separated autonomous part of the form

$$\dot{x}(t) = \int_0^\eta dR(\tau)x(t-\tau) + \int_0^\eta d_s W(t,s)x(t-s) \ (t \ge 0). \tag{2.1}$$

Let us introduce a linear operator $L : C([-\eta,\infty),\mathbf{C}^n) \to C([0,\infty),\mathbf{C}^n)$ by the relation

$$(Lx)(t) = \int_0^\eta d_s W(t,s)x(t-s) \ (x \in C([-\eta,\infty),\mathbf{C}^n)). \tag{2.2}$$

Then (2.1) takes the form

$$\dot{x}(t) = \int_0^\eta dR(\tau)x(t-\tau) + (Lx)(t).$$

It is assumed that for any $u \in L^2([-\eta,\infty),\mathbf{C}^n) \cap C([-\eta,\infty),\mathbf{C}^n)$,

$$\|Lu\|_{L[0,\infty)} \equiv \left(\int_0^\infty \|(Lu)(t)\|^2_{C^n} dt\right)^{1/2} \le q\|u\|_{L[-\eta,\infty)}. \quad (2.3)$$

For example, let L be defined as

$$(Lu)(t) = \sum_{j=1}^m A_j(t)u(t-h_j(t)) \quad (t \ge 0), \quad (2.4)$$

where $A_j(t)$ are bounded on $[0,\infty)$ continuous matrices:

$$q_j = \sup_{t\ge 0} \|A_j(t)\|_{C^n} < \infty, \quad (2.5)$$

and $h_j(t)$ $(j = 1,...,m)$ are differentiable scalar-valued functions with the properties

$$1 - \dot{h}_j(t) \ge b_j, \text{ and } 0 \le h_j(t) \le \eta \ (t \ge 0,\ j = 1,...,m) \quad (2.6)$$

with constants $b_j > 0$.

Proposition 9.2.1 *Let relations (2.5) and (2.6) be fulfilled. Then condition (2.3) holds with*

$$q = \sum_{j=1}^m q_j b_j^{-1/2}. \quad (2.7)$$

Proof: Relation (2.5) gives us the inequality

$$b_j\|u(t-h_j(t))\|^2_{L[0,\infty)} \le \int_0^\infty \|u(t-h_j(t))\|^2_{C^n}(1-\dot{h}_j(t))dt.$$

Thus,

$$b_j\|u(t-h_j(t))\|^2_{L[0,\infty)} \le \int_{-h_j(0)}^\infty \|u(s)\|^2_{C^n} ds = \|u\|^2_{L[0,\infty)} +$$

$$\int_{-h_j(0)}^0 \|u(s)\|^2_{C^n} ds \le \|u\|^2_{L[0,\infty)} + \|u\|^2_{L[-\eta,0)}$$

for $j = 1,...,m$. Therefore

$$\|\sum_{j=1}^m A_j(t)u(t-h_j(t))\|_{L[0,\infty)} \le \sum_{k=1}^m q_j\|u(t-h_j(t))\|_{L[0,\infty)} \le$$

$$\sum_{k=1}^m q_j(b_j)^{-1/2}(\|u\|_{L[0,\infty)} + \|u\|_{L[-\eta,0]})$$

as claimed. □

Recall also that

$$K(p) = \int_0^\eta exp(-ps)dR(s) - Ip,$$

$$\Gamma(K(p)) = \sum_{k=0}^{n-1} \frac{g^k(B(p))}{\sqrt{k!}d^{k+1}(K(p))} \quad (p \in \mathbf{C}), \text{ and } \Gamma_0(K) \equiv \sup_{\omega \in \mathbf{R}} \Gamma(K(\omega i)),$$

where $d(K(p))$ is the smallest modulus of eigenvalues of $K(p)$ (see Sections 8.1 and 8.3). Recall that estimates for $\Gamma_0(K)$ are presented in Section 8.3.

Theorem 9.2.2 *Let all the zeros of* $det K(p)$ *belong to the open left half-plane. In addition, let the conditions (2.3) and*

$$q\Gamma_0(K) < 1 \tag{2.8}$$

hold. Then equation (2.1) is asymptotically stable.

This theorem is a particular case of Theorem 10.3.2 below.

Corollary 9.2.3 *Let all the zeros of* $det K(p)$ *lie in the open left half-plane, and condition (2.6) hold. Moreover, let* $A_j(t)$ *be continuous matrices with the property*

$$\int_0^\infty \|A_j(t)\|_{C^n} dt < \infty \ (j = 1, ..., m). \tag{2.9}$$

Then the equation

$$\dot{x}(t) = \int_0^\eta dR(\tau)x(t-\tau) + \sum_{j=1}^m A_j(t)x(t-h_j(t)) \quad (t \geq 0)$$

is asymptotically stable.

In fact, under condition (2.9) we have $\|A_j(t)\|_{C^n} \to 0$ as $t \to \infty$. So for any $q_j > 0$ there is a $t_0 \geq 0$ such that

$$\sup_{t \geq t_0} \|A_j(t)\|_{C^n} \leq q_j.$$

Taking sufficiently small q_j and applying Proposition 9.2.1 and Theorem 9.2.2 with t_0 instead of 0, we arrive at the result.

Example 9.2.4 *Consider in* $\mathbf{C}^n$ *the system*

$$\dot{x} = Ax + (Lx)(t), \tag{2.10}$$

where A *is a constant Hurwitz matrix,* L *is defined by (2.2) and satisfies condition (2.3).*

Clearly, in the considered case, $K(p) = A - pI$ and the equality $g(B(p)) \equiv g(A)$ is valid. It is easy to see that

$$d(K(iw)) = d(A - Iwi) \geq |\alpha(A)| = \min_{k=1,\ldots,n} |Re\lambda_k(A)|$$

for all real w. Therefore,

$$\Gamma_0(K) = \Gamma(A) \equiv \sum_{k=0}^{n-1} \frac{g^k(A)}{\sqrt{k!}|\alpha(A)|^{k+1}}.$$

If $q\Gamma(A) < 1$, then system (2.10) is asymptotically stable due to Theorem 9.2.2.

Example 9.2.5 *Consider the system*

$$\dot{y}_1(t) + a_{11}y_1(t) + a_{12}y_2(t) + a_{13}y_3(t) + b_1y_1(t-1)$$
$$= \sum_{k=1}^{3} c_{1k}(t)y_k(t) + e_{1k}(t)y_k(t - h(t)),$$
$$\dot{y}_2(t) + a_{22}y_2(t) + a_{23}y_3(t) + b_2y_2(t-1)$$
$$= \sum_{k=1}^{3} c_{2k}(t)y_k(t) + e_{2k}(t)y_k(t - h(t)),$$

$$\dot{y}_3(t) + a_{33}y_3(t) + b_3y_3(t-1) = \sum_{k=1}^{3} c_{3k}(t)y_k(t) + e_{3k}(t)y_k(t - h(t)), \quad (2.11)$$

where $y = (y_1, y_2, y_3)$; a_{kk}, b_k are real numbers, and $c_{jk}(t), e_{jk}(t), h(t)$ are bounded real functions defined on $[0, \infty)$. In addition,

$$1 - \dot{h}(t) \geq b > 0, \; 0 < h(t) \leq 1 \; (t \geq 0). \quad (2.12)$$

Under consideration, we have

$$(Ly)(t) = C(t)y(t) + E(t)y(t - h(t)),$$

where

$$C(t) = (c_{jk}(t))_{j,k=1}^3; \; E(t) = (e_{jk}(t))_{j,k=1}^3.$$

Due to Proposition 9.2.1 and (2.12), condition (2.3) holds with

$$q = q_0 + q_1 b^{-1/2},$$

where

$$q_0 = \sup_{t\geq 0} \|C(t)\|_{C^n}, \; q_1 = \sup_{t\geq 0} \|E(t)\|_{C^n}.$$

Further, clearly $K(p) = -A - Bexp(-p) - pI$, where

$$A = (a_{jk})_{j,k=1}^3; \; B = diag[b_k]_{k=1}^3$$

with

$$a_{jk} = 0 \text{ for } j > k \ (j, k = 1, 2, 3).$$

So $K(p)$ is a triangular matrix, and we have

$$g(B(p)) = \sqrt{a_{12}^2 + a_{13}^2 + a_{23}^2} = g(A).$$

(see relation (1.6) from Section 8.1). Besides,

$$d(K(p)) = \min_{k=1,2,3} |p + a_{kk} + b_k exp(-p)|.$$

Equation $p + a_{kk} + b_k exp(-p) = 0$ is equivalent to the following one:

$$(p + a_{kk})e^p + b_k = 0.$$

Thanks to Theorem 8.12.5, all the roots of that equation have negative real parts if and only if the conditions

$$a_{kk} > -1; \ a_{kk} + b_k > 0; \ b_k < z_k sin \ z_k - acos \ z_k \ (k = 1, 2, 3) \tag{2.13}$$

hold, where z_k is a root of the equation $z = -a_{kk} \ tg \ z \ (0 < z < \pi)$ if $a \neq 0$, and $z_k = \pi/2$ if $a_{kk} = 0$. Put

$$d_0 = \inf_{\omega \in \mathbf{R}} d_0(K(i\omega)) \equiv \inf_{\omega \in \mathbf{R}} \min_{k=1,2,3} |i\omega + a_{kk} + b_k exp(-i\omega)|.$$

Thanks to Theorem 9.2.2, we can assert that if the inequalities (2.13), and

$$(q_0 + q_1 b^{-1/2})(\frac{1}{d_0} + \frac{g(A)}{d_0^2} + \frac{g^2(A)}{2d_0^3}) < 1$$

hold, then system (2.11) is asymptotically stable.

9.3 The freezing method for systems with delay

Consider in $\mathbf{C}^n$ the system

$$\dot{x}(t) = A_0(t)x(t) + A_1(t)x(t-h_1) + ... + A_m(t)x(t-h_m) \ (t \geq 0), \tag{3.1}$$

where $h_k \ : \ 0 = h_0 < h_1 < ... < h_m$ are positive constants, and $A_k(t)$ are variable $n \times n$-matrices with the properties

$$\|A_k(t) - A_k(s)\|_{C^n} \leq q_k \ |t - s|$$

$$(q_k = const \geq 0; \ t, s \geq 0; \ k = 0, ..., m), \tag{3.2}$$

and

$$\sup_{s\geq 0}\sum_{k=0}^{m}\|A_k(s)\|_{C^n}<\infty. \tag{3.3}$$

For a fixed $s \geq 0$, let $K(s,p)$ be the characteristic matrix of equation (3.1). That is,

$$K(s,p) = B(s,p) - pI \text{ where } B(s,p) = \sum_{k=0}^{m} A_k(s)e^{-ph_k} \ (p \in \mathbf{C}).$$

To formulate the result put

$$\Gamma(K(s,p)) = \sum_{k=0}^{n-1}\frac{g^k(B(s,p))}{\sqrt{k!}d^{k+1}(K(s,p))} \quad (p\in\mathbf{C}),$$

where $d(K(s,p))$ is the smallest modulus of eigenvalues of the matrix $K(s,p)$:

$$d(K(s,p)) = \min_{k=1,\dots,n}|\lambda_k(K(s,p))|.$$

Throughout this section it is assumed that all the zeros of $\det K(s,p)$ *for every* $s \geq 0$ *belong to a half-plane* $\{z \in \mathbf{C} : Re\, z \leq -\alpha_0\}$ with a positive α_0. For a non-negative number ν, denote

$$\overline{V}_\nu = \nu + \sup_{s\geq 0}\sum_{k=0}^{m} e^{\nu h_k}\|A_k(s)\|_{C^n},$$

$$\overline{\Gamma}_\nu \equiv \sup_{t\geq 0}\sup_{\omega\in\mathbf{R}}\Gamma(K(t,-\nu+i\omega)), \text{ and } \overline{a}_\nu \equiv 2\sqrt{\overline{V}_\nu\overline{\Gamma}_\nu(\overline{V}_\nu\overline{\Gamma}_\nu+1)}.$$

Now we are in a position to formulate the main result of the section.

Theorem 9.3.1 *Let conditions (3.2), (3.3) be fulfilled. In addition, for a positive number*

$$\nu < \alpha_0, \tag{3.4}$$

let the inequality

$$\tilde{q} \equiv \sum_{k=0}^{m} q_k < \frac{\nu^2}{\overline{a}_\nu} \tag{3.5}$$

hold. Then equation (3.1) is stable.

The proofs are presented in the next section. Let us give an estimate for $\overline{\Gamma}_\nu$. Put

$$\tilde{g}_\nu(B) \equiv \sup_{s\geq 0}\sup_{\omega\in\mathbf{R}} g(B(s,-\nu+i\omega))$$

and

$$d_\nu(K) \equiv \inf_{s\geq 0}\inf_{\omega\in\mathbf{R}} d(K(s,-\nu+i\omega)).$$

One can write

$$\overline{\Gamma}_\nu \le \sum_{k=0}^{n-1} \frac{\tilde{g}_\nu^k(B)}{\sqrt{k!}d_\nu^{k+1}(K)}. \tag{3.6}$$

Due to Proposition 2.1.1

$$g(B(s, -\nu + i\omega)) \le \sqrt{1/2}\sum_{k=0}^{m} e^{\nu h_k} N(A_k(s)exp(-i\omega h_k) - A_k^*(s)exp(i\omega h_k))$$

$$(s \ge 0,\ \omega \in \mathbf{R}).$$

Hence,

$$\tilde{g}_\nu(B) \le \sup_{s\ge 0}[\sqrt{1/2}N(A_0(s) - A_0^*(s)) + \sqrt{2}\sum_{k=1}^{m} e^{\nu h_k} N(A_k(s))]. \tag{3.7}$$

Example 9.3.2 *Let us consider the equation*

$$\ddot{x}(t) + 2(c + sin\beta t)\dot{x}(t) + x(t) - \gamma x(t-1) = 0\ (\beta, \gamma. c = const > 0). \tag{3.8}$$

Besides, $c > 3$, and $\gamma \le 1/3$. Clearly $q_0 = 2\beta, q_1 = 0$ and thus $\tilde{q} = 2\beta$. Rewriting equation (3.8) in the form (3.1), we get $B(s,p) = A_0(s) + A_1e^{-p}$ with

$$A_0(s) = \begin{pmatrix} -2(c + sin\beta s) & -1 \\ 1 & 0 \end{pmatrix} \text{ and } A_1 = \begin{pmatrix} 0 & \gamma \\ 0 & 0 \end{pmatrix}.$$

So

$$K(s,p) = \begin{pmatrix} -2(c + sin\beta s) - p & -1 + \gamma e^{-p} \\ 1 & -p \end{pmatrix}.$$

Consequently,

$$det K(s,p) = p^2 + 2(c + sin\beta s)p + 1 - \gamma e^{-p}.$$

Hence

$$\lambda_{1,2}(K(s,p)) = -c - sin\beta s \pm \sqrt{(c + sin\beta s)^2 - 1 + \gamma e^{-pt}}.$$

Take

$$m_0 = \inf_{t\ge 0}(c + sin\beta t - \sqrt{(c + sin\beta t)^2 - 1/2}\ ) \text{ and } \alpha_0 = min\{m_0, ln\frac{1}{2\gamma}\}.$$

Then for $Rep \ge -\alpha_0$,

$$Re\lambda_{1,2}(K(s,p)) \le -c - sin\beta s + \sqrt{(c + sin\beta s)^2 - 1 + 1/2} \le -\alpha_0 < 0. \tag{3.9}$$

This fact shows that the roots of $det(A_0 + A_1e^{-p} - pI)$ lie in the half-plane $\{p \in \mathbf{C} : Rep < -\alpha_0\}$. Take an arbitrary positive $\nu < \alpha_0$. For instance, one can take $\nu = \alpha_0/2$. We have

$$|\lambda_{1,2}(K(t, -\nu + i\omega))| = |\nu - i\omega - c - sin\beta t \pm \sqrt{(c + sin\beta t)^2 - 1 + \gamma e^{\nu - i\omega}}| \ge d_\nu$$

for all $\omega \in \mathbf{R}$, where

$$d_\nu = \inf_{t\geq 0}(c - \nu + \sin\beta t - \sqrt{(c + \sin\beta t)^2 - 1 + \gamma e^\nu}).$$

According to (3.9), $d_\nu > 0$, since $\nu < \alpha_0$. Further, due to (3.7)

$$g(B(s, -\nu + i\omega)) \leq 2 + \sqrt{2}\gamma e^\nu \ (\omega \in \mathbf{R}).$$

Thus, by virtue of relation (3.6)

$$\overline{\Gamma}_\nu \leq \Gamma_{1\nu} = d_\nu^{-1} + (2 + \sqrt{2}\gamma e^\nu) d_\nu^{-2}.$$

In addition,

$$\overline{V}_\nu = \nu + \sup_{s\geq 0} \|A_0(s)\|_{C^2} + e^\nu \|A_1\|_{C^2} \leq \nu + 2(c+2) + \gamma e^\nu,$$

and

$$\overline{a}_\nu \leq 2\sqrt{\overline{V}_\nu \Gamma_{1\nu}(\overline{V}_\nu \Gamma_{1\nu} + 1)}.$$

Due to Theorem 9.3.1, if for some $\nu < \alpha_0$, the inequality

$$\beta < \frac{\nu^2}{4\sqrt{\overline{V}_\nu \Gamma_{1\nu}(\overline{V}_\nu \Gamma_{1\nu} + 1)}}$$

holds, then equation (3.8) is stable.

9.4 Proof of Theorem 9.3.1

Put

$$L(s)x(t) \equiv A_0(s)x(t) + A_1(s)x(t - h_1) + \ldots + A_m(s)x(t - h_m) \ (t, s \geq 0).$$

For a fixed $s \geq 0$, denote by $G_s(t)$ the Green function of the equation

$$\dot{x}(t) - L(s)x(t) = 0. \tag{4.1}$$

Lemma 9.4.1 *Let the inequality*

$$\|G_s(t)\|_{C^n} \leq C_\nu e^{-\nu t} \ (t, s \geq 0) \tag{4.2}$$

hold with constants $\nu > 0$ and C_ν independent of s. If, in addition, conditions (3.2), (3.3), and

$$\tilde{q} C_\nu < \nu^2 \tag{4.3}$$

are fulfilled, then equation (3.1) is stable.

Proof: Let $x(t)$ be a solution of equation (3.1) with the initial condition

$$x(t) = \Phi(t) \; (-h_m \leq t \leq 0). \tag{4.4}$$

Fix $s \geq 0$ and rewrite (3.1) in the form

$$\dot{x}(t) - L(s)x(t) = (L(t) - L(s))x(t).$$

Setting $\pi(t,s) = (L(t) - L(s))x(t)$, we get

$$\dot{x}(t) - L(s)x(t) = \pi(t,s).$$

A solution of the latter equation with the initial condition (4.4) can be represented as

$$x(t) = \phi_s(t) + \int_0^t G_s(t-\tau)\pi(\tau,s)d\tau \; (t \geq 0), \tag{4.5}$$

where $\phi_s(t)$ is the solution of the homogeneous equation (4.1) with the initial condition (4.4).

Let us use the representation of solutions of homogeneous autonomous systems. Taking into account that $\phi_s(t)$ is a solution of (4.1), we can write according to Lemma 8.6.1

$$\phi_s(t) = G_s(t)\Phi(0) + \sum_{k=1}^{m} A_k(s) \int_{-h_k}^{0} G_s(t-\tau-h_k)\Phi(\tau)d\tau.$$

That representation, (3.3) and (4.2) give

$$||\phi_s(t)||_{C^n} \leq c_1 < \infty \; (c = const; t, s \geq 0).$$

Moreover,

$$c_1 \leq c_2 ||\Phi||_{C([-\eta,0],\mathbf{C}^n)} \; (c_2 = const). \tag{4.6}$$

From (4.2) and (4.5) the inequalities

$$||x(t)||_{C^n} \leq ||\phi_s(t)||_{C^n} + \int_0^t ||G_s(t-\tau)||_{C^n} ||\pi(\tau,s)||_{C^n} d\tau \leq$$

$$c_1 + C_\nu \int_0^t e^{-\nu(t-s)} ||\pi(\tau,s)||_{C^n} d\tau$$

follow. According to (3.2)

$$||\pi(\tau,s)||_{C^n} \leq \sum_{k=0}^{m} ||(A_k(\tau) - A_k(s))x(\tau-h_k)||_{C^n} \leq$$

$$\sum_{k=0}^{m} q_k |s-\tau| ||x(\tau-h_k)||_{C^n}.$$

Take $t = s$. We get

$$\|x(t)\|_{C^n} \leq c_1 + C_\nu \int_0^t e^{-\nu(t-\tau)} \sum_{k=0}^m q_k(t-\tau)\|x(\tau - h_k)\|_{C^n} d\tau =$$

$$c_1 + C_\nu \sum_{k=0}^m q_k \int_{-h_k}^{t-h_k} e^{-\nu(t-z-h_k)}(t - z - h_k)\|x(z)\|_{C^n} dz.$$

Hence

$$\|x(t)\|_{C^n} \leq c_1 + C_\nu \sum_{k=0}^m q_k \int_0^{t-h_k} e^{-\nu(t-z-h_k)}(t - z - h_k)\|x(z)\|_{C^n} dz + c_3,$$

where

$$c_3 = \sup_{t\geq 0} C_\nu \sum_{k=0}^m \int_{-h_k}^0 e^{-\nu(t-z-h_k)} q_k(t - z - h_k)\|\Phi(z)\|_{C^n} dz.$$

Setting $m(x, t_0) = \max_{0\leq t\leq t_0} \|x(t)\|_{C^n}$ for a fixed $t_0 > 0$, we arrive at the relations

$$m(t_0) \leq c_5 + C_\nu m(t_0) \sum_{k=0}^m q_k \int_0^{t-h_k} e^{-\nu(t-z-h_k)}(t - z - h_k) dz \leq$$

$$c_5 + C_\nu m(t_0)\tilde{q} \int_0^\infty e^{-\nu\tau}\tau d\tau = c_5 + C_\nu m(t_0)\tilde{q}\nu^{-2} \ (c_5 = c_1 + c_3).$$

But condition (4.3) implies the inequality

$$m(t_0) \leq c_5(1 - C_\nu \tilde{q}\nu^{-2})^{-1}.$$

Taking into account that t_0 is arbitrary, we arrive at the estimate

$$\sup_{t\geq 0} \|x(t)\|_{C^n} \leq c_5(1 - C_\nu \tilde{q}\nu^{-2})^{-1} < \infty.$$

Clearly,

$$c_3 \leq c_4\|\Phi\|_{C([-\eta,0],\mathbf{C}^n)} < \infty \ (c_4 = const).$$

Since $c_5 = c_1 + c_3$, that inequality and (4.6) yield the required result. □

Proof of Theorem 9.3.1: To obtain estimate (4.2) let us apply Lemma 8.5.1. Clearly, in the case of equation (4.1)

$$V_\nu(R) = \nu + \sum_{k=0}^m e^{\nu h_k}\|A_k(s)\|_{C^n} \leq \overline{V}_\nu \ (s \geq 0),$$

and $\Gamma(K(s, -\nu + i\omega)) \leq \overline{\Gamma}_\nu$ $(s \geq 0)$. Due to Lemma 8.5.1,

$$\|G_s(t)\|_{C^n} \leq \overline{a}_\nu e^{-\nu t} \ (t, s \geq 0).$$

Now Lemma 9.4.1 yields the required result. □

9.5 Integrally small perturbations

9.5.1 Solution estimates

Consider in $\mathbf{C}^n$ the system

$$\dot{x}(t) = A(t)x(t) + B(t)x(t-h) \quad (t \geq 0, h = const > 0), \tag{5.1}$$

where $B(t)$ is a variable matrix. Take the initial condition

$$x(t) = \Phi(t) \text{ for } -h \leq t \leq 0 \tag{5.2}$$

with a given continuous vector-valued function $\Phi(t)$. Let us suppose that the function

$$\alpha_W(t) \equiv \sup_{x \in \mathbf{C}^n} Re \frac{((A(t) + B(t+h))x, x)}{(x, x)}$$

is a Riemann integrable one. Here $(.,.)$ is the scalar product. The aim of the present section is to prove the following

Theorem 9.5.1 *Let the inequality*

$$J_B \equiv \sup_{t \geq 0} \int_t^{t+h} ||B(s)||_{C^n} ds < 1 \tag{5.3}$$

be fulfilled. Then for any solution $x(t)$ of (5.1) with a continuous initial function $\Phi(t)$, the estimate

$$||x(t)||_{C^n} \leq (1 - J_B)^{-1} c_W(\Phi) exp[\int_0^t (\alpha_W(s) + \eta_W(s))ds] \; (t \geq h) \tag{5.4}$$

is valid with the notation

$$\eta_W(s) = (1 - J_B)^{-1} \, ||B(s+h)||_{C^n} \int_s^{s+h} ||A(\tau) + B(\tau + h)||_{C^n} e^{\int_\tau^s \alpha_W(\tau_1) d\tau_1} d\tau,$$

and

$$c_W(\Phi) = ||\Phi(0)||_{C^n} + \int_0^h e^{-\int_0^s \alpha_W(\tau)d\tau} ||B(s)\Phi(s-h)||_{C^n} ds.$$

The proof of that theorem is presented in Subsection 9.5.3

9.5.2 Stability conditions

Theorem 9.5.1 yields.

Corollary 9.5.2 *Under condition (5.3), let the inequality*

$$\overline{\lim}_{t \to \infty} \int_0^t [\alpha_W(s) + \eta_W(s)]ds < \infty \tag{5.5}$$

be fulfilled. Then equation (5.1) is stable.

Let us assume that under the condition $\alpha_W(t) \leq 0$ $(t \geq 0)$, the inequality

$$\tilde{J}_{A+B} \equiv \sup_{t\geq 0} \int_t^{t+h} \|A(s) + B(s+h)\|_{C^n} ds < \infty \text{ and} \tag{5.6}$$

is fulfilled. Then clearly,

$$\Lambda_W \equiv \sup_{t\geq 0} \int_t^{t+h} |\alpha_W(s)| ds < \infty$$

and

$$\int_0^t \eta_W(s) ds \leq \tilde{J}_{A+B}(1-J_B)^{-1} e^{\Lambda_W} \int_0^t \|B(s+h)\|_{C^n} ds.$$

Furthermore, take into account that

$$\overline{\lim}_{t\to\infty} t^{-1} \int_0^t \|B(s+h)\|_{C^n} ds \leq \overline{\lim}_{j\to\infty} (hj)^{-1} \int_0^{(j+1)h} \|B(s+h)\|_{C^n} ds \leq$$

$$\overline{\lim}_{j\to\infty} (hj)^{-1} \sum_{k=0}^{j} \int_{kh}^{(k+1)h} \|B(s+h)\|_{C^n} ds \leq h^{-1} J_B.$$

Now Corollary 9.5.2 implies

Corollary 9.5.3 *Under conditions (5.3) and (5.6), let the inequality*

$$\overline{\lim}_{t\to\infty} t^{-1} \int_0^t \alpha_W(s) ds + \frac{e^{\Lambda_W} J_B \tilde{J}_{A+B}}{h(1-J_B)} \leq 0$$

be fulfilled. Then equation (5.1) is stable.

Now consider the scalar equation

$$\dot{x}(t) = -b(t)x(t-h) \tag{5.7}$$

with a continuous positive function $b(t)$. Denote by z_0 the unique real root of the equation

$$z(1+e^z) = 1.$$

Clearly $0 < z_0 < 1$. Assume that

$$J_B = \sup_{t\geq 0} \int_t^{t+h} b(s) ds < z_0. \tag{5.8}$$

Then $J_B < 1$, and $J_B(1+e^{J_B}) < 1$. since the function $z(1+e^z)$ increases as $z > 0$ increases. Therefore, $J_B e^{J_B} < 1 - J_B$, and

$$(1-J_B)^{-1} J_B e^{J_B} < 1. \tag{5.9}$$

On the other hand, we can write

$$\eta_W(s) = (1 - J_B)^{-1}\, b(s+h) \int_s^{s+h} b(\tau + h) e^{-\int_\tau^s b(\tau_1+h)d\tau_1} d\tau \leq$$

$$(1 - J_B)^{-1}\, b(s+h) J_B e^{J_B}.$$

Relation (5.9) implies

$$\int_0^t [-b(s+h) + J_B e^{J_B}(1 - J_B)^{-1}\, b(s+h)]ds \leq 0.$$

Since in the considered case $\alpha_W(t) = -b(t+h)$, condition (5.5) holds, and according to Corollary 9.5.2 the following result is true:

Corollary 9.5.4 *Let inequality (5.8) be fulfilled. Then equation (5.7) is stable.*

Note that Theorem 9.5.1 can be easily extended to the equation

$$\dot{x}(t) = \sum_{k=0}^{m} A_k(t) x(t - h_k) \;\; (t \geq 0,\; h_0 = 0),$$

where $A_k(t)$ $(k = 0, ..., m)$ are variable matrices, $h_1, ..., h_m$ are positive numbers.

9.5.3 Proof of Theorem 9.5.1

Put $\tilde{A}(t) = A(t) + B(t+h)$, and denote by $U(t,s)$ the evolution operator of the equation

$$\dot{x}(t) = \tilde{A}(t) x(t).$$

Let us suppose that there is a Riemann integrable function $\alpha(t)$ and a positive constant K such that

$$||U(t,s)||_{C^n} \leq K e^{\int_s^t \alpha(\tau)d\tau} \;\; (t, s \geq 0). \tag{5.10}$$

Lemma 9.5.5 *Let the inequalities (5.10) and*

$$\theta \equiv K \sup_{t \geq 0} \int_t^{t+h} ||B(s)||_{C^n} ds < 1$$

be fulfilled. Then for any solution $x(t)$ of (5.1) with a continuous initial function $\Phi(t)$, the estimate

$$||x(t)||_{C^n} \leq (1-\theta)^{-1} c(\Phi) exp[\int_0^t (\alpha(s) + \eta(s))ds] \;\; (t \geq h)$$

is valid with the notation

$$\eta(s) = K(1-\theta)^{-1}\, ||B(s+h)||_{C^n} \int_s^{s+h} ||A(\tau) + B(\tau+h)||_{C^n} e^{\int_\tau^s \alpha(\tau_1)d\tau_1} d\tau,$$

and

$$c(\Phi) = K[\|\Phi(0)\|_{C^n} + \int_0^h e^{-\int_0^s \alpha(\tau)d\tau}\|B(s)\Phi(s-h)\|_{C^n}ds].$$

Proof: For the sake of the simplicity throughout this proof we set $\|.\|_{C^n} = |.|$. Rewrite (5.1) in the form

$$\dot{x}(t) = \tilde{A}(t)x(t) + B(t)x(t-h) - B(t+h)x(t) \ (t \geq 0).$$

This equation is equivalent to the following one:

$$x(t) = U(t,0)x(0) + \int_0^t U(t,s)[B(s)x(s-h) - B(s+h)x(s)]ds \ (t \geq 0).$$

Hence,

$$x(t) = U(t,0)x(0) + \int_0^{t-h} [U(t,s+h) - U(t,s)]B(s+h)x(s)ds -$$

$$\int_{t-h}^t U(t,s)B(s+h)x(s)ds + f_1(t) \ (t \geq h), \tag{5.11}$$

where

$$f_1(t) = \int_0^h U(t,s)B(s)\Phi(s-h)ds.$$

By (5.10) we get

$$|\int_{t-h}^t U(t,s)B(s+h)x(s)ds| \leq K\int_{t-h}^t e^{\int_s^t \alpha(\tau)d\tau}|B(s+h)||x(s)|ds. \tag{5.12}$$

Further, it can be written,

$$\tilde{J}(t) \equiv |\int_0^{t-h} [U(t,s+h) - U(t,s)]B(s+h)x(s)ds| \leq$$

$$\int_0^{t-h} |U(t,s+h) - U(t,s)||B(s+h)||x(s)|ds \ (t \geq h).$$

Take into account that

$$\frac{\partial}{\partial s}U(t,s) = -U(t,s)\tilde{A}(s)$$

(see Section 1.6). Hence by (5.10)

$$|U(t,s+h) - U(t,s)| \leq \int_s^{s+h} |U(t,\tau)\tilde{A}(\tau)|d\tau \leq$$

$$e^{\int_s^t \alpha(\tau_1)d\tau_1} K \int_s^{s+h} e^{\int_\tau^s \alpha(\tau_1)d\tau_1}|\tilde{A}(\tau)|d\tau.$$

Therefore

$$\tilde{J}(t) \leq \int_0^{t-h} e^{\int_s^t \alpha(\tau)d\tau} w(s)|x(s)|ds, \tag{5.13}$$

where

$$w(s) = K|B(s+h)| \int_s^{s+h} e^{\int_\tau^s \alpha(\tau_1)d\tau_1} |\tilde{A}(\tau)|d\tau.$$

In addition, thanks to (5.10)

$$|U(t,0)x(0) + f_1(t)| \leq K[e^{\int_0^t \alpha(\tau)d\tau}|x(0)| + \int_0^h e^{\int_s^t \alpha(\tau)d\tau}|B(s)\Phi(s-h)|ds] =$$

$$e^{\int_0^t \alpha(\tau)d\tau} c(\Phi).$$

Now (5.11), (5.12) and (5.13) imply

$$|x(t)| \leq e^{\int_0^t \alpha(\tau)d\tau} c(\Phi) + \int_0^{t-h} e^{\int_s^t \alpha(\tau)d\tau} w(s)|x(s)|ds +$$

$$K \int_{t-h}^t e^{\int_s^t \alpha(\tau)d\tau} |B(s+h)||x(s)|ds.$$

Put

$$y(t) = |x(t)|e^{-\int_0^t \alpha(\tau)d\tau}. \tag{5.14}$$

Clearly,

$$y(t) \leq c(\Phi) + K\int_{t-h}^t |B(s+h)|y(s)ds + \int_0^{t-h} w(s)y(s)ds \ (t \geq h).$$

Denote

$$\zeta(t) = \max_{0 \leq s \leq t} y(s).$$

Then for a fixed positive $T < \infty$,

$$\zeta(T) \leq c(\Phi) + \theta\zeta(T) + \int_0^{T-h} w(s)\zeta(s)ds.$$

Under the condition $\theta < 1$ this gives

$$\zeta(T) \leq (1-\theta)^{-1}[c(\Phi) + \int_0^T w(s)\zeta(s)ds].$$

Replacing T by t and applying the Gronwall inequality, we get

$$\zeta(t) \leq (1-\theta)^{-1} c(\Phi)\, exp\,[\,(1-\theta)^{-1}\int_0^t w(s)ds].$$

Since $y(t) \leq \zeta(t)$, due to (5.14), we arrive at the required estimate. □

Proof of Theorem 9.5.1: The result is due to Lemma 9.5.5, and the Wazewski inequality

$$\|U(t,s)\|_{C^n} \leq e^{\int_s^t \alpha_W(\tau)d\tau} \ (t, s \geq 0).$$

9.6 The generalized Wazewski and Lozinskii inequalities

In this section $\|.\|_n$ and $\|.\|_{\mathbf{C}^n}$ are the arbitrary and Euclidean norms in $\mathbf{C}^n$, respectively.

Let $A(t), B(t)$ be variable $n \times n$-matrices. The present section is devoted to a generalization of the Wazewski inequality (see Section 4.2) to the equation

$$\dot{x} = A(t)x + B(t)x(h(t)) \ (t \geq 0), \tag{6.1}$$

where $h(t)$ is an increasing differentiable scalar-valued function defined on $[0, \infty)$ and satisfying the inequalities

$$-\eta \leq h(t) \leq t \ (t \geq 0; \ 0 < \eta = const < \infty). \tag{6.2}$$

Let $U(t,s)$ be the evolution operator (the fundamental solution) of the "shorten" equation

$$\dot{v} = A(t)v \ (t \geq 0).$$

Assume that there is a bounded measurable scalar-valued function $a(t) \leq 0$, such that the inequality

$$\|U(t,s)\|_n \leq e^{\int_s^t a(\tau)d\tau} \ (s, t \geq 0) \tag{6.3}$$

holds, and put

$$q(s) = \|B(s)\|_n \ exp[-\int_{h(s)}^s a(\tau)d\tau],$$

where $\psi(z)$ is the function inverse to $h(s)$. That is, if $\tau = h(s)$, then $s = \psi(\tau)$. Now we are in a position to formulate the main result of the section.

Theorem 9.6.1 *Let inequality (6.3) hold. Then any solution $x(t)$ of (6.1) with an initial function Φ satisfies the inequality*

$$\|x(t)\|_n \leq \ z(t) exp[\int_0^t a(\tau)d\tau] \ (t \geq \psi(0)), \tag{6.4}$$

where $z(t)$ is a solution of the scalar equation

$$z(t) = e_0(\Phi) + \int_{\psi(0)}^{t} q(s)z(h(s))ds \ (t \geq \psi(0)) \tag{6.5}$$

with the notation

$$e_0(\Phi) = \|\Phi(0)\|_n + \int_0^{\psi(0)} exp[\int_s^0 \alpha(\tau)d\tau]\|B(s)\|_n \|\Phi(h(s))\|_n ds.$$

This theorem is a particular case of Theorem 10.10.1 below. Further, put

$$\Lambda(t) = \overline{\lim}_{h \to +0} \frac{1}{h}[\|I_{\mathbf{C}^n} + hA(t)\|_n - 1].$$

Corollary 9.6.2 *Let $\Lambda(t)$ be a Riemann integrable function. Then inequality (6.4) holds with $a(t) = \Lambda(t)$.*

Indeed, thanks to the Lozinskii inequality (see Section 4.2), condition (6.3) is valid with $a(t) = \Lambda(t)$. Now the result is due to the latter theorem. So the latter corollary is a generalization of the Lozinskii inequality.

Corollary 9.6.3 *Let the conditions (6.2), and*

$$Re(A(t)g, g) \leq a(t)\|g\|^2_{\mathbf{C}^n} \quad (g \in \mathbf{C}^n)$$

be fulfilled. Then estimate (6.4) holds.

Indeed, this result is due to Theorem 9.6.1 and the Wazewski inequality (see Section 4.2). So the latter corollary is a generalization of the Wazewski inequality.

Since $h(t) \leq t$, we get by equation (6.5)

$$z(t) \leq e_0(\Phi) + \int_0^t \dot{\psi}(s_1)q(\psi(s_1))z(s_1)ds_1.$$

Due to the Gronwall inequality,

$$z(t) \leq e_0(\Phi)exp\{\int_0^t \dot{\psi}(s_1)q(\psi(s_1))ds_1\}.$$

Now Theorem 9.6.1 gives the estimate

$$\|x(t)\|_{\mathbf{C}^n} \leq \ e_0(\Phi) \ exp\{\int_0^t [a(\tau) + \dot{\psi}(\tau)q(\psi(\tau))]d\tau]\} =$$

$$e_0(\Phi) \ exp\{\int_0^t a(\tau)d\tau + \int_{\psi(0)}^{\psi(t)} q(s)ds\} \ (t \geq \psi(0)). \tag{6.6}$$

We thus have derived

Corollary 9.6.4 *Under the hypothesis of Theorem 9.6.1, let the condition*

$$\overline{\lim}_{t\to\infty}\{\int_0^t a(\tau)d\tau + \int_0^{\psi(t)} q(s)ds\} < \infty$$

be fulfilled. Then equation (6.1) is stable

Let the relation

$$\Lambda_1 \equiv \overline{\lim}_{t\to\infty} t^{-1}\{\int_0^t a(\tau)d\tau + \int_0^{\psi(t)} q(s)ds\} < \infty$$

hold. Then due to (6.6) for the Lyapunov exponent of a solution $x(t)$ of equation (6.1) we have the inequality

$$\kappa(x) = \overline{\lim}_{t\to\infty} \frac{ln||x(t)||_{\mathbf{C}^n}}{t} < \Lambda_1.$$

In particular, the condition $\Lambda_1 < 0$ provides asymptotic stability.

Now take $h(t) = t - h_0$, where h_0 is a positive constant. Then Corollary 9.6.4 implies

Corollary 9.6.5 *Under the hypothesis of Theorem 9.6.1, let $h(t) = t - h_0$ and the condition*

$$\int_0^\infty (a(s) + q(s))ds < +\infty$$

be fulfilled. Then equation (6.1) is stable.

9.7 Linear time-variant systems with small delays

9.7.1 Statement of the result

Let us consider the equation

$$\dot{x}(t) = \sum_{k=0}^{m} A_k x(t - h_k) + (Lx)(t) \quad (t \geq 0), \tag{7.1}$$

where A_k $(k = 0, ..., m)$ are constant matrices,

$$0 = h_0 < h_1 < ... < h_m = \eta \equiv const < \infty.$$

The linear operator $L : [0, \infty) \times C([-\eta, \infty), \mathbf{C}^n) \to C([0, \infty), \mathbf{C}^n)$, is defined by (2.2). Again, it is assumed that for every $u \in L^2([-\eta, \infty), \mathbf{C}^n) \cap C([-\eta, \infty), \mathbf{C}^n)$, the inequality

$$||Lu||_{L^2([0,\infty),\mathbf{C}^n)} \leq q||u||_{L^2([-\eta,\infty),\mathbf{C}^n)} \tag{7.2}$$

holds. Put

$$\tilde{A} = \sum_{k=0}^{m} A_k, \Gamma(\tilde{A}) = \sum_{k=1}^{n-1} \frac{g^k(\tilde{A})}{|\alpha(\tilde{A})|^{k+1}\sqrt{k!}}, \text{ and } V_2 = \sum_{k=1}^{m} h_k \|A_k\|_{C^n}.$$

Recall that $\omega(\tilde{A}) = \max_k |Im\lambda_k(\tilde{A})|, \alpha(\tilde{A}) = \max_k Re\lambda_k(\tilde{A})$.

Theorem 9.7.1 *Let $\tilde{A}$ be a Hurwitz matrix and the conditions (7.2) and*

$$[(\alpha^2(\tilde{A}) + \omega^2(\tilde{A}))^{1/2} V_2 + q]\, \Gamma(\tilde{A}) < 1 \tag{7.3}$$

hold. Then equation (7.1) is exponentially stable.

This theorem is a particular case of Theorem 10.7.2 below. If $\tilde{A}$ is a normal (in particular, a Hermitian) matrix, then $g(\tilde{A}) = 0, \Gamma(\tilde{A}) = |\alpha(\tilde{A})|^{-1}$, and condition (7.4) takes the form

$$(\alpha^2(\tilde{A}) + \omega^2(\tilde{A}))^{1/2} V_2 + q < |\alpha(\tilde{A})|.$$

Remark 9.7.2 *Theorem 9.7.1 is exact, and condition (7.3) cannot be improved without additional restrictions.*

In fact, let us consider the system

$$\dot{x}(t) = Ax(t) + qx(t), \tag{7.4}$$

where A is a normal Hurwitz matrix. Under consideration $V_2 = 0$, and (7.3) takes the form

$$q < |\alpha(A)|. \tag{7.5}$$

But for the asymptotic stability of equation (7.4) it is necessary and sufficient that $\alpha(A) + q < 0$. This condition coincides with (7.5).

9.7.2 Example

Let us consider in $\mathbf{C}^n$ the equation

$$\dot{x}(t) = A_0 x(t) + A_1 x(t - \eta) + \int_0^{\eta} T(t, s) x(t - s) ds, \tag{7.6}$$

where the matrix-valued function $T(t, s)$ defined on $[0, \infty) \times [-\eta, 0]$ satisfies the inequality

$$m_T \equiv \sup_{t \geq 0} \int_0^{\eta} \|T(t, s)\|_{C^n} ds < \infty. \tag{7.7}$$

Take into account, that

$$\int_0^{\infty} \| \int_0^{\eta} T(t, s) x(t - s) ds \|_{C^n}^2 dt =$$

$$\int_0^\infty \| \int_0^\eta T(t,s)x(t-s)ds \int_0^\eta T(t,s_1)x(t-s_1)ds_1 \|_{C^n} dt \le$$

$$\int_0^\infty \int_0^\eta \|T(t,s)\|_{C^n} \|x(t-s)\|_{C^n} ds \int_0^\eta \|T(t,s_1)\|_{C^n} \|x(t-s_1)\|_{C^n} ds_1 dt \le$$

$$m_T^2 \sup_{0\le s,s_1\le\eta} \int_0^\infty \|x(t-s)\|_{C^n} \|x(t-s_1)\|_{C^n} dt \le$$

$$m_T^2 \sup_{0\le s,s_1\le\eta} 1/2[\int_0^\infty \|x(t-s)\|_{C^n}^2 + \|x(t-s_1)\|_{C^n}^2 dt].$$

Hence it follows that

$$\int_0^\infty \| \int_0^\eta T(t,s)x(t-s)ds\|_{C^n}^2 dt \le m_T^2 \|x\|_{L[-\eta,\infty)}^2$$

for any $x \in L^2([-\eta,\infty), \mathbf{C}^n) \cap C([-\eta,\infty), \mathbf{C}^n)$. We thus get condition (7.2) with

$$(Lx)(t) = \int_0^\eta T(t,s)x(t-s)ds,$$

and $q = m_T$. Furthermore, in the considered case $\tilde{A} = A_0 + A_1$, $V_2 = \eta\|A_1\|_{C^n}$. Now Theorem 9.7.1 yields the following result. Let the condition

$$[(\alpha^2(\tilde{A}) + \omega^2(\tilde{A}))^{1/2}\, \eta\|A_1\|_{C^n} + m_T]\, \Gamma(\tilde{A}) < 1$$

be valid. Then the equation (7.6) is exponentially stable.

9.8 Notes

Note 9.1 The stability of time-variant linear retarded systems was investigated by many authors. One of the main methods for the analysis of the stability of time-variant systems is connected with the development of the direct Lyapunov method. In particular, Krasovskii (1963) proposed (in the linear and nonlinear case) to use continuous functionals on the relevant space of continuous functions. Nonautonomous linear retarded systems were also analyzed by the Razumikhin-type theorems (Razumikhin, 1960). Many various applications of the Razumikhin-type theorems and the Lyapunov-Krasovskii functionals to linear systems are presented in (Hale, 1977), (Kolmanovskii and Nosov, 1986), (Kolmanovskii and Myshkis, 1992), etc.

Note 9.2 Section 9.2 is based on the paper (Gil', 1994). Theorem 9.3.1 is published in (Gil', 1997). The material of Section 9.5 is taken from (Gil', 1998a). Theorem 9.6.1 is new. Another generalization of the Wazewski inequality is presented in (Kolmanovskii, 1995) among other interesting results. Under the evident restrictions, Corollary 9.5.2 makes the mentioned Kolmanovskii result a bit more precise. Now consider the scalar equation

$$\dot{x}(t) = -b(t)x(t-h(t)) \tag{8.1}$$

with continuous bounded positive functions $b(t), h(t)$. Yoneyama (1987) established the following result ("3/2-stability theorem"). Let the condition

$$\sup_{t\geq 0} h(t) \sup_{t\geq 0} b(t) \leq \frac{3}{2}$$

hold. Then equation (8.1) is stable. Clearly, Corollary 9.5.4 supplements this result when the delay is constant.

Note 9.3 The stability conditions for systems with small delays presented in Section 9.7 are based on the paper (Gil', 1998b). There have been many stability conditions for linear systems with sufficiently small delays usually in the category of "robust stability" cf. (Mori, 1985), (Mori and Kokame, 1989), (Hmamed, 1986), and references therein. Besides, mainly scalar equations or systems with constant discrete commensurable delays are considered while Theorem 9.7.1 allows us to investigate systems containing incommensurable, time-variant and distributed delays.

Note 9.4 Let us mention the very interesting approach to the stability analysis of linear systems which is connected with the invertibility of corresponding functional-differential operators (Kurbatov, 1990).

Some other results on the stability of linear time-variant retarded systems can be found in (Berezanski and Braverman, 1995), (Azbelev *et al.*, 1993), etc.

References for Chapter 9

Azbelev, N. V., Berezansky, L. M., Simonov P.M. and A.V. Chistyakov (1993). Stability of linear systems with time lag (part 4), *Differential Eqs. 29*, 153-160.

Berezansky, L. M. and E. Braverman (1995). Preservation of the exponential stability under perturbations of linear delay impulsive differential equations, *Journal for Analysis and Appl., 14*, 157-174 .

Gil', M.I. (1994). On absolute stability of differential-delay systems, *IEEE, Trans. Automatic Control, 39, N 12*, 2481-2484.

Gil', M.I. (1997). On the freezing method for differential equations with delay, *Analysis, 17*, 25-33.

Gil', M.I. (1998). On generalized Wazewski and Lozinskii inequalities for semilinear abstract differential-delay equations, *Journal of Inequalities and Appl.* (accepted for publication).

Halanay, A. (1966). *Differential Equations. Stability. Oscillations, Time Lag.* Ac. Press, New York.

Hale, J.K. (1977). *Theory of Functional Differential Equations*, Springer-Verlag, New York.

Hale, J.K., Verduyn Lunel, S.M. (1993). *Introduction to Functional Differential Equations*, Springer-Verlag, New York.

Hmamed, A. (1986). On the stability of time-delay systems: new results, *Int. Journal of Control, 43*, 321-324.

Kolmanovskii, V.B. (1995). Applications of differential inequalities for stability of some functional differential equations, *Nonlinear Analysis, TMA, 25, Nos 9-10*, 1017-1028.

Kolmanovskii, V.B. and A.D. Myshkis (1992). *Applied Theory of Functional Differential Equations*, Kluwer Academic Publishers, Dordrecht.

Kolmanovskii, V.B. and V.K. Nosov (1986). *Stability of Functional Differential Equations*, Ac. Press, New York

Krasovskii, N.N. (1963). *Stability of Motion*. Stanford Univ. Press, Stanford.

Kurbatov, V.G. (1990). *Linear Differential-Difference Equations*. Izdatel'stvo Voroneg University, Voroneg. In Russian.

Mori, T. (1985). Criteria for stability of time-delay systems, *IEEE Transactions on Automatic Control, 30*, 158-161.

Mori, T., and Kokame H. (1989). Stability of $\dot{X} = AX(t) + BX(t - \tau)$, *IEEE Transactions on Automatic Control, 34*, 460-462.

Myshkis, A.D. (1972). *Linear Differential Equations with Delaying Argument*, Nauka, Moscow. In Russian.

Razumikhin, B.S. (1960). Applications of the Lyapunov method in the stability of systems with delay. *Autom. and Telemech. 21*, 740-749. In Russian.

Yoneyama, T. (1987). The 3/2-stability theorem for one dimensional delay-differential equations, *J. of Math. Anal. and Appl., 125*, 161-173.

Chapter 10

NONLINEAR SYSTEMS WITH DELAY

In the present chapter nonlinear retarded systems with separated linear parts and continuous nonlinearities are considered. Mainly, the linear parts are time-invariant. Only Section 10.10 deals with nonlinear systems with time-variant linear parts.

In Section 10.1 we introduce the definitions of stability of nonlinear retarded systems.

Section 10.2 is devoted to L^2- estimates for solutions.

Absolute stability is discussed in Section 10.3.

A bound for the region of attraction of the stationary solution is presented in Section 10.4.

Section 10.5 deals with the stability in the first approximation.

The exponential stability of nonlinear retarded systems with separated autonomous linear parts is discussed in Section 10.6. An estimate for the Lyapunov exponents of solutions is also derived.

In Sections 10.7 and 10.8 we investigate nonlinear systems having small delays.

Section 10.9 deals with the stability of nonlinear retarded systems with respect to arbitrary delays.

In Section 10.10 we establish solution estimates and stability conditions for nonlinear systems with time-variant linear parts.

Section 10.11 is devoted to the global feedback stabilization.

Conditions for the input-output stability are derived in Section 10.12.

10.1 Definitions

Let $\eta < \infty$ be a positive number. If $s \in \mathbf{R}, a \geq 0$, and $x \in C([s-\eta, s+a], \mathbf{C}^n)$, then for any $t \in [s, s+a]$ define x_t by the relation $x_t(\theta) \equiv x(\theta+t), \; -\eta \leq \theta \leq 0$.

For some $t_0 \geq 0$, let us consider in $\mathbf{C}^n$ the equation

$$\dot{x}(t) = F(t, x_t) \; (t \geq t_0), \tag{1.1}$$

where the dot means the right derivative, and F continuously maps $[0, \infty) \times C([t_0 - \eta, t_0], \mathbf{C}^n)$ into $\mathbf{C}^n$. For example, we can take $t_0 = 0$ and

$$F(t, x_t) = f(t, x(t), \int_0^\eta T(t, s) x(t - s) ds)$$

where $f : [0, \infty) \times \mathbf{C}^{2n} \to \mathbf{C}^n$ and $T(t, s)$ is a matrix-valued function defined on $[0, \infty) \times [-\eta, 0]$.

Take the initial condition

$$x(t) = \Phi(t) \text{ for } t_0 - \eta \leq t \leq t_0 \tag{1.2}$$

with a given continuous vector-valued function $\Phi(t)$.

A (global) solution $x(t) = x(t, \Phi)$ of the initial problem (1.1), (1.2) with an initial function Φ is an absolutely continuous function which satisfies this equation almost everywhere on $[t_0, \infty)$, and (1.2) holds. The existence of all considered solutions is assumed. Let us give some sufficient solvability conditions for local solutions. Denote $D_\eta = [t_0, \infty) \times C([t_0 - \eta, t_0], \mathbf{C}^n)$.

Lemma 10.1.1 *Suppose $\Phi \in C([t_0 - \eta, t_0], \mathbf{C}^n)$, and let $F : D_\eta \to \mathbf{C}^n$ be a continuous function satisfying in some neighborhood of any point of D_η the Lipschitz condition in the second argument. Then there is $t_\Phi > t_0$, such that problem (1.1), (1.2) has on $[t_0, t_\Phi)$ a unique solution.*

This lemma is a particular case of the well-known Theorem 2.1 from the book (Kolmanovskii and Myshkis, 1992, p.40).

Definition 10.1.2 *Let $F(t, 0) \equiv 0$ for all $t \geq 0$. Then the zero solution of (1.1) is said to be stable if for given $t_0 \geq 0$ and any $\epsilon > 0$ there exists a $\delta(t_0) > 0$, such that the inequality*

$$||\Phi||_{C([t_0-\eta,t_0],\mathbf{C}^n)} \leq \delta(t_0) \text{ implies } ||x(t, \Phi)||_{\mathbf{C}^n} \leq \epsilon \; (t \geq t_0)$$

for a solution $x(t, \Phi)$ with an initial function Φ.

The zero solution of (1.1) is said to be asymptotically stable if it is stable, and there is an open set $B \subseteq C([t_0 - \eta, t_0], \mathbf{C}^n)$, such that $\Phi \in B$ implies $x(t, \Phi) \to 0$ as $t \to \infty$. Besides, B is called the region of attraction of the zero solution.

If the zero solution of (1.1) is asymptotically stable and $B = C([t_0 - \eta, t_0], \mathbf{C}^n)$, then it is globally asymptotically stable.

We will say that the zero solution of (1.1) is exponentially stable if there are positive constants r_0, Λ_0 and l_0, such that for given $t_0 \geq 0$ and an initial function Φ, the condition $||\Phi||_{C([t_0-\eta,t_0],C^n)} \leq r_0$ implies the relation

$$||x(t, \Phi)||_{C^n} \leq l_0 ||\Phi||_{C([t_0-\eta,t_0],C^n)} \, e^{-\Lambda_0 t} \quad (t \geq t_0).$$

The zero solution of (1.1) is globally exponentially stable if it is exponentially stable and the region of attraction coincides with $C([t_0-\eta, t_0], \mathbf{C}^n)$. That is, $r_0 = \infty$.

Definition 10.1.3 *Let $F(t,0) \equiv 0$ for all $t \geq 0$. Then the zero solution of (1.1) is said to be uniformly stable if for any $\epsilon > 0$ there exists a $\delta > 0$ independent of t_0, such that the inequality*

$$||\Phi||_{C([t_0-\eta,t_0],\mathbf{C}^n)} \leq \delta \textit{ implies } ||x(t,\Phi)||_{\mathbf{C}^n} \leq \epsilon \ (t \geq t_0).$$

The zero solution of (1.1) is said to be uniformly asymptotically stable if it is uniformly stable, and there is an open set $B \subseteq C([-\eta,0],\mathbf{C}^n)$, such that $\Phi \in B$ implies $x(t,\Phi) \to 0$ as $t \to \infty$.

One can prove that in the linear case these definitions agree with the definitions from Section 9.1

Recall that sometimes we write $||.||_{L[a,b]}$ instead of $||.||_{L^2([a,b],\mathbf{C}^n)}$.

10.2 L^2- estimates for solutions

10.2.1 Statement of the result

Let us consider in $\mathbf{C}^n$ the equation

$$\dot{x}(t) = \int_0^\eta dR(\tau)x(t-\tau) + F(t,x_t) \ \ (t \geq 0), \tag{2.1}$$

where F continuously maps $[0,\infty) \times C([-\eta,0],\mathbf{C}^n)$ into $\mathbf{C}^n$, and R has a bounded variation. Again take the initial condition

$$x(t) = \Phi(t) \text{ for } -\eta \leq t \leq 0. \tag{2.2}$$

It is assumed that for any $u \in L^2([-\eta,\infty),\mathbf{C}^n) \cap C([-\eta,\infty),\mathbf{C}^n)$ the inequality

$$||F(.,u_t)||_{L[0,\infty)} \equiv [\int_0^\infty ||F(t,u_t)||^2_{\mathbf{C}^n} dt]^{1/2} \leq q||u||_{L[0,\infty)} + M(u) \tag{2.3}$$

holds where $M(.)$ is a continuous (generally nonlinear) functional defined on space $L^2([-\eta,0],\mathbf{C}^n)$.

For instance, let there be constants $b_j > 0$ and $q_j \geq 0$ $(j = 1,...,m < \infty)$, such that the relation

$$||F(t,u_t)||_{C^n} \leq \sum_{k=1}^m q_j||u(t-h_j(t))||_{C^n} \text{ for all } u \in L^2([-\eta,\infty),\mathbf{C}^n) \tag{2.4}$$

holds, where $h_j(t)$ are differentiable scalar-valued functions with the properties

$$1 - \dot{h}_j(t) \geq b_j > 0, \text{ and } 0 \leq h_j(t) \leq \eta \ (t \geq 0, \ j = 1,...,m). \tag{2.5}$$

Proposition 10.2.1 *Let relations (2.4) and (2.5) be fulfilled. Then condition (2.3) holds with*

$$q = \sum_{j=1}^{m} q_j b_j^{-1/2}, \tag{2.6}$$

and

$$M(u) = q[\int_{-\eta}^{0} \|u(t)\|_{C^n}^2 dt]^{1/2} = q\|u\|_{L[-\eta,0]} \; (u \in L^2([-\eta,0],\mathbf{C}^n)).$$

Proof: Relation (2.5) gives us the inequality

$$b_j\|u(t-h_j(t))\|_{L[0,\infty)}^2 \leq \int_0^\infty \|u(t-h_j(t))\|_{C^n}^2(1-h_j(t))dt.$$

Thus,

$$b_j\|u(t-h_j(t))\|_{L[0,\infty)}^2 \leq \int_{-h_j(0)}^{\infty} \|u(s)\|_{C^n}^2 ds = \|u\|_{L[0,\infty)}^2 + \int_{-h_j(0)}^{0} \|u(s)\|_{C^n}^2 ds \leq$$

$$\|u\|_{L[0,\infty)}^2 + \|u\|_{L[-\eta,0]}^2$$

for $j = 1, ..., m$. Therefore by condition (2.4),

$$\|F(.,u_t)\|_{L[0,\infty)} \leq \sum_{k=1}^{m} q_j\|u(t-h_j(t))\|_{L[0,\infty)} \leq \sum_{k=1}^{m} q_j(b_j)^{-1/2}(\|u\|_{L[0,\infty)}+\|u\|_{L[-\eta,0]})$$

as claimed. □

Again introduce the characteristic matrix of the linear part of (2.1):

$$K(p) = \int_0^\eta exp(-ps)dR(s) - Ip \; (p \in \mathbf{C}),$$

and put

$$\Gamma(K(p)) = \sum_{k=0}^{n-1} \frac{g^k(B(p))}{\sqrt{k!}d^{k+1}(K(p))} \; (p \in \mathbf{C}) \text{ and } \Gamma_0(K) \equiv \sup_{\omega\in\mathbf{R}} \Gamma(K(\omega i)),$$

where $d(K(p))$ is the smallest modulus of eigenvalues of $K(p)$ (see Sections 8.1 and 8.3). Now we are in a position to formulate the main result of this section.

Theorem 10.2.2 *Let all the zeros of* $det K(p)$ *lie in the open left half-plane. Let the conditions (2.3) and*

$$q\Gamma_0(K) < 1 \tag{2.7}$$

hold. Then any solution $x(t)$ *of equation (2.1) with a continuous initial function* Φ *belongs to* $L^2([0,\infty),\mathbf{C}^n)$. *Moreover, the bound*

$$\|x\|_{L[0,\infty)} \leq (1-\Gamma_0(K)q)^{-1}(\Gamma_0(K)M(\Phi) + \|\phi\|_{L[0,\infty)}) \tag{2.8}$$

is valid, where $\phi(t)$ is the solution of the equation

$$\phi = \int_0^\eta dR(\tau)\phi(t-\tau)$$

with the initial function Φ.

Due to Lemma 8.6.2 under consideration, ϕ belongs to $L^2([0,\infty), \mathbf{C}^n)$.

10.2.2 Proof of Theorem 10.2.2

Put

$$\tilde{b}_0 = \sup_{\omega \in \mathbf{R}} ||K^{-1}(i\omega)||_{C^n}.$$

Lemma 10.2.3 *Let all the zeros of $\det K(p)$ lie in the open left half-plane. In addition, under condition (2.3), let the inequality*

$$q\tilde{b}_0 < 1 \tag{2.9}$$

hold. Then the estimate

$$||x||_{L[0,\infty)} < (1-\tilde{b}_0 q)^{-1}(\tilde{b}_0 M(\Phi) + ||\phi||_{L[0,\infty)})$$

is valid for any solution $x(t)$ of equation (2.1) with the initial condition (2.2).

Proof: For a positive $a < \infty$ let us introduce the scalar-valued function by the relations $\chi_a(t) = 1$ $(0 \leq t \leq a)$ and $\chi_a(t) = 0$ $(t > a)$. Consider the equation

$$\dot{u}(t) + \int_0^\eta dR(t-s)u(s) = \chi_a(t)F(t, u_t) \quad (t \geq 0) \tag{2.10}$$

with the initial function Φ. Clearly, any solution $w_a(t)$ of (2.10) for $t \geq a$ coincides with a solution ϕ of the homogeneous equation considered on $[a, \infty)$ with the corresponding initial condition. Therefore, according to Lemma 8.6.2, $w_a \in L^2([0,\infty), \mathbf{C}^n)$. Rewrite equation (2.10) in the form

$$\dot{u}(t) + \int_0^\eta dR(t-s)u(s) = f(t) \tag{2.11}$$

with $f(t) = \chi_a(t)F(t, u_t)$. Lemma 8.3.3 and restriction (2.3) imply

$$||w_a||_{L[0,\infty)} \leq \tilde{b}_0||F(.,(w_a)_t)||_{L[0,\infty)} + ||\phi||_{L[0,\infty)} \leq$$

$$\tilde{b}_0(q||w_a||_{L[0,\infty)} + M(w_a)) + ||\phi||_{L[0,\infty)}.$$

But by (2.2) $M(w_a) = M(\Phi)$. Now condition (2.9) ensures the inequality

$$||w_a||_{L[0,\infty)} \leq (1-\Gamma_0(K)q)^{-1}(\Gamma_0(K)M(\Phi) + ||\phi||_{L[0,\infty)}).$$

Passing at the left part of that inequality to the limit as $a \to \infty$, and taking into account that $w_a(t)$ for $t \leq a$ coincides with a solution $x(t)$ of (2.1), we arrive at the result. □

The assertion of Theorem 10.2.2 immediately follows from Lemmata 10.2.3 and 8.3.4. □

10.3 Absolute stability

10.3.1 Statement of the result

Definition 10.3.1 *We will say that the zero solution of equation (2.1) is absolutely stable in the class of nonlinearities satisfying the inequality*

$$\|F(.,u_t)\|_{L[0,\infty)} \equiv \left[\int_0^\infty \|F(t,u_t(t))\|^2_{C^n}dt\right]^{1/2} \leq q\|u\|_{L[-\eta,\infty)} \tag{3.1}$$

for any $u \in L^2([-\eta,\infty),\mathbf{C}^n) \cap C([-\eta,\infty),\mathbf{C}^n)$, *if under (3.1), the zero solution of (2.1) is globally asymptotically stable. Moreover, there is a positive constant* N *independent of the specific form of the function* F, *such that the inequality*

$$\|x(t)\|_{C^n} \leq N\|\Phi\|_{C([-\eta,0],\mathbf{C}^n)} \quad (t \geq 0)$$

holds for any solution of (2.1) with the initial condition (2.2).

Theorem 10.3.2 *Let all the zeros of* $\det K(p)$ *lie in the open left half-plane, and condition (2.7) hold. Then the zero solution of equation (2.1) is absolutely stable in the class of nonlinearities (3.1).*

Example 10.3.3 *Consider in* $\mathbf{C}^n$ *the following system*

$$\dot{x} = Ax + F(t,x_t), \tag{3.2}$$

where A *is a constant Hurwitz matrix, and* F *satisfies condition (3.1).*

Clearly, in the considered case $K(p) = A - pI$, and the equality $g(K(p)) \equiv g(A)$ is valid. It is easy to see that

$$d(K(iw)) = d(A - Iwi) \geq |\alpha(A)| = \min_{k=1,\dots,n} |Re\lambda_k(A)|$$

for all real w. Therefore,

$$\Gamma_0(K) = \Gamma(A) \equiv \sum_{k=0}^{n-1} \frac{g^k(A)}{\sqrt{k!}|\alpha(A)|^{k+1}}.$$

If $q\Gamma(A) < 1$, then the zero solution of system (3.2) is absolutely stable in the class of nonlinearities (3.1) due to Theorem 10.3.2. If A is a Hermitian matrix, then $g(A) = 0$ and $\Gamma_0(K) = 1/|\alpha(A)|$.

Example 10.3.4 *Consider the system*

$$\begin{aligned}
\dot{y}_1(t) + a_{11}y_1(t) + a_{12}y_2(t) + a_{13}y_3(t) + b_1y_1(t-1) &= f_1(y(t),y(t-1),t),\\
\dot{y}_2(t) + a_{22}y_2(t) + a_{23}y_3(t) + b_2y_2(t-1) &= f_2(y(t),y(t-1),t),\\
\dot{y}_3(t) + a_{33}y_3(t) + b_3y_3(t-1) &= f_3(y(t),y(t-1),t),
\end{aligned} \tag{3.3}$$

where $y = (y_1, y_2, y_3)$, and a_{kk}, b_k are real numbers. Besides, the real scalar valued functions f_j defined on $\mathbf{R}^6 \times [0, \infty)$ satisfy the conditions

$$f_j^2(u, h, t) \leq q_0^2(u_1^2 + u_2^2 + u_3^2)/3 + q_1^2(h_1^2 + h_2^2 + h_3^2)/3 \tag{3.4}$$

for every $u = (u_k), h = (h_k) \in \mathbf{R}^3, t \geq 0$. In the considered case $K(p) = -A - Bexp(-p) - pI$, where

$$A = (a_{jk})_{j,k=1}^3; \; B = diag[b_k]_{k=1}^3 \text{ with } a_{jk} = 0 \text{ for } j > k \; (j, k = 1, 2, 3).$$

So $K(p)$ is a triangular matrix, and we have

$$g(K(p)) = \sqrt{a_{12}^2 + a_{13}^2 + a_{23}^2} = g(A).$$

(see Example 2.1.3). Besides,

$$d(K(p)) = \min_{k=1,2,3} |p + a_{kk} + b_k exp(-p)|.$$

Equation $p + a_{kk} + b_k exp(-p) = 0$ is equivalent to the following one:

$$(p + a_{kk})exp(p) + b_k = 0.$$

Thanks to Theorem 8.12.5, all roots of that equation have negative real parts if and only if the conditions

$$a_{kk} > -1; \; a_{kk} + b_k > 0; \; b_k < z_k sin\, z_k - acos\, z_k \; (k = 1, 2, 3) \tag{3.5}$$

hold, where z_k is a root of the equation $z = -a_{kk}\, tg\, z$ $(0 < z < \pi)$ if $a \neq 0$, and $z_k = \pi/2$ if $a_{kk} = 0$. Put

$$d_0 = \inf_{\omega \in \mathbf{R}} d_0(K(i\omega)) \equiv \inf_{\omega \in \mathbf{R}} \min_{k=1,2,3} |i\omega + a_{kk} + b_k exp(-i\omega)|.$$

Due to Proposition 10.2.1 and (3.4), condition (3.1) is satisfied with $q = q_0 + q_1$.

Thanks to Theorem 10.3.2, we can assert that if the inequalities (3.5), and

$$(q_0 + q_1)\left(\frac{1}{d_0} + \frac{g(A)}{d_0^2} + \frac{g^2(A)}{2d_0^3}\right) < 1$$

hold, then the zero solution of system (3.3) is absolutely stable in the class of nonlinearities (3.4).

10.3.2 Proofs

Lemma 10.3.5 *Let all the zeros of $detK(p)$ lie in the open left half-plane, and conditions (3.1), and (2.7) hold. Then the zero solution of (2.1) is globally* *asymptotically stable, and any solution of (2.1) with an initial function Φ is subject to the estimate*

$$\|x(t)\|_{C^n} \leq \theta(K, \Phi, q) \quad (t \geq 0) \tag{3.6}$$

with the notation

$$\theta(K, \Phi, q) \equiv \frac{1 + V(R)\Gamma_0(K)}{1 - q\Gamma_0(K)}[2\|\Phi(0)\|_{C^n} + \|\Phi\|_{L[-\eta,0]} \sqrt{2(V(R)+q)}\,].$$

Proof: For brevity in the present proof, we set $\Gamma_0(K) = \Gamma_0$, and

$$\|y\|_M = \|y(t)\|_{C([0,\infty),\, C^n)}$$

for every continuous vector-valued function y bounded on $[0,\infty)$. Under (3.1) using Theorem 10.2.2, we get the inequality

$$\|x\|_{L[0,\infty)} \leq (1 - \Gamma_0 q)^{-1}(\Gamma_0 q\|\Phi\|_{L[-\eta,0]} + \|\phi\|_{L[0,\infty)}). \tag{3.7}$$

Furthermore, take into account that

$$\|\int_0^\eta dR(s)x(t-s)\|^2_{L[0,\infty)} = \int_0^\infty \|\int_0^\eta dR(s)x(t-s)\|^2_{C^n} dt \leq$$

$$\leq (\int_0^\eta \|dR(s)\|_{C^n})^2 \sup_{-\eta\leq s\leq 0} \int_0^\infty \|x(t-s)\|^2_{C^n} dt = V^2(R)\ (\|x\|^2_{L[0,\infty)} + \|\Phi\|^2_{L[-\eta,0]}).$$

According to (2.1) and (3.1), this relation yields

$$\|\dot{x}(t)\|_{L[0,\infty)} = \|\int_0^\eta dR(\tau)x(t-\tau)\|_{L[0,\infty)} + \|F(t, x_t)\|_{L[0,\infty)} \leq$$

$$(V(R) + q)\ (\|x\|_{L[0,\infty)} + \|\Phi\|_{L[-\eta,0]}).$$

Moreover, by Lemma 8.4.4, $\|x(t)\|_{C^n} \to 0$ as $t \to \infty$, and

$$\|x\|_M \leq \sqrt{2\|x\|_{L[0,\infty)}\|\dot{x}\|_{L[0,\infty)}}.$$

Consequently,

$$\|x\|_M \leq \sqrt{2(V(R)+q)(\|x\|_{L[0,\infty)} + \|\Phi\|_{L[-\eta,0]})\|x\|_{L[0,\infty)}} \leq$$

$$\sqrt{2(V(R)+q)}(\|x\|_{L[0,\infty)} + \|\Phi\|_{L[-\eta,0]}).$$

Now (3.7) gives

$$\|x\|_M \leq \sqrt{2(V(R)+q)}[(1 - q\Gamma_0)^{-1}(\|\phi\|_{L[0,\infty)} + q\Gamma_0\|\Phi\|_{L[-\eta,0]}) +$$

$$\|\Phi\|_{L[-\eta,0]}] = \sqrt{2(V(R)+q)}(1 - q\Gamma_0)^{-1}[\|\phi\|_{L[0,\infty)} + \|\Phi\|_{L[-\eta,0]}].$$

Lemma 8.6.2 implies

$$||x||_M \leq \frac{\sqrt{2(V(R)+q)}}{1-q\Gamma_0}[W_0(K)||\Phi(0)||_{C^n} + (\Gamma_0 V(R)+1)||\Phi||_{L[-\eta,0]}]. \quad (3.8)$$

But according to (2.7),

$$\sqrt{2(V(R)+q)}W_0(K) = 2\sqrt{(V(R)+q)\Gamma_0(K)(V(R)\Gamma_0(K)+1)}$$

$$\leq 2(V(R)\Gamma_0(K)+1).$$

Thus, by (3.8) $||x||_M \leq \theta(K,\Phi,q)$, as claimed. □

The assertion of Theorem 10.3.2 follows from the latter lemma and the Schwarz inequality

10.4 A bound for the region of attraction

10.4.1 Statement of the result

Let us consider in $\mathbf{C}^n$ the equation

$$\dot{x}(t) = \int_0^\eta dR(\tau)x(t-\tau) + F(t, x(t-h_1(t)), ..., x(t-h_m(t))) \ (t \geq 0), \quad (4.1)$$

where $h_j(t)$ $(j = 1, ..., m)$ are differentiable scalar-valued functions with the properties

$$1 - \dot{h}_j(t) \geq b_j > 0, \text{ and } 0 \leq h_j(t) \leq \eta \ (t \geq 0, \ j = 1, ..., m) \quad (4.2)$$

again, and F continuously maps $[0,\infty) \times \mathbf{C}^{nm}$ into $\mathbf{C}^n$. Take the initial condition (2.2) and recall that $\Omega(r) \equiv \{h \in \mathbf{C}^n : ||h||_{C^n} \leq r\}$ for a positive number $r \leq \infty$. It is assumed that there are positive constants q_j $(j = 1, ..., m)$, such that

$$||F(t, y_1, ..., y_m)||_{C^n} \leq \sum_{j=1}^m q_j ||y_j||_{C^n} \text{ for all } y_j \in \Omega(r) \ (j = 1, ..., m, \ t \geq 0). \quad (4.3)$$

Put

$$q = \sum_{j=1}^m q_j b_j^{-1/2}.$$

Remember that $\theta(K, \Phi, q)$ is defined in the previous section.

Theorem 10.4.1 *Let all the zeros of* $\det K(p)$ *lie in the open left half-plane. In addition, let conditions (4.2), (4.3) and (2.7) hold. Then the zero solution* *of (4.1) is asymptotically stable. Moreover, any initial function Φ satisfying the inequality*

$$\theta(K, \Phi, q) < r, \tag{4.4}$$

belongs to the region of attraction of the zero solution, and the bound

$$||x(t)||_{C^n} \leq \theta(K, \Phi, q) \quad (t \geq 0) \tag{4.5}$$

is true for the corresponding solution $x(t)$ of equation (4.1).

The proof is presented in the next subsection.

Now assume that F satisfies the following conditions: there are positive constants l_j and $p_j > 1$, such that the inequality

$$||F(t, y_1, ..., y_m)||_{C^n} \leq \sum_{j=1}^{m} l_j ||y_j||_{C^n}^{p_j} \text{ for all } y_j \in \mathbf{C}^n \ (j = 1, ..., m; \ t \geq 0) \tag{4.6}$$

is true. Take some positive $r < \infty$. Then condition (4.6) implies the inequality (4.3) with

$$q_j = l_j r^{p_j - 1}.$$

So

$$q = q(r) = \sum_{j=1}^{m} l_j r^{p_j - 1} b_j^{-1/2}.$$

If r is sufficiently small, then condition (2.7) holds. Since all the zeros of $det K(p)$ are in the open left half-plane, due to Theorem 10.4.1, the zero solution of equation (4.1) under the restrictions (4.2) and (4.6) is asymptotically stable. Moreover, condition (4.4) gives a bound for the region of attraction.

10.4.2 Proofs

We have divided the proof of Theorem 10.4.1 into a series of lemmata

Lemma 10.4.2 *Let the assumptions of Theorem 10.4.1 hold with $\Omega(r) = \mathbf{C}^n$. That is, $r = \infty$. Then estimate (4.5) is valid for any solution $x(t)$ of equation (4.1) with the initial function Φ.*

Proof: By Proposition 10.2.1, we have

$$||F(., u)||_{L[0,\infty)} \equiv [\int_0^\infty ||F(t, u(t - h_1(t)), ..., u(t - h_m(t)))||_{C^n}^2 dt]^{1/2}$$

$$\leq q(||u||_{L[0,\infty)} + ||u||_{L[-\eta,0]})$$

with an arbitrary $u \in L^2([-\eta, \infty), \mathbf{C}^n)$. Now Lemma 10.3.5 implies the result.

Lemma 10.4.3 *Let a continuous mapping $F : [0,\infty) \times \mathbf{C}^{nm} \to \mathbf{C}^n$ satisfy condition (4.3) with a $r < \infty$. Then for any $\epsilon > 0$, there is a function*

$$F_1 : [0,\infty) \times \mathbf{C}^{nm} \to \mathbf{C}^n$$

with the properties

$$\|F_1(t, y_1, ..., y_m)\|_{C^n} \le \sum_{j=1}^{m} (q_j + \epsilon)\|y_j\|_{C^n} \text{ for all } y_j \in \mathbf{C}^n, \tag{4.7}$$

and

$$F_1(t, y_1, ..., y_m) = F(t, y_1, ..., y_m) \text{ for } y_j \in \Omega(r) \ (j = 1, ..., m;\ t \ge 0). \tag{4.8}$$

Proof: In accordance with the classical Urysonh lemma (see e.g. (Gemignani, 1972, p. 108)), for any $\delta > 0$ there exists a continuous function f acting from $\mathbf{C}^{nm}$ into $[0, 1]$ such that $f(y_1, ..., y_m) = 1$ for all $y_j \in \Omega(r)$ $(j = 1, ..., m)$ and $f(y_1, ..., y_m) = 0$ for all $y_j \in \mathbf{C}^n$ with

$$\|y_j\|_{C^n} \ge r + \delta \ (j = 1, ..., m).$$

Put

$$F_1(t, y_1, ..., y_m) = f(y_1, ..., y_m)F(t, y_1, ..., y_m).$$

Taking into account that F is continuous and choosing sufficiently small $\delta > 0$, we obtain the result. □

Proof of Theorem 10.4.1: In the case $r = \infty$, the theorem is valid due to Lemma 10.4.2. Let $r < \infty$. Consider the equation

$$\dot{x} = \int_0^\eta dR(\tau)x(t-\tau) + F_1(t, x(t - h_1(t)), ..., x(t - h_m(t))) \ (t \ge 0), \tag{4.9}$$

where the vector-valued function F_1 defined on $[0,\infty) \times \mathbf{C}^{nm}$ satisfies conditions (4.7) and (4.8). The function F_1 can be constructed due to Lemma 10.4.3. Thus, for a solution of (4.9) Lemma 10.4.2 gives estimate (4.5). Condition (4.4) and estimate (4.5) guarantee that the solution of (4.9) lies in $\Omega(r)$. But according to (4.8), it is simultaneously the solution of (4.1). This proves the result. □

10.5 Stability in the first approximation

Again consider in $\mathbf{C}^n$ equation (4.1). For a vector $y = (y_1, ..., y_m) \in \mathbf{C}^{mn}$ with $y^k \in \mathbf{C}^n$, put naturally

$$\|y\|_{C^{mn}} = \|(y_1, ..., y_m)\|_{C^{mn}} \equiv (\|y_1\|^2_{C^n} + ... + \|y_m\|^2_{C^n})^{1/2}.$$

We will say that (4.1) *is a quasilinear equation* if F continuously maps $[0,\infty)\times \mathbf{C}^{nm}$ into $\mathbf{C}^n$ with the property

$$\frac{\|F(t,y_1,...,y_m)\|_{C^n}}{\|(y_1,...,y_m)\|_{C^{mn}}} \to 0 \text{ as } \|(y_1,...,y_m)\|_{C^{mn}} \to 0 \tag{5.1}$$

uniformly in t.

Theorem 10.5.1 *Let condition (5.1) hold, and all the zeros of* $\det K(p)$ *lie in the open left half-plane. Then the zero solution of (4.1) is asymptotically stable.*

Proof: From condition (5.1) it follows that condition (4.3) holds for any $r<\infty$. Moreover, $q_j = q_j(r) \to 0$ as $r \to 0$. Take sufficiently small r in such a way that condition (2.7) is fulfilled. Now the result is due to Theorem 10.4.1. □

10.6 Exponential stability of nonlinear retarded systems

10.6.1 Statement of the result

In the present section exponential stability conditions and an estimate for the Lyapunov exponents of solutions are derived for a class of nonlinear retarded systems.

Let us consider in $\mathbf{C}^n$ the equation

$$\dot{x}(t) = \int_0^\eta dR(\tau)x(t-\tau) + F(t,x,x(t-h_1),...,x(t-h_m)) \ (t\geq 0), \tag{6.1}$$

where h_j $(j=1,...,m)$ are positive numbers, such that

$$\max_{k=1,...,m} h_k \leq \eta < \infty; \tag{6.2}$$

the matrix-valued function $R(\tau)$ has a bounded variation, and F continuously maps $[0,\infty)\times \mathbf{C}^{n(m+1)}$ into $\mathbf{C}^n$. It is assumed that for a positive number $r\leq\infty$, there are constants q_j $(j=0,...m)$, such that

$$\|F(t,y_0,y_1,...,y_m)\|_{C^n} \leq \sum_{j=0}^m q_j\|y_j\|_{C^n} \text{ for all } y_j\in\Omega(r) \ (j=0,1,...,m;\ t\geq 0). \tag{6.3}$$

Recall that $\Omega(r)=\{h\in \mathbf{C}^n : \|h\|_{C^n}\leq r\}$. Let $K(p)$ be the characteristic matrix of the linear part of (6.1). *Throughout the present section it is assumed that all the zeros of* $\det K(p)$ *belong to a half-plane* $\{z\in \mathbf{C} : Re z \leq -\alpha_0\}$ with a positive α_0.

Again put

$$\Gamma(K(p)) = \sum_{k=0}^{n-1} \frac{g^k(B(p))}{\sqrt{k!}d^{k+1}(K(p))} \quad (p \in \mathbf{C})$$

(see Section 8.1). Further, for a number $\nu \geq 0$, set

$$\Gamma_\nu(K) \equiv \sup_{\omega \in \mathbf{R}} \Gamma(K(-\nu + \omega i)), \quad V_\nu(R) \equiv \nu + \int_0^\eta e^{\nu s} \|dR(s)\|_{C^n},$$

and

$$a_\nu(K) \equiv 2\sqrt{V_\nu(R)\Gamma_\nu(K)[V_\nu(R)\Gamma_\nu(K) + 1]}.$$

The aim of this section is to prove the following

Theorem 10.6.1 *Let condition (6.3) be fulfilled. In addition, for a positive*

$$\nu < \alpha_0, \tag{6.4}$$

let the relation

$$\tilde{q}_\nu \equiv \sum_{k=0}^{m} q_k e^{\nu h_k} < \frac{\nu}{a_\nu(K)} \tag{6.5}$$

hold. Then the zero solution of (6.1) is exponentially stable. Moreover, any initial function Φ satisfying the inequality

$$T_\nu(\Phi, K) \equiv \|\Phi(0)\|_{C^n} + V_\nu(R) \int_{-\eta}^0 e^{\nu\tau} \|\Phi(\tau)\|_{C^n} d\tau \leq \frac{r}{(1 + \tilde{q}_\nu)a_\nu(K)}, \tag{6.6}$$

belongs to the region of attraction of the zero solution, and the corresponding solution $x(t)$ is subject to the estimate

$$\|x(t)\|_{C^n} \leq \tilde{b}_\nu exp\,[-(\nu - \tilde{q}_\nu a_\nu(K))t] \ (t \geq 0) \tag{6.7}$$

with

$$\tilde{b}_\nu = a_\nu T_\nu(\Phi, K)(1 + \tilde{q}_\nu V_\nu^{-1}(R)).$$

The proofs are presented in Subsection 10.6.3

Corollary 10.6.2 *Let condition (6.3) hold with $\Omega(r) = \mathbf{C}^n$. That is, $r = \infty$. Then under conditions (6.4) and (6.5) the zero solution of (6.1) is globally exponentially stable, and estimate (6.7) holds for any solution of (6.1).*

Let us give some estimates for $\Gamma_\nu(K)$. Put

$$v_\nu(B) \equiv \sup_{\omega \in \mathbf{R}} g(B(-\nu + i\omega))$$

and

$$d_\nu(K) \equiv \inf_{-2V_\nu(R) \leq \omega \leq 2V_\nu(R)} d(K(-\nu + i\omega)).$$

Below we will check that

$$\inf_{-\infty\le\omega\le\infty} d(K(-\nu+i\omega)) = d_\nu(K). \tag{6.8}$$

Consequently, if $d_\nu(K)>0$, then

$$\Gamma_\nu(K) \le \sum_{k=0}^{n-1} \frac{v_\nu^k(B)}{\sqrt{k!}\, d_\nu^{k+1}(K)}.$$

Thanks to the definition of $g(A)$, it can be written

$$g(B(-\nu+i\omega)) \le N(B(-\nu+i\omega)) \le \sqrt{n}||B(-\nu+i\omega)||_{C^n} \le$$

$$\sqrt{n}\int_0^\eta e^{\nu s}||dR(s)||_{C^n}.$$

Consider the case

$$B(p) = \sum_{k=0}^m A_k exp(-ph_k)\ (h_0=0), \tag{6.9}$$

where $A_k (k=1,...,m)$ are constant matrices. Due to the relation

$$g(A) \le \sqrt{1/2}N(A-A^*)$$

for a matrix A (see Proposition 2.1.1), we have

$$g(B(-\nu+i\omega)) \le \sqrt{1/2}\sum_{k=0}^m e^{\nu h_k} N(A_k exp(-i\omega h_k) - A_k^* exp(i\omega h_k)) \le$$

$$\sqrt{1/2}N(A_0-A_0^*) + \sqrt{2}\sum_{k=1}^m e^{\nu h_k} N(A_k)\ (\omega\in\mathbf{R}).$$

It is simple to see that in the case (6.9) we have

$$V_\nu(R) = \nu + \sum_{k=0}^m ||A_k||_{C^n} e^{\nu h_k}.$$

Below we also use the following

Lemma 10.6.3 *Let positive numbers a, b and h satisfy the conditions*

$$bh<1 \text{ and } 0 \le \frac{bh(1-ha)}{1-bh} \le \pi. \tag{6.10}$$

Then

$$\min_{\omega\in\mathbf{R}} |i\omega + a + be^{-ih\omega}| = a+b.$$

10.6.2 Example

Consider the system

$$\dot{x}_1(t) + a_{11}x_1(t) + a_{12}x_2(t) + b_1x_2(t-h) = f_1(x(t), x(t-h)),$$

$$\dot{x}_2(t) + a_{22}x_2(t) + a_{21}x_1(t) + b_2x_2(t-h) = f_2(x(t), x(t-h)), \qquad (6.11)$$

where $x(t) = (x_1(t), x_2(t))$, and a_{jk}, b_j, $(j, k = 1, 2)$ are given real numbers. Besides, a_{11}, a_{22} and b_2 are positive, and the scalar valued functions f_j $(j = 1, 2)$ defined on $\mathbf{C}^4$ satisfy the following conditions: there are positive constants l_0, l_1, such that

$$\sqrt{|f_1(u,z)|^2 + |f_2(u,z)|^2} \leq l_0\|u\|_{C^2}^{p_0} + l_1\|z\|_{C^2}^{p_1} \; (p_0, p_1 > 1; \; t \geq 0) \qquad (6.12)$$

for every $u, z \in \mathbf{C}^2$. In addition, $b_2h < 1$. Rewrite the considered system in the form

$$\dot{x}(t) + A_0x(t) + A_1x(t-h) = F(x(t), x(t-h))$$

with the triangular matrices

$$A_0 = \begin{pmatrix} a_{11} & a_{12} \\ 0 & a_{22} \end{pmatrix}, A_1 = \begin{pmatrix} 0 & b_1 \\ 0 & b_2 \end{pmatrix},$$

and

$$F(x(t), x(t-h)) = \begin{pmatrix} f_1(x(t), x(t-h)) \\ -a_{21}x_1 + f_2(x(t), x(t-h)) \end{pmatrix}.$$

Clearly, $K(p) = -A_0 - A_1exp(-ph) - pI$. Since $K(p)$ is a triangular matrix,

$$d(K(p)) = \min\{|p + a_{22} + b_2e^{-hp}|, \; |a_{11} + p|\}.$$

Assume that for a $\nu > 0$, relations

$$\nu < a_{jj} \; (j = 1, 2), \; hb_2e^{h\nu} < 1, \; 0 \leq \frac{b_2he^{h\nu}(1 - h\,(a_{22} - \nu))}{1 - b_2he^{h\nu}} \leq \pi \qquad (6.13)$$

hold. Then, by Lemma 10.6.3

$$\min_{\omega \in \mathbf{R}} |i\omega - \nu + a_{22} + b_2e^{h\nu}e^{-ih\omega}| = a_{22} - \nu + b_2e^{h\nu}.$$

Clearly,

$$\min_{\omega \in \mathbf{R}} |i\omega - \nu + a_{11}| = a_{11} - \nu.$$

These relations show that roots of the characteristic function lie in the half-plane $Rez < -\alpha_0$ with

$$\alpha_0 = \min\{a_{22} + b_2, \; a_{11}\}.$$

Put

$$d_\nu = \min\{a_{22} - \nu + b_2e^{h\nu}, \; a_{11} - \nu\}.$$

Thus, $d(K(-\nu+i\omega)) \geq d_\nu$ $(\omega \in \mathbf{R})$. Moreover, since the characteristic matrix is triangular, we can write

$$g(B(-\nu+i\omega)) = g(A_0 + A_1 e^{\nu h - ih\omega}) = |a_{12} + b_1 e^{\nu h - ih\omega}| \leq v_\nu,$$

where

$$v_\nu = |a_{12}| + |b_1| e^{\nu h}.$$

(see Example 2.1.3). Consequently, $\Gamma_\nu(K) \leq \Gamma_{1\nu}(K)$ where

$$\Gamma_{1\nu}(K) = d_\nu^{-1} + v_\nu d_\nu^{-2}.$$

In addition,

$$a_\nu(K) \leq \tilde{a}_{1\nu}(K) \equiv 2\sqrt{V_\nu(R)\Gamma_{1\nu}(K)[V_\nu(R)\Gamma_{1\nu}(K)+1]}$$

with

$$V_\nu(R) = \nu + ||A_0||_{C^2} + e^{\nu h}||A_1||_{C^2}.$$

Further, according to (6.12), for a given positive $r < \infty$, condition (6.3) holds with $m = 1$,

$$q_0 = r^{p_0 - 1} l_0 + |a_{21}| \text{ and } q_1 = l_1 r^{p_1 - 1}.$$

Now Theorem 10.6.1 implies

Proposition 10.6.4 *For a positive ν, let relations (6.13) hold. Then under (6.12) for any sufficiently small r, the relation*

$$\tilde{q}_\nu \equiv r^{p_0-1} l_0 + |a_{21}| + l_1 r^{p_1-1} e^{\nu h} < \frac{\nu}{\tilde{a}_{1\nu}(K)}.$$

holds, and therefore, the zero solution of system (6.11) is exponentially stable. Moreover, any initial function $\Phi(t)$ $(-h \leq t \leq 0)$ satisfying the inequality

$$||\Phi(0)||_{C^2} + V_\nu(R)\int_{-h}^{0} e^{\nu\tau}||\Phi(\tau)||_{C^2} d\tau \leq \frac{r}{(1+\tilde{q}_\nu)\tilde{a}_{1\nu}(K)},$$

belongs to the region of attraction of the zero solution, and the corresponding solution is subject to the estimate

$$||x(t)||_{C^2} \leq const\, e^{-(\nu - \tilde{q}_\nu \tilde{a}_{1\nu}(K))t} \quad (t \geq 0).$$

10.6.3 Proofs

Lemma 10.6.5 *Under condition (6.4) a solution $\phi(t)$ of the homogeneous equation*

$$\dot{\phi} = \int_0^\eta dR(\tau)\phi(t-\tau)$$

with an initial function Φ is subject to the inequality

$$||\phi(t)||_{C^n} \leq a_\nu(K) T_\nu(\Phi, K) e^{-\nu t} \quad (t \geq 0).$$

Proof: By virtue of Lemmata 8.5.1 and 8.6.1,

$$\|\phi(t)\|_{C^n} \le \|G(t)\|_{C^n}\|\Phi(0)\|_{C^n}+$$

$$\int_0^\eta \|dR(s)\|_{C^n} \int_{-\eta}^0 \|G(t-\tau-s)\|_{C^n}\|\Phi(\tau)\|_{C^n}d\tau \le$$

$$a_\nu e^{-\nu t}\|\Phi(0)\|_{C^n} + \int_0^\eta \|dR(s)\|_{C^n} \int_{-\eta}^0 a_\nu e^{-\nu(t-\tau-s)}\|\Phi(\tau)\|_{C^n}d\tau.$$

Consequently,

$$\|\phi(t)\|_{C^n} \le a_\nu e^{-\nu t}[\|\Phi(0)\|_{C^n} + \int_0^\eta \|dR(s)\|_{C^n} \int_{-\eta}^0 e^{\nu(\tau+s)}\|\Phi(\tau)\|_{C^n}d\tau].$$

Hence the result immediately follows. □

Lemma 10.6.6 *Assume that inequality (6.3) holds with $\Omega(r) = \mathbf{C}^n$. That is, $r = \infty$. Then under conditions (6.4) and (6.5), estimate (6.7) is valid for any solution $x(t)$ of equation (6.1) with an initial function Φ.*

Proof: It is simple to check that equation (6.1) is equivalent to the following one:

$$x(t) = \phi(t) + \int_0^t G(t-s)F(s, x(s), x(s-h_1), ..., x(s-h_m))ds,$$

where ϕ is the same as above. Due to (6.3),

$$\|x(t)\|_{C^n} \le \|\phi(t)\|_{C^n} + \int_0^t \|G(t-s)\|_{C^n} \sum_{j=0}^m q_j\|x(s-h_j)\|_{C^n}ds.$$

Now Lemmata 8.5.1 and 10.6.5 yield

$$\|x(t)\|_{C^n} \le a_\nu T_\nu(\Phi, K)e^{-\nu t} + \int_0^t a_\nu e^{-\nu(t-s)} \sum_{j=0}^m q_j\|x(s-h_j)\|_{C^n}ds.$$

Hence,

$$\|x(t)\|_{C^n} \le a_\nu e^{-\nu t}[T_\nu(\Phi, K) + \sum_{j=0}^m q_j \int_{-h_j}^{t-h_j} e^{\nu(z+h_j)}\|x(z)\|_{C^n}dz].$$

Therefore

$$\|x(t)\|_{C^n} \le a_\nu e^{-\nu t}[T_\nu(\Phi, K) + \tilde{q}_\nu \int_{-\eta}^t e^{z\nu}\|x(z)\|_{C^n}dz].$$

According to (6.2) this gives

$$\|x(t)\|_{C^n} \leq a_\nu e^{-\nu t}[T_\nu(\Phi, K) + \tilde{q}_\nu \int_0^t e^{z\nu}\|x(z)\|_{C^n} dz +$$

$$\tilde{q}_\nu \int_{-\eta}^0 e^{z\nu}\|\Phi(z)\|_{C^n} dz].$$

Put $\psi(t) = e^{\nu t}\|x(t)\|_{C^n}$, and take into account that

$$\int_{-\eta}^0 e^{z\nu}\|\Phi(z)\|_{C^n} dz \leq T_\nu(\Phi, K)V_\nu^{-1}(R).$$

Then

$$\psi(t) \leq a_\nu[T_\nu(\Phi, K)(1 + \tilde{q}_\nu V_\nu^{-1}(R)) + \tilde{q}_\nu \int_0^t \psi(z)dz].$$

By the Gronwall lemma,

$$\psi(t) \leq a_\nu T_\nu(\Phi, K)(1 + \tilde{q}_\nu V_\nu^{-1}(R)) exp\, [\tilde{q} a_\nu t] \; (t \geq 0).$$

This proves the result. □

Proof of Theorem 10.6.1: In the case of $r = \infty$, the theorem is valid due to Lemma 10.6.5. Consider the equation

$$\dot{x} = \int_0^\eta dR(\tau)x(t-\tau) + F_1(t, x(t), x(t-h_1), \ldots, x(t-h_m)) \; (t \geq 0), \quad (6.14)$$

where the vector-valued function F_1 defined on $[0, \infty) \times (\mathbf{C}^n)^{m+1}$ satisfies conditions (4.7) and (4.8). The function F_1 can be constructed due to Lemma 10.4.3. Thus, for a solution of equation (6.14) Lemma 10.6.5 gives estimate (6.7). Condition (6.6) and estimate (6.7) guarantee that the solution of (6.14) lies in $\Omega(r)$. But according to (4.8) it is simultaneously the solution of (6.1). This proves the result. □

Equality (6.8) follows from Lemma 8.3.2, since $K_\nu(p)$ is the characteristic function of equation

$$\dot{y}(t) = \nu y(t) + \int_0^\eta e^{\nu t} dR(\tau)x(t-\tau).$$

Proof of Lemma 10.6.3. Consider the function

$$J(\omega) \equiv |i\omega + a + be^{-ih\omega}|^2 = (\omega + b\, sin\, h\omega)^2 + (a + b\, cos\, h\omega)^2 =$$

$$\omega^2 + a^2 + b^2 + 2b\omega sin\, h\omega + 2ab\, cosh\omega.$$

Under (6.10), we have $1 \geq ah$, and

$$J'(\omega) = 2[\omega(1 + bhcos\, h\omega) + b(1-ah)\, sin\, h\omega] \geq 0 \; (0 \leq \omega h \leq \pi).$$

and

$$J'(\omega) \geq 2[h^{-1}\pi(1-bh) - b(1-ah)] \geq 0 \ (\omega h \geq \pi).$$

But $J(0) = (a+b)^2$ and $J(\omega) = J(-\omega)$. This proves the result. □

10.7 Nonlinear systems with small delays

10.7.1 Statement of the result

Let us consider in $\mathbf{C}^n$ the equation

$$\dot{x}(t) = \sum_{k=0}^{m} A_k x(t-h_k) + F(t, x_t) \quad (t \geq 0), \tag{7.1}$$

where A_k $(k = 0, ..., m)$ are constant matrices,

$$0 = h_0 < h_1 < ... < h_m = \eta \equiv const < \infty$$

are numbers, and F continuously maps $[0, \infty) \times C([-\eta, 0], \mathbf{C}^n)$ into $\mathbf{C}^n$.

Again it is assumed that for every $v \in L^2([-\eta, \infty), \mathbf{C}^n) \cap C([-\eta, \infty), \mathbf{C}^n)$, the inequality

$$||F(., v_t)||_{L^2([0,\infty),\mathbf{C}^n)} \leq q||v||_{L^2([-\eta,\infty),\mathbf{C}^n)} \tag{7.2}$$

holds.

Definition 10.7.1 *We will say that the zero solution of equation (7.1) is absolutely exponentially stable in the class of nonlinearities (7.2) if there exist positive constants N and ϵ independent of the specific form of the function F, such that for any solution $x(t)$ of (7.1) with a continuous initial function Φ, the inequality*

$$||x(t)||_{C^n} \leq N exp\,[-\epsilon t] ||\Phi||_{C([-\eta,0],C^n)} \ (t \geq 0)$$

is fulfilled.

Put

$$\tilde{A} = \sum_{k=0}^{m} A_k, \ V_2 = \sum_{k=1}^{m} h_k ||A_k||_{C^n}. \tag{7.3}$$

Recall that $\omega(\tilde{A}) = \max_k |Im\lambda_k(\tilde{A})|$, $\alpha(\tilde{A}) = \max_k Re\lambda_k(\tilde{A})$, and

$$\Gamma(\tilde{A}) = \sum_{k=1}^{n-1} \frac{g^k(\tilde{A})}{|\alpha(\tilde{A})|^{k+1}\sqrt{k!}}.$$

Throughout the present section, it is assumed that $\tilde{A}$ is a Hurwitz matrix.

Theorem 10.7.2 *Let the condition*

$$[(\alpha^2(\tilde{A}) + \omega^2(\tilde{A}))^{1/2}\ V_2 + q]\ \Gamma(\tilde{A}) < 1 \tag{7.4}$$

hold. Then the zero solution of equation (7.1) is absolutely exponentially stable in the class of nonlinearities (7.2).

If $\tilde{A}$ is a normal (in particular, Hermitian) matrix, then $g(\tilde{A}) = 0, \Gamma(\tilde{A}) = |\alpha(\tilde{A})|^{-1}$, and condition (7.4) takes the form

$$[(\alpha^2(\tilde{A}) + \omega^2(\tilde{A}))^{1/2}\ V_2 + q]\ |\alpha(\tilde{A})|^{-1} < 1. \tag{7.5}$$

Remark 10.7.3 *Theorem 10.7.2 is exact, and condition (7.4) cannot be improved without additional restrictions.*

In fact, let us consider the system

$$\dot{x}(t) = Ax(t) + F(t, x(t)), \tag{7.6}$$

where A is a normal Hurwitz matrix, and F continuously maps $[0, \infty) \times \mathbf{C}^n$ into $\mathbf{C}^n$ with the property

$$||F(t, h)||_{C^n} \leq q||h||_{C^n}\ (h \in \mathbf{C}^n;\ t \geq 0). \tag{7.7}$$

Under consideration $V_2 = 0, g(A) = 0$, and (7.4) takes the form

$$q|\alpha(A)|^{-1} < 1. \tag{7.8}$$

Take $F(t, h) = qh$ ($h \in \mathbf{C}$). Then for the asymptotic stability of equation (7.6) it is necessary and sufficient that $\alpha(A) + q < 0$. But this condition coincides with (7.8).

10.7.2 Example

Let us consider in $\mathbf{C}^n$ the equation

$$\dot{x}(t) = A_0x(t) + A_1x(t - \eta) + f(t, x(t), \int_0^\eta T(t, s)x(t - s)ds), \tag{7.9}$$

where the function $f : [0, \infty) \times \mathbf{C}^{2n} \to \mathbf{C}^n$ has the property

$$||f(t, h, w)||_{\mathbf{C}^n} \leq q_0||h||_{C^n} + q_1||w||_{C^n}\ (h, w \in \mathbf{C}^n;\ t \geq 0), \tag{7.10}$$

and the matrix-valued function $T(t, s)$ defined on $[0, \infty) \times [-\eta, 0]$ satisfies the inequality

$$m_T \equiv \sup_{t \geq 0} \int_0^\eta ||T(t, s)||_{C^n} ds < \infty. \tag{7.11}$$

Taking into account that

$$\int_0^\infty \|\int_0^\eta T(t,s)x(t-s)ds\|^2_{C^n}dt \le m_T^2\|x\|^2_{L[-\eta,\infty)}$$

(see Subsection 9.7.2), we easily get the inequality

$$[\int_0^\infty \|f(t,u(t),\int_0^\eta T(t,s)u(t-s)ds)\|^2_{C^n}dt]^{1/2} \le$$

$$q_0\|u\|_{L[0,\infty)} + q_1 m_T\|u\|_{L[-\eta,\infty)}$$

for any $u \in L^2([-\eta,\infty),\mathbf{C}^n) \cap C([-\eta,\infty),\mathbf{C}^n)$. Hence it follows that (7.2) holds with

$$F(t,u_t) = f(t,u(t),\int_0^\eta T(t,s)u(t-s)ds), \text{ and } q = q_0 + q_1 m_T. \tag{7.12}$$

Clearly,

$$\tilde{A} = A_0 + A_1,\ V_2 = \eta\|A_1\|_{C^n}. \tag{7.13}$$

Now Theorem 10.7.2 yields the following result. Let $\tilde{A}, q$ and V_2 be defined by (7.12) and (7.13). Let condition (7.4) hold. Then the zero solution of equation (7.9) is absolutely exponentially stable in the class of nonlinearities satisfying conditions (7.10), (7.11).

10.8 Proof of Theorem 10.7.2

Throughout the present section, for brevity put $\|.\|_{L^2([0,\infty),\mathbf{C}^n)} = \|.\|_L$,

$$\|y(i\omega)\|_{\hat{L}} \equiv [\frac{1}{2\pi}\int_{-\infty}^\infty \|y(i\omega)\|^2_{C^n}d\omega]^{1/2}\ (y \in L^2(\mathbf{R},\mathbf{C}^n)).$$

Denote

$$c_0(\tilde{A}) = \sup_{\omega\in\mathbf{R}} \|(i\omega I - \tilde{A})^{-1}\|_{C^n}, \text{ and } c_1(\tilde{A}) = \sup_{\omega\in\mathbf{R}} \|\omega(i\omega I - \tilde{A})^{-1}\|_{C^n}.$$

Let us consider the equation

$$\dot{x}(t) = \sum_{k=0}^m A_k x(t-h_k) + f(t)\ \ (t \ge 0). \tag{8.1}$$

Lemma 10.8.1 *Under the condition $c_1(\tilde{A})V_2 < 1$, let $f \in L^2([0,\infty),\mathbf{C}^n)$. Then for any solution $x(t)$ of (8.1) with a continuous initial function Φ, the bound*

$$\|x\|_L \le [1 - c_1(\tilde{A})V_2]^{-1}c_0(\tilde{A})\|f\|_L + m_1\|\Phi\|_{C[-\eta,0]} \tag{8.2}$$

is valid with a constant m_1 independent of Φ.

Proof: Rewrite (8.1) in the form

$$\dot{x}(t) = \tilde{A}x(t) + \sum_{k=1}^{m} A_k(x(t-h_k) - x(t)) + f(t) \ (t \geq 0). \tag{8.3}$$

The Laplace transform of $x(t-h_k)$ is $\tilde{x}(\lambda)e^{-h_k\lambda} + v_k(\lambda)$ where $\tilde{x}(\lambda)$ is the Laplace transform of $x(t)$, and

$$v_k(\lambda) = \int_0^{h_k} e^{-t\lambda} x(t-h_k)dt = \int_0^{h_k} e^{-t\lambda}\Phi(t-h_k)dt.$$

Thus the application of the Laplace transformation to equation (8.3) gives

$$\lambda\tilde{x}(\lambda) - x(0) = \tilde{A}\tilde{x}(\lambda) + \sum_{k=1}^{m} A_k[\tilde{x}(\lambda)(e^{-h_k\lambda} - 1) + v_k(\lambda)] + \tilde{F}(\lambda) \ (\lambda \in \mathbf{C}),$$

where $\tilde{F}(\lambda)$ is the Laplace transform of the function $f(t)$. Hence,

$$\tilde{x}(\lambda) = (\lambda I - \tilde{A})^{-1}[\sum_{k=1}^{m} A_k(e^{-h_k\lambda} - 1)x(\lambda) + \tilde{F}(\lambda)] + e(\lambda) \ (\lambda \in \mathbf{C}), \tag{8.4}$$

where

$$e(\lambda) = (\lambda I - \tilde{A})^{-1}(x(0) + \sum_{k=1}^{m} A_k v_k(\lambda)).$$

Since $|1 - e^{-hi\omega}| \leq h|\omega| \ (\omega \in \mathbf{R}, h > 0)$, we have

$$||(\tilde{A} - i\omega I)^{-1} \sum_{k=1}^{m} A_k(1 - e^{-h_k i\omega})\tilde{x}(i\omega)||_{\tilde{L}} \leq$$

$$||\omega(\tilde{A} - i\omega I)^{-1}||_{C^n} \sum_{k=1}^{m} h_k ||A_k||_{C^n} ||\tilde{x}(i\omega)||_{\tilde{L}} \leq$$

$$c_1(\tilde{A})V_2||\tilde{x}(i\omega)||_{\tilde{L}} \ (\omega \in \mathbf{R}). \tag{8.5}$$

Due to (8.4) and (8.5), it can be written

$$||\tilde{x}(i\omega)||_{\tilde{L}} \leq V_2 c_1(\tilde{A})||\tilde{x}(i\omega)||_{\tilde{L}} + c_0(\tilde{A})||\tilde{F}||_{\tilde{L}} + ||e(i\omega)||_{\tilde{L}}.$$

Hence, taking into account that $c_1(\tilde{A})V_2 < 1$, we get

$$||\tilde{x}(i\omega)||_{\tilde{L}} \leq (1 - V_2 c_1(\tilde{A}))^{-1}(c_0(\tilde{A})||\tilde{F}||_{\tilde{L}} + ||e(i\omega)||_{\tilde{L}}). \tag{8.6}$$

The simple calculations show that

$$||v_k(\omega i)||_{C^n} \leq h_k ||\Phi||_{C[-\eta,0]} \ (\omega \in \mathbf{R}),$$

and

$$\|e(i\omega)\|_{\tilde{L}} \leq \|(i\omega I - \tilde{A})^{-1}\|_{\tilde{L}}(\|x(0)\|_{C^n}+$$

$$\sum_{k=1}^{m} h_k\|A_k\|_{C^n} \sup_{\omega\in\mathbf{R}} \|v_k(\omega i)\|_{C^n}) \leq$$

$$\|(i\omega I - \tilde{A})^{-1}\|_{\tilde{L}}(\|\Phi(0)\|_{C^n} + V_2\|\Phi\|_{C[-\eta,0]}).$$

Hence,

$$\|e(i\omega)\|_{\tilde{L}} \leq \|(i\omega I - \tilde{A})^{-1}\|_{\tilde{L}}(1 + V_2)\|\Phi\|_{C[-\eta,0]}.$$

Relation (8.6) yields

$$\|\tilde{x}(i\omega)\|_{\tilde{L}} \leq (1 - V_2 c_1(\tilde{A}))^{-1} c_0(\tilde{A})\|\tilde{F}\|_{\tilde{L}} + m_1\|\Phi\|_{C[-\eta,0]}.$$

Now due to the Parseval equality, we arrive at the required result. □

Lemma 10.8.2 *Under condition (7.2), let the inequality*

$$c_1(\tilde{A})V_2 + c_0(\tilde{A})q < 1 \tag{8.7}$$

be fulfilled. Then for any solution $x(t)$ of (7.1) with a continuous initial function $\Phi(t)$, the bound

$$\|x\|_L \leq m_0[1 - (c_1(\tilde{A})V_2 + qc_0(\tilde{A}))]^{-1}\|\Phi\|_{C[-\eta,0]}$$

is valid with a constant m_0.

Proof: Rewrite (7.1) in the form (8.1) with $f(t) = F(t, x_t(t))$, where x is the solution of (7.1). Combining the previous lemma with condition (7.2), we easily get

$$\|x\|_L \leq [1 - c_1(\tilde{A})V_2]^{-1}c_0(\tilde{A})\|f\|_L + m_1\|\Phi\|_{C[-\eta,0]} \leq$$

$$[1 - c_1(\tilde{A})V_2]^{-1}c_0(\tilde{A})q(\|x\|_L + \|\Phi\|_{L[-\eta,0]}) + m_1\|\Phi\|_{C[-\eta,0]}.$$

Consequently,

$$\|x\|_L \leq [1 - c_1(\tilde{A})V_2]^{-1}c_0(\tilde{A})q\|x\|_L + m_2\|\Phi\|_{C[-\eta,0]}$$

with a constant m_2. Condition (8.7) implies

$$[1 - c_1(\tilde{A})V_2]^{-1}c_0(\tilde{A})q < 1.$$

Thus,

$$\|x\|_L \leq (1 - c_0(\tilde{A})q[1 - c_1(\tilde{A})V_2]^{-1})^{-1}m_2\|\Phi\|_{C[-\eta,0]} =$$

$$= (1 - c_0(\tilde{A})q - c_1(\tilde{A})V_2)^{-1}m_0\|\Phi\|_{C[-\eta,0]}.$$

As claimed. □

Lemma 10.8.3 *The relations* $c_0(\tilde{A}) \leq \Gamma(\tilde{A})$, *and*

$$c_1(\tilde{A}) \leq \Gamma(\tilde{A})[\alpha^2(\tilde{A}) + \omega^2(\tilde{A})]^{1/2}$$

are true.

Proof: By Lemma 8.1.1, we have the inequality

$$\|(\tilde{A} - I\lambda)^{-1}\|_{C^n} \leq \sum_{k=0}^{n-1} \frac{g^k(\tilde{A})}{\sqrt{k!}\rho^{k+1}(\tilde{A}, \lambda)} \text{ for all regular } \lambda, \tag{8.8}$$

where $\rho(\tilde{A}, \lambda)$ is the distance between the spectrum of $\tilde{A}$ and a complex point λ. Since $\tilde{A}$ is a Hurwitz matrix,

$$\rho(\tilde{A}, i\omega) \geq |\alpha(\tilde{A})| \ (\omega \in \mathbf{R}), \tag{8.9}$$

and, therefore, $c_0(\tilde{A}) \leq \Gamma(\tilde{A})$. Further, we have

$$|\lambda_k(\tilde{A}) - i\omega|^2 \geq \alpha^2(\tilde{A}) + (\omega_k - \omega)^2$$

for any eigenvalue $\lambda_k(\tilde{A})$ of $\tilde{A}$ with $Im\lambda_k(\tilde{A}) = \omega_k$. Hence simple calculations show that

$$\max_{\omega \in \mathbf{R}} \omega^2 |\lambda_k(\tilde{A}) - i\omega|^{-2} \leq \frac{\alpha^2(\tilde{A}) + \omega_k^2}{\alpha^2(\tilde{A})}.$$

Thus

$$\omega^2 \rho^{-2}(\tilde{A}, \omega) \leq \frac{\alpha^2(\tilde{A}) + \omega^2(\tilde{A})}{\alpha^2(\tilde{A})} \ (\omega \in \mathbf{R}).$$

Therefore according to (8.8) and (8.9),

$$\|\omega(\tilde{A} - i\omega I)^{-1}\| \leq \sum_{k=0}^{n-1} \frac{|\omega| g^k(\tilde{A})}{\sqrt{k!}\rho^{k+1}(\tilde{A}, \omega)} \leq$$

$$(\alpha^2(\tilde{A}) + \omega^2(\tilde{A}))^{1/2} \sum_{k=0}^{n-1} \frac{g^k(\tilde{A})}{\sqrt{k!}|\alpha(\tilde{A})|^{k+1}} = (\alpha^2(\tilde{A}) + \omega^2(\tilde{A}))^{1/2}\Gamma(\tilde{A}) \ (\omega \in \mathbf{R}).$$

As claimed. □

Lemmas 10.8.2 and 10.8.3 yield

Corollary 10.8.4 *Let conditions (7.2) and (7.4) hold. Then for any solution* $x(t)$ *of (7.1) with an initial function* $\Phi(t)$, *the bound*

$$\|x\|_L \leq m_0 \left[1 - \Gamma(\tilde{A})(V_2(\alpha^2(\tilde{A}) + \omega^2(\tilde{A}))^{1/2} + q)\right]^{-1} \|\Phi\|_{C[-\eta, 0]} \ (m_0 = const)$$

is valid.

Proof of Theorem 10.7.2: From (7.1) it follows the inequality

$$||\dot{x}||_L \leq \sum_{k=0}^{m} ||A_k||_{C^n} ||x(t-h_k)||_L + ||F||_L \leq V_1(||x||_L + ||x||_{L[-\eta,0]}) + ||F||_L,$$

where $F = F(t, x_t)$,

$$V_1 = \sum_{k=0}^{m} ||A_k||_{C^n}.$$

Due to (7.2) this gives

$$||\dot{x}||_L \leq (q + V_1)(||x||_L + ||\Phi||_{L[-\eta,0]}).$$

Now Corollary 10.8.4 implies the relation

$$||\dot{x}||_L \leq N_1 ||\Phi||_{C[-\eta,0]} \ (N_1 = const).$$

By virtue of Lemma 8.4.4 we get

$$||x||_{C([0,\infty);\mathbf{C}^n)} \leq \sqrt{2||x||_L ||\dot{x}||_L} \leq N_2 ||\Phi||_{C[-\eta,0]} \ (N_2 = const). \tag{8.10}$$

Furthermore, substitute in (7.1) the relation

$$x(t) = x_\epsilon(t) e^{-\epsilon t} \tag{8.11}$$

with a small enough $\epsilon > 0$. Applying our reasoning above, we can assert according to (8.10) that

$$||x_\epsilon||_{C([-\eta,\infty);\mathbf{C}^n)} \leq N_\epsilon ||\Phi||_{C[-\eta,0]} \ (N_\epsilon = const).$$

Hence, (8.11) yields the absolute exponential stability. □

10.9 Stability with respect to arbitrary delay

Let us consider in $\mathbf{C}^n$ the equation

$$\dot{x} = Ax + F(t, x(t-h_1), ..., x(t-h_m))$$
$$(t \geq 0;\ h_1, ..., h_m = const > 0), \tag{9.1}$$

where A is a constant Hurwitz $n \times n$-matrix, F maps $[0,\infty) \times \mathbf{C}^{mn}$ into $\mathbf{C}^n$ with the following property: there are positive numbers $q_0, ..., q_m$, such that

$$||F(t, y_1, ..., y_m)|| \leq \sum_{k=1}^{m} q_k ||y_k|| \text{ for all } y_k \in \mathbf{C}^n \ (k = 1, ..., m). \tag{9.2}$$

Again put

$$q = \sum_{k=1}^{m} q_k \text{ and } \Gamma(A) = \sum_{k=0}^{n-1} \frac{g^k(A)}{\sqrt{k!}|\alpha(A)|^{k+1}}.$$

Theorem 10.9.1 *Let the condition $q\Gamma(A) < 1$ be fulfilled. Then the zero solution of equation (9.1) is absolutely stable in the class of nonlinearities (9.2).*

Proof: Let us apply Theorem 10.3.2 to equation (9.1). In the considered case, clearly, $\Gamma_0(K) = \Gamma(A)$. Moreover, due to Proposition 10.2.1 condition (9.2) implies condition (3.1). Now Theorem 10.3.2 yields the result. □

10.10 Systems with nonautonomous linear parts

10.10.1 Solution estimates

Let $A(t)$ be a variable matrix. Consider in $\mathbf{C}^n$ the equation

$$\dot{x} = A(t)x + F(t, x(h(t)))\ (t \geq 0),\ x(t) = \Phi(t)\ (-\eta \leq t \leq 0), \tag{10.1}$$

where $h(t)$ is an increasing differentiable scalar-valued function defined on $[0, \infty)$ and satisfying the inequalities

$$-\eta \leq h(t) \leq t\ (t \geq 0;\ 0 < \eta = const < \infty). \tag{10.2}$$

In addition, F maps $[0, \infty) \times \mathbf{C}^n$ into $\mathbf{C}^n$.

Let $U(t, s)$ be the evolution operator (fundamental solution) of the "shorten" equation

$$\dot{v} = A(t)v\ (t \geq 0).$$

Assume that there are bounded measurable scalar-valued functions $a(t) \leq 0$ and $w(t)$, such that the inequalities

$$\|U(t, s)\|_{\mathbf{C}^n} \leq Me^{\int_s^t a(\tau)d\tau}\ (M = const;\ t, s \geq 0) \tag{10.3}$$

and

$$\|F(t, h)\|_n \leq w(t)\|h\|_n\ (h \in \mathbf{C}^n,\ t \geq 0) \tag{10.4}$$

hold. Here $\|.\|_n$ is an arbitrary norm in $\mathbf{C}^n$.

Denote by $\psi(z)$ the function inverse to $h(s)$. That is, if $\tau = h(s)$, then $s = \psi(\tau)$. Put

$$q(s) = w(s)\ exp[-\int_{h(s)}^{s} a(\tau)d\tau].$$

Now we are in a position to formulate the main result of the section.

Theorem 10.10.1 *Let inequalities (10.3) and (10.4) hold. Then any solution $x(t)$ of (10.1) with an initial function Φ satisfies the inequality*

$$\|x(t)\|_n \leq\ z(t)exp[\int_0^t a(\tau)d\tau]\ (t \geq \psi(0)), \tag{10.5}$$

where $z(t)$ is a solution of the scalar equation

$$z(t) = e_0(\Phi) + M \int_{\psi(0)}^{t} q(s)z(h(s))ds \ (t \geq \psi(0)) \tag{10.6}$$

with the notation

$$e_0(\Phi) = M\|\Phi(0)\|_n + M \int_0^{\psi(0)} exp[\int_s^0 a(\tau)d\tau]w(s)\|\Phi(h(s))\|_n ds.$$

The proof is presented in Subsection 10.10.3. Further, put

$$\Lambda(t) = \overline{\lim}_{h\to+0}\frac{1}{h}[\|I_{\mathbf{C}^n} + hA(t)\|_n - 1].$$

Corollary 10.10.2 *Let $\Lambda(t)$ be a Riemann integrable function, and let condition (10.4) hold. Then inequality (10.5) is valid with $a(t) = \Lambda(t)$ and $M = 1$.*

Indeed, due to the Lozinskii inequality the condition (10.3) holds with $a(t) = \Lambda(t)$ and $M = 1$ (see Section 4.2). Now the result is due to Theorem 10.10.1.

Corollary 10.10.3 *Let the conditions (10.4) with the Euclidean norm $\|.\|_n = \|.\|_{C^n}$, and*

$$Re(A(t)g, g) \leq a(t)\|g\|^2_{C^n} \quad (g \in \mathbf{C}^n)$$

be fulfilled. Then inequality (10.5) is valid and $M = 1$.

Indeed, due to the Wazewski inequality the condition (10.3) holds with $M = 1$ (see Section 4.2). Now the result is due to Theorem 10.10.1.

10.10.2 Stability conditions

Since $h(t) \leq t$, we get by equation (10.6)

$$z(t) \leq e_0(\Phi) + M \int_0^t \dot{\psi}(s_1)q(\psi(s_1))z(s_1)ds_1.$$

Due to the Gronwall inequality,

$$z(t) \leq e_0(\Phi)exp\{M \int_0^t \dot{\psi}(s_1)q(\psi(s_1))ds_1\}.$$

Consequently, Theorem 10.10.1 gives the estimate

$$\|x(t)\|_n \leq e_0(\Phi)\ exp\{\int_0^t [a(\tau) + M\dot{\psi}(\tau)q(\psi(\tau))]d\tau]\} =$$

$$e_0(\Phi)\ exp\{\int_0^t a(\tau)d\tau + M \int_{\psi(0)}^{\psi(t)} q(s)ds\} \ (t \geq \psi(0)). \tag{10.7}$$

We thus have derived

Corollary 10.10.4 *Under the hypothesis of Theorem 10.10.1, let the condition*

$$\overline{\lim}_{t\to\infty}\{\int_0^t a(\tau)d\tau + M\int_0^{\psi(t)} q(s)ds\} < \infty$$

be fulfilled. Then the zero solution of equation (10.4) is globally stable.

Let the relation

$$\Lambda_1 \equiv \overline{\lim}_{t\to\infty}t^{-1}\{\int_0^t a(\tau)d\tau + M\int_0^{\psi(t)} q(s)ds\} < \infty$$

hold. Then due to (10.7) for the Lyapunov exponent of a solution $x(t)$ of equation (10.4) we have the inequality

$$\kappa(x) = \overline{\lim}_{t\to\infty}\frac{ln||x(t)||_n}{t} < \Lambda_1.$$

In particular, the condition $\Lambda_1 < 0$ provides the global exponential stability.

Now take $h(t) = t - h_0$, where h_0 is a positive constant. Then inequality (10.7) implies

Corollary 10.10.5 *Under the hypothesis of Theorem 10.10.1, let $h(t) = t - h_0$ and the condition*

$$\overline{\lim}_{t\to\infty}t^{-1}\int_0^t (a(s) + Mq(s))ds < 0$$

be fulfilled. Then the zero solution of equation (10.4) is globally exponentially stable.

10.10.3 Proof of Theorem 10.10.1

Equation (10.1) is equivalent to the following one:

$$x(t) = U(t,0)x(0) + \int_0^t U(t,s)F(s,x(s))ds.$$

Due to relations (10.3) and (10.4) that equation yields

$$||x(t)||_n \leq Mexp[\int_0^t \alpha(\tau)d\tau]||x(0)||_n +$$

$$M\int_0^t exp[\int_s^t \alpha(\tau)d\tau]w(s)||x(h(s))||_n ds.$$

Therefore

$$||x(t)||_n \leq Mexp[\int_0^t \alpha(\tau)d\tau](||\Phi(0)||_n +$$

$$\int_0^{\psi(0)} exp[\int_s^0 \alpha(\tau)d\tau]w(s)\|\Phi(h(s))\|_n ds +$$

$$M\int_{\psi(0)}^t exp[\int_s^0 \alpha(\tau)d\tau]w(s)\|x(h(s))\|_n ds) =$$

$$exp[\int_0^t \alpha(\tau)d\tau](e_0(\Phi) + M\int_{\psi(0)}^t w(s)\|x(h(s))\|_n exp[\int_s^0 \alpha(\tau)d\tau]ds)$$

for $t \geq \psi(0)$. Put

$$y(t) = \|x(t)\|_n \; exp[-\int_0^t \alpha(\tau)d\tau]. \tag{10.8}$$

Then

$$y(t) \leq e_0(\Phi) + M\int_{\psi(0)}^t exp[\int_s^{h(s)} \alpha(\tau)d\tau]w(s)y(h(s))ds =$$

$$e_0(\Phi) + M\int_{\psi(0)}^t q(s)y(h(s))ds.$$

Or

$$y(t) \leq e_0(\Phi) + M\int_0^{h(t)} \dot{\psi}(s_1)q(\psi(s_1))y(s_1)ds_1.$$

Set

$$K(t,s) = M\dot{\psi}(z)q(\psi(z)) \text{ for } 0 \leq s \leq h(t);\; K(t,s) = 0 \text{ for } h(t) < s \leq t.$$

Then

$$y(t) \leq e_0(\Phi) + \int_0^t K(t,s)y(s)ds \; (t \geq 0).$$

By Lemma 1.7.1 we get the relation $y(t) \leq z_1(t)$, where $z_1(t)$ is a solution of the equation

$$z_1(t) = e_0(\Phi) + \int_0^t K(t,s)z_1(s)ds = e_0(\Phi) + \int_0^{h(t)} \dot{\psi}(s_1)q(\psi(s_1))z_1(s_1)ds_1.$$

Clearly, this relation is equivalent to equation (10.6). Now (10.8) implies the required estimate (10.5). □

10.11 Global feedback stabilization

10.11.1 Statement of the result

Let us consider in $\mathbf{C}^n$ system

$$\dot{x}(t) = A_0x(t) + A_1x(t-h) + F(t,x,x(t-h)) + Bu(t)$$

$$(u(t) \in \mathbf{C}^m, t \geq 0), \tag{11.1}$$

where A_j $(j = 0, 1)$ are constant $n \times n$-matrices, B is a constant $n \times m$-matrix, $h = const > 0$, and F continuously maps $[0, \infty) \times \mathbf{C}^n \times \mathbf{C}^n$ into $\mathbf{C}^n$. It is assumed that there are constants q_j $(j = 0, 1)$, such that

$$\|F(t, z_0, z_1)\|_{C^n} \leq q_0\|z_0\|_{C^n} + q_1\|z_1\|_{C^n} \text{ for all } z_j \in \mathbf{C}^n \ (j = 0, 1;\ t \geq 0). \tag{11.2}$$

Let us take the feedback law

$$u(t) = Lx(t), \tag{11.3}$$

where L is a constant $m \times n$ matrix. Then (11.1) takes the form

$$\dot{x}(t) = A_0x(t) + A_1x(t-h) + F(t, x(t), x(t-h)) + BLx(t) \ (t \geq 0). \tag{11.4}$$

Definition 10.11.1 *System (11.1) is said to be globally stabilizable (at $x = 0$) by means of the feedback law (11.3) if there is a matrix L, such that the zero solution of (11.4) is globally asymptotically stable.*

Put $\tilde{A} = A_0 + A_1$. The pair $(\tilde{A}, B)$ *is stabilizable* if there exists a matrix $L \in \mathbf{C}^{m \times n}$, such that $\tilde{A} + BL$ is a Hurwitz matrix.

Let the pair $(\tilde{A}, B)$ be stabilizable. Then there is a matrix L, such that

$$\alpha(\tilde{A} + BL) < 0. \tag{11.5}$$

Put

$$\Gamma(\tilde{A} + BL) \equiv \sum_{k=0}^{n-1} \frac{g^k(\tilde{A} + BL)}{\sqrt{k!}|\alpha(\tilde{A} + BL)|^{k+1}},$$

and recall that $\omega(\tilde{A}) = \max_k |Im\lambda_k(\tilde{A})|$, $\alpha(\tilde{A}) = \max_k Re\lambda_k(\tilde{A})$.

Theorem 10.11.2 *Let condition (11.2) hold, and $(\tilde{A}, B)$ be a stabilizable pair. In addition, let there exist a matrix L satisfying under (11.5) the inequality*

$$\Gamma(\tilde{A} + BL)[h\ \|A_1\|_{C^n}\ (\alpha^2(\tilde{A} + BL) + \omega^2(\tilde{A} + BL))^{1/2} + q_0 + q_1] < 1.$$

Then (11.1) is globally stabilizable by means of the feedback law (11.3).

This result easily follows from Theorem 10.7.2 and Proposition 10.2.1.

10.11.2 Example

Let us consider the equation

$$\ddot{x}(t) - a_1\dot{x}(t) - x(t) - a_2x(t-h) + f(x(t), x(t-h)) = bu(t), \tag{11.6}$$

where a_1, a_2, b are positive constants, the scalar-valued function f defined on $\mathbf{R}^2$ satisfies the relation

$$|f(x_0, x_1)| \leq q_0|x_0| + q_1|x_1| \ (x_0, x_1 \in \mathbf{R}). \tag{11.7}$$

Rewrite this equation in the form (11.1) with

$$A_0 = \begin{pmatrix} a_1 & 1 \\ 1 & 0 \end{pmatrix}, \; A_1 = \begin{pmatrix} 0 & a_2 \\ 0 & 0 \end{pmatrix}, \text{ and } B = \begin{pmatrix} b \\ 0 \end{pmatrix}.$$

In the present case $||A_1|| = a_2$, and $(\tilde{A}, B)$ is a stabilizable pair. Take $L = -(l_1, l_2)$ with constants $l_1 > a_1/b, l_2 > (1 + a_2)/b$. Then

$$\tilde{A} + BL = A_0 + A_1 + BL = \begin{pmatrix} -c_1 & -c_2 \\ 1 & 0 \end{pmatrix}$$

where

$$c_1 = bl_1 - a_1 > 0, \; c_2 = bl_2 - 1 - a_2 > 0.$$

We have

$$\alpha(\tilde{A} + BL) = \alpha_0(c_1, c_2) \equiv Re\{-c_1/2 + \sqrt{c_1^2/4 - c_2}\},$$

and $\omega(\tilde{A} + BL) = \omega_0(c_1, c_2)$ where

$$\omega_0(c_1, c_2) \equiv \sqrt{c_2 - c_1^2/4} \text{ if } c_2 > c_1^2/4; \; \omega_0(c_1, c_2) = 0 \text{ if } c_2 \leq c_1^2/4.$$

According to Example 2.1.4, it can be written $g(\tilde{A} + BL) \leq 1 + c_2$. Therefore,

$$\Gamma(\tilde{A} + BL) = \frac{1}{|\alpha_0(c_1, c_2)|} + \frac{g(\tilde{A} + BL)}{|\alpha_0(c_1, c_2)|^2} \leq G(c_1, c_2),$$

where

$$G(c_1, c_2) \equiv \frac{1}{|\alpha_0(c_1, c_2)|} + \frac{1 + c_2}{|\alpha_0(c_1, c_2)|^2}.$$

If for some $c_1, c_2 > 0$, we have the inequality

$$G(c_1, c_2) \left[ha_2(\omega_0^2(c_1, c_2) + \alpha_0^2(c_1, c_2))^{1/2} + q_1 + q_0\right] < 1,$$

then due to Theorem 10.11.2 system (11.6) under condition (11.7) is globally stabilizable.

10.12 Input-output stability of nonlinear retarded systems

10.12.1 Statement of the result

Let us consider in $\mathbf{C}^n$ the system

$$\dot{x}(t) - \int_0^\eta dR(\tau)x(t - \tau) - F(t, x_t) = Mu(t), \tag{12.1}$$

$$y(t) = Cx(t) + Du(t) \quad (t \geq 0), \tag{12.2}$$

Here $0 < \eta = const < \infty$; M, C, D are constant matrices; the $n \times n$-matrix-valued function $R(\tau)$ has a bounded variation; function F maps $[0, \infty) \times C([-\eta, 0), \mathbf{C}^n)$ into $\mathbf{C}^n$, $u : [0, \infty) \to \mathbf{C}^{m_1}$ $(m_1 \leq n)$ is the input, $y : [0, \infty) \to \mathbf{C}^{m_2}$ $(m_2 \leq n)$ is the output. All inputs below are assumed to be bounded for finite $t \geq 0$.

Let us introduce the following

Definition 10.12.1 *System (12.1), (12.2) is said to be input-output L^2-stable, if the conditions*

$$x(t) = 0 \text{ for } -\eta \leq t \leq 0,$$

and $u \in L^2([0, \infty), \mathbf{C}^{m_1})$ imply $y(t) \in L^2([0, \infty), \mathbf{C}^{m_2})$. System (12.1) is input-output L^2-stable with finite gain and zero bias if it is L^2-stable and, in addition, there exists a constant γ_2, such that

$$||y(t)||_{L^2([0,\infty),\mathbf{C}^{m_2})} \leq \gamma_2 ||u||_{L^2([0,\infty),\mathbf{C}^{m_1})}. \tag{12.3}$$

Further, let $K(p)$ be the characteristic matrix-valued function of the linear part of (12.1). It is assumed that all the zeros of $det K(p)$ belong to the open left half-plane. Again put

$$\Gamma_0(K) \equiv \sup_{\omega \in \mathbf{R}} \Gamma(K(\omega i)) \text{ with } \Gamma(K(p)) = \sum_{k=0}^{n-1} \frac{g^k(B(p))}{\sqrt{k!} d^{k+1}(K(p))},$$

where $d(K(p))$ $(p \in \mathbf{C})$ is the smallest modulus of eigenvalues of the matrix $K(p)$. Remember that simple estimates for $\Gamma_0(K)$ are presented in Section 8.3. The aim of this section is to prove the following

Theorem 10.12.2 *Let the conditions (7.2) and*

$$q\Gamma_0(K) < 1 \tag{12.4}$$

hold. Then system (12.1), (12.2) is input-output L^2 - stable with finite gain and zero bias. Moreover, the constant γ_2 in (12.3) can be defined by the equality

$$\gamma_2 = ||C||(1 - q\Gamma_0(K))^{-1}\Gamma_0(K)||M|| + ||D||.$$

The proofs are presented in Subsection 10.12.3.

10.12.2 Example

Consider in $\mathbf{C}^n$ the system

$$dx/dt - Ax - F(t, x_t) = Mu(t),$$

$$y = Cx + Du, \tag{12.5}$$

where A is a constant Hurwitz constant matrix, and F satisfies condition (7.2).

As it is shown in Example 10.3.3, under consideration $\Gamma_0(K) \leq \Gamma(A)$. So, if $q\Gamma(A) < 1$, then the system (12.5) under condition (7.2) is input-output L^2 - stable with finite gain and zero bias due to Theorem 10.12.2

10.12.3 Proof of Theorem 10.12.2

For a fixed solution $x(t)$ of (12.1) with the zero initial condition, put

$$f(t) = F(t, x_t) + Mu(t).$$

Then (12.1) takes the form

$$\dot{x} = \int_0^\eta dR(s)x(t-s) + f(t) \ (t \geq 0).$$

Due to Lemma 8.2.1 a solution of (12.1) with the zero initial condition admits the representation

$$x(t) = -\frac{1}{2\pi}\int_{-\infty}^{\infty} [K(is)]^{-1} exp(ist) f(is) ds,$$

where $\hat{f}(p)$ is the Laplace transform of $f(t)$. Denote by $F(p)$ and $u(p)$ the Laplace transforms of $F(t, x_t)$ and of $Mu(t)$. respectively. Thanks to Parseval's equality $\dot{u} \in L^2((-i\infty, i\infty), \mathbf{C}^n)$, and the equality

$$||x||^2_{L[0,\infty)} = \frac{1}{2\pi}\int_{-\infty}^{\infty} ||[K(is)]^{-1} f(is)||^2_{C^n} ds$$

holds. Therefore,

$$||x||^2_{L[0,\infty)} = \frac{1}{2\pi}\int_{-\infty}^{\infty} ||[K(is)]^{-1}(F(is) + u(is))||^2_{C^n} ds.$$

Hence,

$$||x||_{L[0,\infty)} \leq [\frac{1}{2\pi}\int_{-\infty}^{\infty} ||[K(is)]^{-1} F(is)||^2_{C^n} ds]^{1/2} +$$

$$[\frac{1}{2\pi}\int_{-\infty}^{\infty} ||[K(is)]^{-1} u(is) ds||^2_{C^n} ds]^{1/2}.$$

This implies

$$||x||_{L[0,\infty)} \leq \tilde{b}[\frac{1}{2\pi}\int_{-\infty}^{\infty} ||F(is)||^2_{C^n} ds]^{1/2} + \nu[\frac{1}{2\pi}\int_{-\infty}^{\infty} ||\dot{u}(is) ds||^2_{C^n} ds]^{1/2},$$

where

$$\tilde{b} = sup\{||K^{-1}(i\omega)||_{C^n} : -\infty < \omega < \infty\}.$$

Due to Parseval's equality we can write

$$||x||_{L[0,\infty)} \leq \nu(||F||_{L[0,\infty)} + ||Mu||_{L[0,\infty)}).$$

Due to Lemma 8.3.4 $\tilde{b} \leq \Gamma_0(K)$. So by virtue of (7.2) we obtain

$$||x||_{L[0,\infty)} \leq \Gamma_0(K)(q||x||_{L[0,\infty)} + ||Mu||_{L[0,\infty)}).$$

Now condition (12.4) provides the inequality

$$||x||_{L[0,\infty)} \leq (1 - q\Gamma_0(K))^{-1}\Gamma_0(K)||Mu||_{L[0,\infty)}.$$

According to (12.2) it can be written

$$||y||_{L^2([0,\infty),\mathbf{C}^{m_2})} \leq ||Cx||_{L^2([0,\infty),\mathbf{C}^{m_2})} + ||Du||_{L^2([0,\infty),\mathbf{C}^{m_2})} \leq$$

$$||C||||x||_{L[0,\infty)} + ||D||||u||_{L^2([0,\infty),\mathbf{C}^{m_1})}.$$

This proves the result. □

10.13 Notes

Note 10.1 As it was mentioned, one of the basic methods in stability theory of nonlinear retarded systems is the Lyapunov functionals one. The most general results here are due to Krasovskii (1963) and Razumikhin (1960). The following excellent books are devoted to a development of the direct Lyapunov method: (Hale and Lunel, 1993), (Kolmanovskii and Nosov, 1986), (Kolmanovskii and Myshkis, 1992), (Gopasalmy, 1992), etc.

Note 10.2 The contents of Sections 10.2 and 10.3 are based on the paper (Gil', 1994). Note that the absolute stability of differential-delay systems has been discussed by many authors (see (Razvan, 1983), (Likhtarnikov and Yakubovich, 1982, 1983a, 1983b), (Kolmanovskii and Nosov, 1986) and references given therein). Usually, the frequency criteria for ordinary equations are extended. In particular, in (Likhtarnikov and Yakubovich, 1983a), a system consisting of the equations

$$\dot{x} = \sum_{j=0}^{k}[A_j x(t - \tau_j) + b_j\psi(t - \tau_j)] + f^0(t), \tag{13.1}$$

and the inequality

$$\int_0^{T_j} G[x(t), x(t-\tau_1), ..., x(t-\tau_k), \psi(t), ..., \psi(t-\tau_k)]dt \geq -\gamma$$

is investigated. Here $x(t), \psi(t)$ are vector-valued functions, A_j, b_j indicate constant matrices, $G[.]$ is a Hermitian form of its arguments,

$$0 = \tau_0 < \tau_1 < ... < \tau_k = \tau$$

are lags, $T_j \to \infty$ as $j \to \infty$, $f^0 \in L^2([0,\infty), \mathbf{C}^n)$, and γ is a positive constant. The very useful quadratic criteria for the absolute stability of this system are established. It is clear that the class of linear parts of systems (2.1) is wider than (13.1). The absolute stability criterion presented in Section 10.3 improves the well-known criteria in the case of systems with many nonlinearities.

Note 10.3 The results presented in Section 10.4 are adapted from (Gil', 1995, Chapter 8) and (Gil', 1997). Theorem 10.8.1 on stability in the first approximation is a trivial corollary of Theorem 10.4.1. Other theorems on stability in the first approximation can be found for example in (Kolmanovskii and Nosov, 1986, p. 75).

Note 10.4 Section 10.6 is based on the paper (Gil', 1998a). Theorem 10.7.2 about the absolute stability of systems with small delays is new. There have been many stability conditions for nonlinear systems with sufficiently small delays in recent years usually in the category of "robust stability" for time-delay systems (e.g. (Wang et al., 1987), (Shyu and Yan, 1993), and references therein).

Note 10.5 Theorem 10.8.1 is a trivial corollary of Theorem 10.3.2. Section 10.9 is based on the paper (Gil', 1998 b). Theorem 10.11.1 is due to the paper (Gil' and Ailon, 1998a). The contents of Section 10.12 are based on the paper (Gil' and Ailon, 1998b). Other approaches to the problem of the stability and stabilizability of retarded systems are presented in (Burton, 1985), (Corduneanu, 1973), (Corduneanu, 1990), (Diekmann et al., 1995), etc.

References for Chapter 10

Burton, T.A. (1985). *Stability and and Periodic Solutions of Ordinary and Functional Differential Equations*, Academic Press, Orlando.

Corduneanu C. (1973). *Integral Equations and Stability of Feedback Systems.* Acad. Press, New York-London.

Corduneanu C. (1990). *Integral Equations and Applications*, Cambridge Univers. Press, Cambridge.

Diekmann O., van Gils S., Lunel S.V., and Walther H.-O. (1995). *Delay Equations*, Springer-Verlag, New York.

Gemignani, M. G. (1972). *Elementary Topology.* Addison-Wesley Publishing Company, Massachusetts.

Gil', M.I. (1994). On absolute stability of differential-delay systems, *IEEE, Trans. Automatic Control, 39, N 12*, 2481-2484.

Gil', M. I. (1995). *Norm Estimations for Operator-valued Functions and Applications.* Marcel Dekker, Inc. New York.

Gil', M.I. (1997). Explicit stability conditions of nonlinear systems with delay. *Nonlinear Analysis. Theory, Methods and Applications, 30, No 5*, 2629-2639.

Gil', M.I. (1998a). On exponential stability of nonlinear systems with delaying argument. *Nonlinear Analysis. Theory, Methods and Applications, 31, N 5/6*, 755-764

Gil', M.I. (1998b). On generalized Wazewski and Lozinskii inequalities for semilinear abstract differential-delay equations, *Journal of Inequalities and Appl.* (accepted for publication).

Gil', M.I. and A. Ailon (1998a). On global feedback stabilization of nonlinear retarded systems, *Dynamics of Continuous, Discrete and Impulsive Systems* (accepted for publication).

Gil', M.I. and A. Ailon. (1998b). On input-output stability of nonlinear retarded systems, *IEEE Trans. of Circuits and Systems, Part I.* (accepted for publication).

Gopasalmy, K. (1992). *Stability and Oscillations in Delay Differential Equations of Population Dynamics.* Kluwer Academic Publishers, Dordrecht.

Halanay, A. (1966). *Differential Equations. Stability. Oscillations, Time lag.* Ac. Press, New York.

Hale, J. K. (1977). *Theory of Functional Differential Equations.* Springer-Verlag, New York.

Hale, J. K., Verduyn Lunel, S. M. (1993). *Introduction to Functional Differential Equations.* Springer-Verlag, New York.

Kolmanovskii, V. B. (1996). Applications of differential inequalities for stability of some functional differential equations. *Nonlinear Analysis, TMA, 25, Nos 9-10*, 1017-1028.

Kolmanovskii, V. B. and A. D. Myshkis (1992). *Applied Theory of Functional Differential Equations.* Kluwer Academic Publishers, Dordrecht.

Kolmanovskii V.B. and V.K. Nosov (1986). *Stability of Functional Differential Equations.* Ac. Press, New York.

Krasovskii, N.N. (1963). *Stability of Motion.* Stanford Univ. Press, Stanford.

Likhtarnikov, A.L., and V.A. Yakubovich (1982). Abstract criteria for absolute stability of nonlinear systems with respect to the linear output and their applications (part 1), *Sibirian Mathematical Journal, 23, N4*, 103-121.

Likhtarnikov A.L., and V.A. Yakubovich (1983a). Abstract criteria for absolute stability of nonlinear systems with respect to the linear output and their applications (part 2), *Sibirian Mathematical Journal, 24 , N5*, 129-148.

Likhtarnikov, A.L. and V.A. Yakubovich (1983b). Absolutely Stability of Nonlinear Systems. Appendix to book: V. Razvan, *Absolute Stability of Systems with Delay*, Nauka, Moscow, 1983. In Russian.

Razumikhin, B. S. (1960). Application of the Lyapunov method in the stability of systems with delay, *Autom. and Telemech. 21*, 740-749.

Razvan, V. (1975). *Stabilitea Absoluta a Sistemelor.* Automate cu Intirziere Ed. Acad. Rep. Soc. Romania, Bucharest.

Shyu, K.K and Yan, J.J. (1993). Robust stability of uncertain time-delay systems, and its stabilization by variable structure control. *Int. Journal of Control, 57, N1*, 237-246.

Wang S.S, Chen, B.S., and Lin, T.P. (1987). Robust stability of uncertain time-delay systems. *Int. Journal of Control, 46, N3*, 963-976.

Chapter 11

LINEAR NEUTRAL TYPE SYSTEMS

In the present chapter we consider linear neutral type functional differential systems.

In Section 11.1 we introduce the notions of characteristic matrices and characteristic functions of linear autonomous neutral type systems.

L^2-norm estimates for solutions of nonhomogeneous time-invariant neutral type equations are presented in Section 11.2.

Section 11.3 is devoted to the representation of solutions by virtue of the Laplace transformation.

Estimates for characteristic matrices of neutral type autonomous systems are derived in Section 11.4.

Solution estimates for nonhomogeneous difference equations are presented in Section 11.5.

Bounds for roots of characteristic functions of neutral type systems are discussed in Sections 11.6.

L^2-norm estimates for derivatives of solutions of time-invariant neutral type equations are presented in Section 11.7.

Conditions for the asymptotic stability of neutral time-invariant type are established in Section 11.8.

The stability of linear neutral time-variant type is investigated in Section 11.9.

11.1 Characteristic functions of autonomous neutral systems

Let $R(\tau)$ and $W(\tau)$ be $n \times n$-matrix-valued functions defined on a segment $[0, \eta]$ $(0 < \eta < \infty)$, continuous on the left and having bounded variations.

Consider in $\mathbf{C}^n$ the autonomous homogeneous neutral type functional differential system

$$\dot{x}(t) - \int_0^\eta dW(\tau)\dot{x}(t-\tau) = \int_0^\eta dR(\tau)x(t-\tau) \ (t \geq 0) \tag{1.1}$$

with the initial condition

$$x(t) = \Phi(t) \text{ for } -\eta \leq t \leq 0, \tag{1.2}$$

where $\Phi(t)$ is a given differentiable vector-valued function.

The following scalar equation

$$\dot{x}(t) - c\dot{x}(t-\eta) = ax(t) + bx(t-\eta) \tag{1.3}$$

with real numbers a, b, c is an example of equation (1.1). Assume that there is a differentiable scalar-function x defined on $[-\eta, \infty)$ and satisfying (1.3) for all $t \geq 0$. In addition, let (1.2) hold with a scalar-valued function Φ. Then that solution in general has a discontinuous derivative at $k\eta$, $k = 0, 1, 2, \ldots$. In fact, suppose

$$\dot{\Phi}(0) \neq c\dot{\Phi}(-\eta) + a\Phi(0) + b\Phi(-\eta).$$

From equation (1.3), this implies $\dot{x}(\eta)$ is discontinuous if $c \neq 0$. The same reasoning applies for all values $k\eta$, $k = 2, \ldots$. On the other hand, if the sewing condition

$$\dot{\Phi}(0) = c\dot{\Phi}(-\eta) + a\Phi(0) + b\Phi(-\eta)$$

is fulfilled, then the solution has a continuous derivative. In the case of equation (1.1) the sewing condition takes the form

$$\dot{\Phi}(0) - \int_0^\eta dW(\tau)\dot{\Phi}(-\tau) = \int_0^\eta dR(\tau)\Phi(-\tau). \tag{1.4}$$

A solution of equation (1.1) on $[0, \infty)$ with an initial function Φ is a continuously differentiable function $x(t) : [-\eta, \infty) \to \mathbf{C}^n$ satisfying equation (1.1) for all $t \geq 0$. Besides, condition (1.2) holds.

Denote by $C^1([a, b]; \mathbf{C}^n)$ the space of continuously differentiable vector-valued functions defined on a segment $[a, b]$ and equipped with the norm

$$||u||_{C^1([a,b];\mathbf{C}^n)} = [||u||^2_{C([a,b];\mathbf{C}^n)} + ||\dot{u}||^2_{C([a,b];\mathbf{C}^n)}]^{1/2} \ (u \in C^1([a, b]; \mathbf{C}^n)).$$

Here $||.||_{\mathbf{C}^n}$ is the Euclidean norm. For brevity sometimes we write

$$||\cdot||_{C^1[a,b]}, \ ||\cdot||_{C[a,b]}, \ ||\cdot||_{L[a,b]}$$

instead of

$$||\cdot||_{C^1([a,b];\mathbf{C}^n)}, \ ||\cdot||_{C([a,b];\mathbf{C}^n)} \text{ and } ||u||_{L^2([a,b];\mathbf{C}^n)},$$

respectively.

The existence and uniqueness of solutions of equation (1.1) for all $t \geq 0$, and any initial function $\Phi \in C^1([-\eta, 0]; \mathbf{C}^n)$ satisfying the sewing condition (1.4) follows from Theorem 12.1.1 below.

Further, denote

$$K_W(p) = I - \int_0^\eta exp(-ps)dW(s).$$

and

$$K_R(p) = \int_0^\eta exp(-ps)dR(s) \quad (p \in \mathbf{C}).$$

Recall that I is the unit matrix. The matrix-valued function $K_0(p)$ defined by the expression

$$K_0(p) \equiv K_R(p) - pK_W(p) =$$

$$\int_0^\eta exp(-ps)dR(s) + p[\int_0^\eta exp(-ps)dW(s) - I] \quad (p \in \mathbf{C})$$

is called *the characteristic matrix*, and $det K_0(p)$ is called *the characteristic function (the characteristic determinant)* of equation (1.1).

Evidently, the system

$$\dot{x} - \sum_{k=1}^{m_1} B_k \dot{x}(t - l_k) = \sum_{j=0}^{m} A_j x(t - h_j) \tag{1.5}$$

with constant matrices A_j, B_k, and constant delays $0 = h_0 < h_1 < ... < h_m$, and $0 < l_1 < ... < l_{m_1}$ can be written in the form (1.1). Besides,

$$K_R(p) = A_0 + A_1 e^{-ph_1} + ... + A_m e^{-ph_m},$$

and

$$K_W(p) = I - B_1 e^{-pl_1} - ... - B_{m_1} e^{-pl_{m_1}}.$$

So the characteristic matrix is

$$K_0(p) = p(B_1 e^{-pl_1} + ... + B_{m_1} e^{-pl_{m_1}} - I) + A_0 + A_1 e^{-ph_1} + ... + A_m e^{-ph_m}.$$

Introduce the matrix

$$\int_0^\eta |dW(s)| \equiv \lim \sum_{k=0}^{N-1} |W(s_{k+1}) - W(s_k)| \text{ as } \max_k |s_{k+1} - s_k| \to 0,$$

where $0 = s_0 < ... < s_N = \eta$ is a partiating of $[-\eta, 0]$, and $|A|$ denotes the matrix whose entries are the modulus of the matrix A. The same sense has the symbol $|h|$ for a vector h. For example, in the case (1.5) we have

$$\int_0^\eta |dW(s)| = \sum_{k=1}^{m_1} |B_k|.$$

Recall that the quantity $g(A)$ of a matrix A is introduced in Section 2.1.

Lemma 11.1.1 *Let the spectral radius ρ_W of the matrix $\int_0^\eta |dW(s)|$ be less than one:*

$$\rho_W \equiv r_s(\int_0^\eta |dW(s)|) < 1. \tag{1.6}$$

Then with the notation

$$\tilde{\Gamma}_W \equiv \sum_{k=0}^{n-1} \frac{g^k(\int_0^\eta |dW(s)|)}{\sqrt{k!}(1-\rho_W)^{k+1}}, \quad c_W \equiv \sup_{\omega \in \mathbf{R}} ||K_W^{-1}(i\omega)||_{C^n}, \tag{1.7}$$

the relation

$$\sup_{\omega \in \mathbf{R}} ||K_W^{-1}(\alpha + i\omega)||_{C^n} \leq \tilde{\Gamma}_W, \tag{1.8}$$

holds for any $\alpha \geq 0$. In particular, $c_W \leq \tilde{\Gamma}_W$. Moreover, there is a norm $||.||_{an}$ in $\mathbf{C}^n$, such that

$$V_a(W) \equiv || \int_0^\eta |dW(s)| ||_{an} < 1. \tag{1.9}$$

Proof: Put

$$M_W(\alpha + i\omega) = \int_0^\eta e^{(-\alpha - i\omega)s} dW(s).$$

It is simple to check that

$$|M_W(\alpha + i\omega)| \leq \int_0^\eta |dW(s)| \; (\alpha \geq 0, \; \omega \in \mathbf{R}). \tag{1.10}$$

Hence, the spectral radius of the matrix $M_W(\alpha + i\omega)$ satisfies the inequality

$$r_s(M_W(\alpha + i\omega)) \leq r_s(\int_0^\eta |dW(s)|) = \rho_W.$$

Moreover,

$$|(I - M_W(\alpha + i\omega))h| \geq (I - \int_0^\eta |dW(s)|) \; |h| \; (h \in \mathbf{C}^n).$$

We thus get

$$||(I - M_W(\alpha + i\omega))h||_{C^n} \geq ||(I - \int_0^\eta |dW(s)|)|h|||_{C^n},$$

and

$$||(I - M_W(\alpha + i\omega))^{-1}||_{C^n} \leq ||(I - \int_0^\eta |dW(s)|)^{-1}||_{C^n}.$$

Due to Lemma 8.1.1 we can write

$$||(I - \int_0^\eta |dW(s)|)^{-1}||_{C^n} \leq \tilde{\Gamma}_W.$$

Since $K_W(\alpha+i\omega) = I - M_W(\alpha+i\omega)$, we get the inequality (1.8). Furthermore, it is well-known that for any matrix A and any $\epsilon > 0$, there is a norm $||.||_{an}$ such that $||A||_{an} < r_s(A) + \epsilon$ where $r_s(A)$ is the spectral radius of A, cf. (Stewart and Sun, 1990, p. 72). This result and (1.6) provides the inequality (1.9). As claimed. □

Lemma 11.1.2 *Let inequality (1.6) be valid. Then the condition*

$$c_0 \equiv \sup_{\omega \in \mathbf{R}} ||K_0^{-1}(i\omega)||_{C^n} < \infty \tag{1.11}$$

holds, provided $\det K_0(p)$ *has no roots on the imaginary axis.*

Proof: Due to (1.6) we have

$$||K_0^{-1}(i\omega)||_{C^n} = ||[i\omega K_W(i\omega) - K_R(i\omega)]^{-1}||_{C^n} =$$

$$||[i\omega I - K_W^{-1}(i\omega)K_R(i\omega)]^{-1}K_W^{-1}(i\omega)||_{C^n} \le$$

$$c_W||[i\omega I - K_W^{-1}(i\omega)K_R(i\omega)]^{-1}||_{C^n}.$$

Clearly, $||K_R(i\omega)||_{C^n} \le V(R)$ $(\omega \in \mathbf{R})$. Recall that $V(R)$ is the variation of R in the Euclidean norm. So for $|\omega| \ge \omega_0 \equiv 2c_W V(R)$ we get

$$||[i\omega I - K_W^{-1}(i\omega)K_R(i\omega)]^{-1}||_{C^n} \le (|\omega| - c_W V(R))^{-1} \le (c_W V(R))^{-1}.$$

Since for each $|\omega| \le \omega_0$ function $K_0(i\omega)$ has no zeros, and it is continuous, we arrive at the relation (1.11). □

11.2 L^2-estimates for solutions of autonomous systems

Let us consider the nonhomogeneous equation

$$\dot{y}(t) - \int_0^\eta dW(s)\dot{y}(t-s) = \int_0^\eta dR(s)y(t-s) + f(t) \text{ for } t \ge 0 \tag{2.1}$$

with a given continuous function f. Take the zero initial condition

$$y(t) = 0, \ -\eta \le t \le 0. \tag{2.2}$$

Solutions of that equation are defined as in the homogeneous case (see the previous section). Besides, the sewing condition (1.4) is automatically satisfied. Recall that c_0 is defined by (1.11).

Lemma 11.2.1 *Let $det K_0(p)$ have no roots in the closed right half-plane. In addition, let conditions (2.2) and (1.6) be fulfilled. Then for any continuous $f \in L^2([0,\infty), \mathbf{C}^n)$ the solution $y(t)$ of (2.1) is in $L^2([0,\infty), \mathbf{C}^n)$. Moreover, that solution can be represented by the equality*

$$y(t) = -\frac{1}{2\pi}\int_{-\infty}^{\infty}[K_0(i\omega)]^{-1}exp(i\omega t)\tilde{f}(i\omega)d\omega, \tag{2.3}$$

where $\tilde{f}(p)$ is the Laplace transform of $f(t)$. In addition, the estimate

$$||y||_{L[0,\infty)} \leq c_0||f||_{L[0,\infty)} \tag{2.4}$$

is valid.

Proof: Integrating (2.1) and using relation (5.4) below, we have

$$\zeta(t) \leq (1 - V_a(W))^{-1}[V_{an}(R)\int_0^t \zeta(\tau)d\tau + \int_0^t |||f(s)|||_{an}ds] \tag{2.5}$$

with the notation

$$V_{an}(R) = \int_0^{\eta} ||dR(s)||_{an}.$$

Thus, according to the Gronwall lemma, we can apply the Laplace transformation to equation (2.1). It gives

$$\tilde{y}(p) = K_0^{-1}(p)\tilde{f}(p).$$

where $\tilde{y}(p)$ is the Laplace transform of the solution. Consequently (2.3) holds, and relation (1.8) yields

$$\int_{-\infty}^{\infty} ||\tilde{y}(iw)||^2_{C^n}dw \leq c_0^2\int_{-\infty}^{\infty} ||\tilde{f}(iw)||^2_{C^n}dw.$$

Exploiting Parseval's equality, we get the inequality (2.4). Hence the result easily follows. □

Consider now the homogeneous equation (1.1) with the initial condition (1.2)

Lemma 11.2.2 *Let $det K_0(p)$ have no roots in the closed right half-plane, and condition (1.6) be fulfilled. Then there is a constant M_0, such that the estimate*

$$||x||_{L[0,\infty)} \leq M_0||\Phi(s)||_{C^1[-\eta,0]} \tag{2.6}$$

is valid for any solution $x(t)$ of equation (1.1) with a differentiable initial function Φ.

Proof: Introduce a function $w(t)$ by the equalities

$$w(t) = \Phi(t) \text{ for } -\eta \leq t \leq 0, \text{ and } w(t) = e^{-\nu t}\Phi(0) \text{ for } t \geq 0 \tag{2.7}$$

with some $\nu > 0$. Substitute the equality $x(t) = w(t) + y(t)$ in (1.1). We have

$$\frac{d}{dt}[y(t) - \int_0^\eta dW(\tau)y(t-\tau)] = \int_0^\eta dR(\tau)y(t-\tau) + f(t),$$

where

$$f(t) = -\dot{w}(t) + \int_0^\eta dW(\tau)\dot{w}(t-\tau) + \int_0^\eta dR(\tau)w(t-\tau). \tag{2.8}$$

In addition,

$$\int_0^\eta dW(\tau)\dot{w}(t-\tau) = \int_0^\eta dW(\tau)\dot{\Phi}(t-\tau) \text{ for } 0 \leq t \leq \eta,$$

and

$$\int_0^\eta dW(\tau)\dot{w}(t-\tau) = -\int_0^\eta dW(\tau)\nu e^{-\nu t}\Phi(0), \ \iota \geq \eta.$$

Similarly,

$$\int_0^\eta dR(\tau)w(t-\tau) = \int_0^\eta dR(\tau)\Phi(t-\tau) \text{ for } 0 \leq t \leq \eta,$$

and

$$\int_0^\eta dR(\tau)w(t-\tau) = \int_0^\eta dR(\tau)e^{-\nu t}\Phi(0), \ t \geq \eta.$$

Simple calculations show that,

$$||\dot{w}||_{L[0,\infty)} \leq \sqrt{\nu/2}||\Phi(0)||_{C^n}, \tag{2.9}$$

and

$$||\int_0^\eta dW(\tau)\dot{w}(t-\tau)||_{L[0,\infty)} \leq V(W)||\dot{w}||_{L[-\eta,\infty)} \leq$$

$$V(W)[||\dot{\Phi}||_{L[-\eta,0)} + \sqrt{\nu/2}||\Phi(0)||_{C^n}].$$

The symbols $V(R), V(W)$ denote the variations in the Euclidean norm of R and W, respectively. Furthermore,

$$||\int_0^\eta dR(\tau)w(t-\tau)||_{L[0,\infty)} \leq V(R)||w||_{L[-\eta,\infty)} \leq$$

$$V(R)[||\Phi||_{L[-\eta,0]} + \frac{1}{\sqrt{2\nu}}||\Phi(0)||_{C^n}].$$

Then according to (2.8) we get

$$||f||_{L[0,\infty)} \leq M_1||\Phi(s)||_{C^1[-\eta,0]}, \tag{2.10}$$

where

$$M_1 = \sqrt{\nu/2} + \sqrt{V^2(W) + \nu/2} + \sqrt{V^2(R) + \frac{1}{2\nu}}.$$

Lemma 11.2.1 asserts that $y \in L^2([0, \infty); \mathbf{C}^n)$, and

$$||y||_{L[0,\infty)} \leq c_0 M_1 ||\Phi(s)||_{C^1[-\eta,0]}. \tag{2.11}$$

Further, obviously,

$$||w||_{L[0,\infty)} \leq \frac{1}{\sqrt{2\nu}} ||\Phi(0)||_{C^n}. \tag{2.12}$$

Now inequalities $||x||_{L[0,\infty)} \leq ||y||_{L[0,\infty)} + ||w||_{L[0,\infty)}$ and (2.11) yield the stated result. □

11.3 Representation of solutions of homogeneous systems

Lemma 11.3.1 *Let all the roots of* $det K_0(p)$ *lie in the open left half-plane. In addition, let condition (1.6) hold. Then any solution* $x(t)$ *of equation (1.1) with a differentiable initial function* $\Phi(t)$ *admits the representation*

$$x(t) = -\frac{1}{2\pi} \int_{-\infty}^{\infty} e^{ti\omega} K_0^{-1}(i\omega) J_0(i\omega, \Phi) d\omega \ (t \geq 0), \tag{3.1}$$

where

$$J_0(p, \Phi) = \Phi(0) - K_W(p)\Phi(0) + \int_0^\eta dR(\tau) \int_{-\tau}^0 e^{-(s+\tau)p} \Phi(s) ds +$$

$$\int_0^\eta dW(\tau) \int_{-\tau}^0 e^{-(s+\tau)p} \dot{\Phi}(s) ds.$$

Proof: According to Lemma 11.2.2, we can apply the Laplace transformation to equation (1.1). Take into account that

$$\int_0^\infty e^{-pt} \dot{x}(t) dt = -x(0) + p\tilde{x}(p) \ (Re p \geq 0), \tag{3.2}$$

where $\tilde{x}(p)$ is the Laplace transform of $x(t)$. Further,

$$\int_0^\infty e^{-pt} \int_0^\eta dR(\tau) x(t-\tau) dt = \int_0^\eta dR(\tau) \int_0^\infty e^{-pt} x(t-\tau) dt =$$

$$\int_0^\eta dR(\tau) \int_{-\tau}^\infty e^{-(s+\tau)p} x(s) ds = J_R(p) + \int_0^\eta e^{-p\tau} dR(\tau) \tilde{x}(p), \tag{3.3}$$

where

$$J_R(p) = \int_0^\eta dR(\tau) \int_{-\tau}^0 e^{-(s+\tau)p}\Phi(s)ds.$$

Similarly,

$$\int_0^\infty e^{-pt} \int_0^\eta dW(\tau)\dot{x}(t-\tau)dt = \int_0^\eta dW(\tau) \int_0^\infty e^{-pt}\dot{x}(t-\tau)dt =$$

$$\int_0^\eta dW(\tau) \int_{-\tau}^\infty e^{-p(s+\tau)}\dot{x}(s)ds = \int_0^\eta dW(\tau)[\int_{-\tau}^0 e^{-p(s+\tau)}\dot{\Phi}(s)ds+$$

$$\int_0^\infty e^{-p(s+\tau)}\dot{x}(s)ds] = J_W(p) + \int_0^\eta e^{-p\tau}dW(\tau)\int_0^\infty e^{-ps}\dot{x}(s)ds =$$

$$J_W(p) + K_W(p)(p\tilde{x}(p) - x(0)),$$

where

$$J_W(p) = \int_0^\eta dW(\tau) \int_{-\tau}^0 e^{-(s+\tau)p}\dot{\Phi}(s)ds.$$

But

$$J_0(p,\Phi) = x(0) - K_W(p)x(0) + J_R(p) + J_W(p).$$

So taking into account (3.2) and (3.3), we get

$$-K_0(p)\tilde{x}(p) = J_0(p,\Phi).$$

Consequently,

$$\tilde{x}(i\omega) = -K_0^{-1}(i\omega)J_0(i\omega,\Phi). \tag{3.4}$$

Now the inverse Laplace transformation yields the required result. □

Since a solution of equation (2.1) with a nonzero initial function is the sum of solutions of the problems (1.1), (1.2) and (2.1), (2.2), due to Lemma 11.2.1 and the latter lemma we get

Theorem 11.3.2 *Let all the roots of* $\det K_0(p)$ *lie in the open left half-plane and condition (1.6) hold. Then any solution* $y(t)$ *of equation (2.1) with a differentiable initial function* Φ, *and a continuous* $f \in L^2([0,\infty), \mathbf{C}^n)$ *admits the representation*

$$y(t) = -\frac{1}{2\pi}\int_{-\infty}^\infty e^{ti\omega}K_0^{-1}(i\omega)[J_0(i\omega,\Phi) + \tilde{f}(i\omega)]d\omega \ (t \geq 0).$$

11.4 Norm estimates for characteristic matrices

Recall that the quantity $g(.)$ is introduced in Section 2.1. Put

$$\Gamma(K_0(p)) = \sum_{k=0}^{n-1} \frac{g^k(K_0(p))}{\sqrt{k!}d^{k+1}(K_0(p))} \quad (p \in \mathbf{C}),$$

where $d_0(K(p))$ is the smallest modulus of eigenvalues of the matrix $K_0(p)$:

$$d(K_0(p)) = \min_{k=1,\dots,n} |\lambda_k(K_0(p))|.$$

Due to proposition 2.1.1

$$g(Ae^{i\theta} + zI) = g(A) \tag{4.1}$$

for every matrix A, and numbers $\theta \in \mathbf{R}, z \in \mathbf{C}$. We can write $g(K_0(p)) = g(B_0(p))$, where

$$B_0(p) = p\int_0^\eta exp(-ps)dW(s) + \int_0^\eta exp(-ps)dR(s) = K_0(p) + Ip \quad (p \in \mathbf{C}). \tag{4.2}$$

Lemma 11.4.1 *Let p be a regular point of $K_0(p)$ (that is, $det K_0(p) \neq 0$). Then the estimate*

$$\|[K_0(p)]^{-1}\|_{C^n} \leq \Gamma(K_0(p))$$

is valid.

Proof: For any invertible $n \times n$-matrix A, Lemma 8.1.1 yields

$$\|A^{-1}\|_{C^n} \leq \sum_{k=0}^{n-1} \frac{g^k(A)}{\sqrt{k!}d^{k+1}(A)}, \tag{4.3}$$

where $d(A)$ is the smallest modulus of eigenvalues of A. Hence, the result follows. □

If $K_0(p)$ is a normal matrix, then $g(K_0(p)) = 0$ (see Section 2.1), and

$$\|[K_0(p)]^{-1}\|_{C^n} \leq d^{-1}(K_0(p)).$$

for any regular p. Put

$$\Gamma_0(K_0) \equiv \sup_{\omega \in \mathbf{R}} \Gamma(K_0(i\omega)). \tag{4.4}$$

The previous lemma and Lemma 11.2.1 imply

Corollary 11.4.2 *Let $det K_0(p)$ have no roots in the closed right plane, and let condition (1.6) be fulfilled. Then a solution $y(t)$ of equation (2.1) under the zero initial condition can be represented by equality (2.3). Moreover, the estimate*

$$||y||_{L[0,\infty)} \leq \Gamma_0(K_0)||f||_{L[0,\infty)}$$

is true.

According to the definition of $g(A)$ it can be directly written as

$$g(K_0(i\omega)) \leq |\omega| N(\int_0^\eta exp(-i\omega s)dW(s)) + N(\int_0^\eta exp(-i\omega s)dR(s)) \leq$$

$$|\omega| \int_0^\eta N(dW(s)) + \int_0^\eta N(dR(s)) \ (\omega \in \mathbf{R}).$$

Recall that $N(.)$ is the Frobenius norm.

Let us derive more precise estimates for $g(K_0(i\omega))$ in the case

$$K_0(p) = -p(I - B_1 exp(-pl_1)) + \sum_{j=1}^{m} A_j exp(-ph_j), \tag{4.5}$$

where A_j, B_1 are constant matrices, and $l_1 > 0, h_j \geq 0$ are numbers, assuming that the relation

$$B_1^* e^{i\theta} = -B_1 e^{-i\theta} \tag{4.6}$$

holds with some real θ. For instance, if B_1 is Hermitian, then (4.6) holds with $\theta = \pi/2$. If it is skew-symmetric, then (4.6) holds with $\theta = 0$. Exploiting Proposition 2.1.1, we get the relation

$$g(A) \leq \sqrt{1/2} N(e^{-i\tau} A^* - e^{i\tau} A) \tag{4.7}$$

for every matrix A, and a number $\tau \in \mathbf{R}$. Consequently,

$$g(K_0(i\omega)) \leq \sqrt{1/2} N(K_0^*(i\omega)e^{-i\tau} - K_0(i\omega)e^{i\tau}) \leq$$

$$\sqrt{1/2}[N(\omega B_1^* exp(-i(\omega l_1 + \tau)) + \omega B_1 exp(i(\omega l_1 + \tau)) +$$

$$\sum_{j=1}^{m} N(A_j^* exp(-i\tau - i\omega h_j) - A_j exp(i\omega h_j + i\tau))].$$

Take $\tau = \theta - l_1\omega$. Then according to (4.6) we obtain,

$$g(K_0(i\omega)) \leq \sqrt{2} \sum_{j=1}^{m} N(A_j) \ (\omega \in \mathbf{R}). \tag{4.8}$$

11.5 Solution estimates for difference equations

Let us consider in $\mathbf{C}^n$ the difference equation

$$z(t) - \int_0^\eta dW(s)z(t-s) = f(t) \quad (t \geq 0) \tag{5.1}$$

with the zero initial condition

$$z(t) = 0, \ -\eta \leq t \leq 0. \tag{5.2}$$

Here $f(t) : [0, \infty) \to \mathbf{C}^n$ is a given continuous vector-valued function. W has a bounded variation.

A solution of the problem (5.1), (5.2) is a continuous function $z : [-\eta, \infty) \to \mathbf{C}^n$ satisfying (5.1) for all $t \geq 0$. Besides, (5.2) holds.

Lemma 11.5.1 *Let the condition (1.6) be fulfilled. Then for any continuous $f \in L^2([0,\infty), \mathbf{C}^n)$, the problem (5.1), (5.2) has a unique solution $z(t)$. Moreover, the inequality*

$$\|z\|_{L[0,\infty)} \leq c_W \|f\|_{L[0,\infty)} \tag{5.3}$$

is valid where c_W is defined by (1.7).

Proof: Due to (1.8) all the zeros of $det\, K_W$ are in the open left half-plane, and the following integral converges

$$z_0(t) = (2\pi)^{-1} \int_{-\infty}^{\infty} e^{i\omega t} K_W(i\omega)\tilde{f}(i\omega)d\omega,$$

where $\tilde{f}(i\omega)$ is the the Laplace transform of $f(t)$. Clearly, this integral satisfies (5.1). So $z_0(t) = z(t)$, and

$$\tilde{z}(i\omega) = K_W^{-1}(i\omega)\tilde{f}(i\omega),$$

is the Laplace transform of $z(t)$. Therefore,

$$\|\tilde{z}(i\omega)\|_{L(\mathbf{R})} \leq c_W \|f(i\omega)\|_{L(\mathbf{R})}.$$

Now the Parseval equality implies the required estimate (5.3). That estimate provides the uniqueness. □

Combining Lemma 11.5.1 with Lemma 11.1.1, we get the following result.

Theorem 11.5.2 *Let all the zeros of $det K_W(p)$ be in the left half-plane, and condition (1.6) hold. Then equation (5.1) with the zero initial condition (5.2) has a unique solution $z(t)$. Moreover, the inequality*

$$\|z\|_{L[0,\infty)} \leq \tilde{\Gamma}_W \|f\|_{L[0,\infty)}$$

is valid, where $\tilde{\Gamma}_W$ is defined by (1.7).

Consider the function

$$K_W(p) = I - B_1 e^{-pl_1} - \dots - B_{m_1} e^{-pl_{m_1}}$$

with constant matrices B_k and constant delays $0 < l_1 < \dots < l_{m_1}$. In this case

$$\int_0^\eta |dW(s)| = \tilde{B}.$$

where

$$\tilde{B} = |B_1| + |B_2| + \dots + |B_{m_1}|.$$

If the spectral radius $r_s(\tilde{B}) < 1$, then

$$\tilde{\Gamma}_W = \sum_{k=0}^{n-1} \frac{g^k(\tilde{B})}{\sqrt{k!}(1 - r_s(\tilde{B}))^{k+1}}.$$

11.6 Estimates for eigenvalues of characteristic matrices

The following result allows us to replace the minimum of $d(K_0(i\omega))$ on the whole axis by the minimum on a finite interval.

Lemma 11.6.1 *Let condition (1.6) be fulfilled. Then with the notation*

$$\overline{d}(K_0) \equiv \inf_{-\infty \le \omega \le \infty} d(K_0(i\omega)); \quad V_1 \equiv c_W(V(R) + \|K_R(0)\|_{C^n}),$$

the relations

$$\overline{d}(K_0) = \inf_{-V_1 \le \omega \le V_1} d(K_0(i\omega)) \le \|K_R(0)\|_{C^n}$$

hold.

Proof: In the framework of this proof to simplify the notation we write $\|.\|$ instead of $\|.\|_{C^n}$. Furthermore, for any fixed real ω and an eigenvalue $\lambda_j(K_0(i\omega))$ of matrix $K_0(i\omega)$ it can be written

$$\lambda_j(K_0(i\omega))h_j(i\omega) = K_0(i\omega)h_j(i\omega) = [i\omega K_W(i\omega) + K_R(i\omega)]h_j(i\omega), \qquad (6.1)$$

where $h_j(\omega)$ is the eigenvector of $K_0(i\omega)$ with $\|h_j(\omega)\| = 1$. The trivial calculations show that $\|K_R(i\omega)\| \le V(R)$ $(\omega \in \mathbf{R})$. Thus

$$|\lambda_j(K_0(i\omega))| \ge |\omega|\|K_W(i\omega)h_j(i\omega)\| - \|K_R(i\omega)\| \ge$$

$$|\omega|\|K_W(i\omega)h_j(i\omega)\| - V(R).$$

It is easy to check that

$$\|K_W(i\omega)h_j(i\omega)\| \geq \frac{1}{\|K_W^{-1}(i\omega)\|} \geq c_W^{-1}.$$

Therefore,

$$|\lambda_j(K_0(i\omega))| \geq c_W^{-1}|\omega| - V(R) \ (\omega \in \mathbf{R}). \tag{6.2}$$

On the other hand, thanks to equality (6.1), it can be written $|\lambda_j(K_0(0))| \leq \|K_R(0)\|$. That is,

$$\min_{-\infty \leq \omega \leq \infty} |\lambda_j(K_0(i\omega))| \leq \|K_R(0)\|.$$

But for

$$|\omega| \geq c_W(V(R) + \|K_R(0)\|) = V_1,$$

relation (6.2) gives the inequality $|\lambda_j(K_0(i\omega))| \geq \|K_R(0)\|$. That is the minimum is attained for $|\omega| \leq V_1$. As claimed. □

Thanks to the latter lemma and Lemma 11.1.1 we can assert

Corollary 11.6.2 *Let the condition (1.6) be fulfilled. Then the relation*

$$\overline{d}(K_0) = \inf_{|\omega| \leq V_0} d(K_0(i\omega))$$

is true with the notation $V_0 \equiv \tilde{\Gamma}_W(V(R) + \|K_R(0)\|_{C^n})$.

Furthermore, assume that

$$\overline{g}(K_0) \equiv \sup_{-\infty \leq \omega \leq \infty} g(K_0(i\omega)) < \infty.$$

Then

$$\Gamma_0(K_0) \leq G_0 \equiv \sum_{k=0}^{n-1} \frac{\overline{g}^k(K_0)}{\overline{d}^{k+1}(K_0)\sqrt{k!}}. \tag{6.3}$$

According to (4.8) the condition $\overline{g}(K_0) < \infty$ holds in the case (4.5), (4.6).

11.7 L^2-estimates for derivatives of solutions

Lemma 11.7.1 *Let all the roots of* $\det K_0(p)$ *lie in the open left half-plane, and let condition (1.6) hold, and* $f \in L^2([0,\infty), \mathbf{C}^n)$. *Then the derivative of a solution* $y(t)$ *of equation (2.1) with the zero initial condition is in* $L^2([0,\infty), \mathbf{C}^n)$. *Moreover, the inequality*

$$\|\dot{y}\|_{L[0,\infty)} \leq c_W(V(R)\|y\|_{L[0,\infty)} + \|f\|_{L[0,\infty)})$$

is valid.

Proof: Put

$$z(t) = \dot{y}(t) - \int_0^\eta dW(\tau)\dot{y}(t-\tau).$$

Then by (2.1) we easily get

$$\|z\|_{L[0,\infty)} \leq V(R)\|y\|_{L[0,\infty)} + \|f\|_{L[0,\infty)}.$$

Thanks to Lemma 11.5.1,

$$\|\dot{y}\|_{L[0,\infty)} \leq c_W\|z\|_{L[0,\infty)} \leq c_W(V(R)\|y\|_{L[0,\infty)} + \|f\|_{L[0,\infty)})$$

as claimed. □

Now consider the homogeneous problem (1.1), (1.2) under condition (1.6). Let a function w be defined by relations (2.7). Substitute the equality $x(t) = w(t) + y(t)$ into (1.1). Then y satisfies equation (2.1) with the zero initial condition, and f defined by (2.8). By virtue of inequalities (2.10), (2.11) and the latter lemma we get

$$\|\dot{y}\|_{L[0,\infty)} \leq c_W(V(R)\|y\|_{L[0,\infty)} + \|f\|_{L[0,\infty)}) \leq M_2\|\Phi(s)\|_{C^1([-\eta,0];\mathbf{C}^n)}$$

($M_2 = const$). Hence taking into account inequality (2.9), we arrive at the relation

$$\|\dot{x}\|_{L[0,\infty)} \leq m_0\|\Phi(s)\|_{C^1([-\eta,0];\mathbf{C}^n)} \tag{7.1}$$

with a constant m_0. So we have derived the following result.

Lemma 11.7.2 *Let all the roots of* $\det K_0(p)$ *lie in the open left half-plane, and condition (1.6) hold. Then the derivative of a solution* $x(t)$ *of equation (1.1) with a differentiable initial function* Φ *is in* $L^2([0,\infty),\mathbf{C}^n)$. *Moreover, inequality (7.1) is valid.*

11.8 Stability of autonomous systems

Equation (1.1) is said to be stable, if there is a constant $C > 0$, such that

$$\|x(t)\|_{C^n} \leq C\|\Phi\|_{C^1([-\eta,0],\mathbf{C}^n)} \ (t \geq 0)$$

for any solution x with a continuously differentiable initial function Φ.

Equation (1.1) is *asymptotically stable* if it is stable and any solution $x(t)$ has the property $x(t) \to 0$ as $t \to \infty$.

Thanks to Lemma 8.4.4, one can write

$$\|x(t)\|^2_{C^n} \leq [2\int_t^\infty \|x(s)\|^2_{C^n}ds \int_t^\infty \|\dot{x}(s)\|^2_{C^n}ds]^{1/2} \ (t \geq 0).$$

For any solution $x(t)$ of the problem (1.1), (1.2) under condition (1.6) the integrals in that relation converge due to Lemma 11.2.2 and Lemma 11.7.2. Therefore, $||x(t)||_{C^n} \to 0$ as $t \to \infty$. Moreover, according to these lemmata

$$||x||_{C[0,\infty)} \leq [2\int_0^\infty ||x(s)||^2_{C^n} ds \int_0^\infty ||\dot{x}(s)||^2_{C^n} ds]^{1/2} \leq m_1||\Phi(s)||_{C^1[-\eta,0]}.$$

Thus we have established the following

Theorem 11.8.1 *Let all the roots of* $det K_0(p)$ *lie in the open left half-plane, and condition (1.6) be fulfilled. Then equation (1.1) is stable.*

11.9 Stability of linear time-variant systems

Let us consider in $\mathbf{C}^n$ the nonautonomous neutral type system

$$\dot{x}(t) = \int_0^\eta d_\tau W_v(t,\tau)\dot{x}(t-\tau) + \int_0^\eta d_\tau R_v(t,\tau)x(t-\tau), \tag{9.1}$$

where $W_v(t,\tau)$ and $R_v(t,\tau)$ are matrix-valued functions defined on $[0,\infty) \times [-\eta, 0]$, having bounded variations in τ for each $t \geq 0$, and continuous in t.

The sewing condition under (1.2) is

$$\dot{\Phi}(0) - \int_0^\eta d_\tau W_v(0,\tau)\dot{\Phi}(-\tau) = \int_0^\eta d_\tau R_v(0,\tau)\Phi(-\tau).$$

The definitions of solutions, stability and asymptotic stability for equation (9.1) are the same as in Sections 11.1 and 11.8.

We will say that equation (9.1) is uniformly stable if for any $t_0 \geq 0$ there is a constant N independent of t_0 such that

$$||x(t)||_{\mathbf{C}^n} \leq N||x||_{C^1[t_0-\eta,t_0]} \ (t \geq t_0)$$

for any solution of (9.1). Equation (9.1) *is uniformly asymptotically stable* if it is uniformly stable and $x(t) \to 0$ as $t \to \infty$.

To derive stability conditions, let us consider in $\mathbf{C}^n$ the following equation with the separated time-invariant part:

$$\dot{x}(t) - \int_0^\eta dW(\tau)\dot{x}(t-\tau) - \int_0^\eta dR(\tau)x(t-\tau) =$$

$$\int_0^\eta d_\tau W(t,\tau)\dot{x}(t-\tau) + \int_0^\eta d_\tau R_v(t,\tau)x(t-\tau), \tag{9.2}$$

where W, W_v, R and R_v are the same above. For example, the following system can be written in the form (9.2):

$$x(t) - \sum_{k=1}^{m_1}[B_k\dot{x}(t-l_k) + A_k x(t-h_k)] = \int_{-\eta}^{0}(B(t,s)\dot{x}(t+s) + A(t,s)x(t+s))ds$$

$$+\sum_{k=1}^{m_2}[\tilde{B}_k(t)\dot{x}(t-\tilde{l}_k(t)) + \tilde{A}_k(t)x(t-\tilde{h}_k(t))],$$

where $l_k > 0, h_k \geq 0$ are constant delays, A_k, B_k $(k = 1, ..., m_1)$ are constant matrices, $\tilde{h}_j(t), \tilde{l}_j(t) \leq \eta$ are variable delays,

$$\tilde{A}_j(t), \tilde{B}_j(t)\ (j = 1, ..., m_2), A(t,s), B(t,s)$$

are variable matrices.

The existence and uniqueness of solutions of equation (9.2) for all $t \geq 0$ under the sewing condition are assumed, although a priory estimates obtained below allow us easily to prove the solution existence. Note that some sufficient solvability conditions follow from Theorem 12.1.1 below.

Introduce operators $L_1, L_0 : C([-\eta,\infty), \mathbf{C}^n) \to C([0,\infty), \mathbf{C}^n)$ by the relations

$$(L_1 h)(t) \equiv \int_0^{\eta} d_\tau W_v(t,\tau)h(t-\tau),$$

and

$$(L_0 h)(t) \equiv \int_0^{\eta} d_\tau R_v(t,\tau)h(t-\tau),\ (t \geq 0),$$

for any $h \in C([-\eta,\infty], \mathbf{C}^n)$. Then (9.2) can be written in the form

$$\dot{x}(t) - \int_0^{\eta} dW(\tau)\dot{x}(t-\tau) - \int_0^{\eta} dR(\tau)x(t-\tau) = (L_1\dot{x})(t) + (L_0 x)(t). \quad (9.3)$$

Lemma 11.9.1 *Let the conditions*

$$q_0 = \sup_{t\geq 0} variation(R_v(t,.)) \equiv \sup_{t\geq 0}\int_0^{\eta} ||d_\tau R_v(t,\tau)||_{C^n} < \infty$$

and

$$q_1 = \sup_{t\geq 0} variation(W_v(t,.)) \equiv \sup_{t\geq 0}\int_0^{\eta} ||d_\tau W_v(t,\tau)||_{C^n} < \infty$$

be valid. Then the inequalities

$$||L_1 u||_{L[0,\infty)} \equiv [\int_0^{\infty} ||(L_1 u)(s)||^2_{C^n} ds]^{1/2} \leq q_1 ||u||_{L[-\eta,\infty)},$$

$$||L_0 u||_{L[0,\infty)} \leq q_0 ||u||_{L[-\eta,\infty)} \quad (9.4)$$

hold for any $u \in C([-\eta,\infty), \mathbf{C}^n) \cap L^2([-\eta,\infty), \mathbf{C}^n)$.

The proof is left to the reader.

For example, let

$$(L_0u)(t) = \sum_{j=1}^{m} A_j(t)u(t - h_j(t)) \text{ and}$$

$$(L_1u)(t) = \sum_{k=1}^{m_1} B_k(t)u(t - l_k(t)), \tag{9.5}$$

where $A_k(t), B_j(t)$ are uniformly bounded variable matrices:

$$q_{0,k} = \sup_{t\geq 0} \|A_k(t)\|_{C^n} < \infty \text{ and } q_{1,j} = \sup_{t\geq 0} \|B_j(t)\|_{C^n} < \infty$$

$$(j = 1, ..., m_1;\ k = 1, ..., m) \tag{9.6}$$

In addition, $h_j(t)$, $l_k(t)$ are differentiable scalar-valued functions with the properties

$$1 - \dot{h}_j(t) \geq b_{0,j},\quad 1 - \dot{l}_k(t) \geq b_{1,k}, \text{ and } 0 \leq h_j(t), l_k(t) \leq \eta, \tag{9.7}$$

where $b_{0,j}$, $b_{1,k}$ are positive numbers; $t > 0$, $j = 1, ..., m$; $k = 1, ..., m_1$.

Lemma 11.9.2 *Under conditions (9.6), (9.7) operators L_1 and L_0 defined by (9.5) satisfy conditions (9.4) with*

$$q_0 = \sum_{j=1}^{m} \frac{q_{0,j}}{\sqrt{b_{0,j}}}, \text{ and } q_1 = \sum_{k=1}^{m_1} \frac{q_{1,k}}{\sqrt{b_{1,k}}}. \tag{9.8}$$

Proof: The relation (9.7) gives

$$b_{0j}\|u(t - h_j(t))\|^2_{L[0,\infty)} \leq \int_0^\infty \|u(t - h_j(t))\|^2_{C^n}(1 - \dot{h}_j(t))dt =$$

$$\int_{-h_j(0)}^{\infty} \|u(t)\|^2_{C^n} dt \leq \|u\|_{L[-\eta,\infty)}$$

for $j = 1, ..., m$. Take into account that by (9.6)

$$\|L_0u\|_{L[0,\infty)} \leq \sum_{k=1}^{m_1} q_{0k}\|u(t - h_k(t))\|_{L[0,\infty)}.$$

Thus,

$$\|L_0u\|_{L[0,\infty)} \leq \sum_{j=1}^{m} q_{0k}(b_{0j})^{-1/2}\|u\|_{L[-\eta,\infty)} = q_0\|u\|_{L[-\eta,\infty)}.$$

I. e. we arrive at the stated result for L_0. Similarly, the inequality for L_1 can be proved. □

Recall that the quantities $\Gamma_0(K_0)$ and $\tilde{\Gamma}_W$ are introduced in Sections 11.4 and 11.5, respectively, and $V(R)$ is the variation of the function $R(s)$ in the Euclidean norm.

Theorem 11.9.3 *Let all the zeros of det $K_0(p)$ lie in the open left half-plane. In addition, let the conditions (1.6), (9.4),*

$$\tilde{\Gamma}_w q_1 < 1, \text{ and } \tilde{\Gamma}_w q_1 + \Gamma_0(K_0)[q_0(1 - \tilde{\Gamma}_w q_1) + q_1(V(R) + q_2)] < 1$$

be fulfilled. Then equation (9.2) is uniformly stable.

This theorem is a particular case of Theorem 12.2.3 below. Replacing the zero by a sufficiently large $t_0 > 0$ in the hypothesis of the previous theorem we easily arrive at the following result.

Corollary 11.9.4 *Let all the zeros of det $K_0(p)$ lie in the open left half-plane. In addition, let the conditions (1.6) and*

$$\int_0^\eta \|d_\tau R_v(t,\tau)\|_{C^n} \to 0, \int_0^\eta \|d_\tau W_v(t,\tau)\|_{C^n} \to 0 \text{ as } t \to \infty \tag{9.9}$$

be valid. Then equation (9.2) is asymptotically stable.

Consider for example the system

$$\dot{x}(t) - \int_0^\eta dW(\tau)\dot{x}(t-\tau) - \int_0^\eta dR(\tau)x(t-\tau)$$

$$= \sum_{k=1}^{m_2} [B_k(t)\dot{x}(t - \tilde{l}_k(t)) + A_k(t)x(t - \tilde{h}_k(t))],$$

where $\tilde{h}_j(t), \tilde{l}_j(t) \le \eta$ are variable delays with properties (9.7), and $A_j(t), B_j(t)$ ($j = 1, ..., m_2$) are continuous matrices with the properties

$$\|A_k(t)\|_{C^n} \to 0, \ \|B_k(t)\|_{C^n} \to 0 \text{ as } t \to \infty.$$

Then according to Corollary 11.9.4 the considered system is asymptotically stable provided all the zeros of *det* $K_0(p)$ lie in the open left half-plane and (1.6) holds.

Example 11.9.5 *Consider in $\mathbf{C}^n$ the system*

$$\dot{x} - A_0 x = \sum_{k=1}^{m_1} B_k(t)\dot{x}(t - l_k(t)) + \sum_{j=1}^{m} A_j(t)x(t - h_j(t)) \ (t \ge 0), \tag{9.10}$$

where A_0 is a constant Hurwitz matrix, $A_j(t), B_k(t)$ ($j = 1, ..., m$; $k = 1, ...m_1$) are variable matrices satisfying (9.6), $h_j(t), l_k(t)$ are variable delays with the properties (9.7). Under consideration $K_W(p) = I$ and $K_0(p) = A_0 - pI$. By virtue of (4.1) $g(K_0(p)) \equiv g(A_0)$. It is easy to see that $d(K_0(iw)) = d(A_0 - Iwi) \ge |\alpha(A_0)|$ for all real w. Therefore,

$$\Gamma_0(K_0) \le \Gamma(A_0) \equiv \sum_{k=0}^{n-1} \frac{g^k(A_0)}{\sqrt{k!}|\alpha(A)|^{k+1}}.$$

Moreover, $W \equiv 0$, $\tilde{\Gamma}_W = 1$, $V(R) = \|A_0\|_{C^n}$. Due to Lemma 11.9.2, condition (9.4) holds with q_0, q_1 defined by (9.8). If, in addition, the inequalities $q_1 < 1$, and

$$q_1 + \Gamma(A_0)[q_0(1 - q_1) + q_1(\|A_0\|_{C^n} + q_0)] < 1$$

are fulfilled, then equation (9.10) is stable due to Theorem 11.9.3.

11.10 Notes

Note 11.1 About other definitions of solutions see Section 12.6.

Note 11.2. Many real processes involve small delays which are often neglected in the process of mathematical modeling. Sometimes this leads to false conclusions. Consider, for instance, the equation

$$\dot{x} + x = a[\dot{x}(t - h) + x(t - h)]$$

with $a > 1$. If $h = 0$, then this equation is asymptotically stable. The characteristic function $D(z) = (z + 1)(1 - ae^{-hz})$ has the following zeros with positive real parts if $h > 0$:

$$z_k = h^{-1}(ln\ a + 2\pi ik),\ i^2 = -1,\ k = 0, \pm 1, \pm 2, \ldots.$$

These roots lead to the instability of the equation.

Note 11.3 Numerous investigations have been devoted to the zeros of the characteristic functions of neutral equations with discrete delays. In particular, it was shown that if the delays are commensurable, then asymptotic roots z_k of such functions belong to certain chains, and $lim\ Imz_k = \pm\infty$ as $k \to \infty$. The real parts of these roots on each chain are either uniformly bounded, or tend to $-\infty$.

If the delays are incommensurable, then there are, in addition to these chains, finitely many strips parallel to the imaginary axis containing certain zeros of the characteristic functions. The real parts of these zeros densely fill intervals of the real axis corresponding to these strips. For more details see (Kolmanovskii and Myshkis, 1992, p. 135), (Bellman and Cooke, 1963), and references therein.

Note 11.4 Theorems 8.12.1-8.12.4 of Pontryagin are valid for neutral systems without changes. The frequency criteria of Michailov and Nyquist (see Section 8.12) are not valid for neutral systems cf. (Kolmanovskii and Nosov, 1986, p. 119).

Note 11.5 The material of Sections 11.2, 11.4-11.6 is based on the paper (Gil', 1993), (see also (Gil', 1995, Chapter 2)). The contents of Sections 11.7-11.9 are adapted from the paper (Gil', 1996).

Due to lack of space, many areas of the theory of stability of linear neutral type functional differential systems have not been mentioned. Many very interesting relevant results can be found in following works: (Hale and Lunel, 1993), (Kolmanovskii and Myshkis, 1992), (Kolmanovskii and Nosov, 1986),

(Bellman and Cooke, 1963), (Burton, 1985), (Salamon, 1984), (Achmerov et al., 1981), (Fiagbedzi, 1994), (Burnes et al., 1984), etc.

References for Chapter 11

Achmerov, R.R., Kamenskii, M.I., Potapov, A.C., Rodkina, A.E. and Sadovskii, B.N. (1981). Theory of equations of neutral type. *Matematicheskii Analiz., 19, Itogi Nauki i Techniki. VINITI*, 55-126. In Russian.

Bellman, R. and K.L. Cooke (1963). *Differential-Difference Equations*. Academic Press, New York.

Burnes, C.I., Spong, M.W., and Tarn, T.J. (1984). A several complex variable approach to feedback stabilization of linear neutral differential-delay systems. *Mathematical System Theory, 17*, 97-133.

Burton, T.A. (1985). *Stability and and Periodic Solutions of Ordinary and Functional Differential Equations*, Academic Press, Orlando.

Fiagbedzi, Y.A. (1994). Feedback stabilization of neutral systems via the transformation technique. *Int. Journal of Control, 59, N 6*, 1579-1589.

Gil', M. I. (1993). Estimates for norms of matrix-valued functions, *Linear and Multilinear Algebra, 35*, 65-73.

Gil', M. I. (1995). *Norm Estimations for Operator-valued Functions and Applications*. Marcel Dekker, Inc. New York.

Gil', M. I. (1996). On absolute stability of neutral type functional differential systems, *Nonlinear Studies, 3, N1*, 59-72.

Hale, J. (1977). *Theory of Functional Differential Equations*, Springer- Verlag, New York.

Hale, J. K., Verduyn Lunel, S. M. (1993). *Introduction to Functional Differential Equations*. Springer-Verlag, New York.

Kolmanovskii, V.B. and A.D. Myshkis. (1992). *Applied Theory of Functional Differential Equations*, Kluwer Academic Publishers, Dordrecht.

Kolmanovskii, V.B. and V.K. Nosov (1986). *Stability of Functional Differential Equations*. Ac. Press, New York.

Pontryagin, L. S. (1955). On zeros of some elementary transcendental functions, *Transactions of AMS, Series 2, v.1*, 96-110.

Salamon, D. (1984). *Control and Observation of Neutral Systems*. Pitman Advanced Publishing Program, London.

Stewart, G. W. and Sun Ji-guang (1990). *Matrix Perturbation Theory*. Academic Press, New York.

Chapter 12

NONLINEAR NEUTRAL TYPE FUNCTIONAL DIFFERENTIAL SYSTEMS

In the present chapter nonlinear neutral type functional differential systems with separated linear parts and continuous nonlinearities are considered.

In Section 12.1 we introduce the stability definitions for nonlinear neutral type systems.

Absolute stability is discussed in Sections 12.2 and 12.3.

Section 12.4 deals with the stability in the first approximation.

12.1 Preliminaries and definitions

Let us consider in $\mathbf{C}^n$ the functional differential equation of the neutral type

$$\dot{x}(t) = F(t, x_t, \dot{x}_t)$$

$$(0 < \eta = const < \infty;\ x_t(\theta) = x(t+\theta), \dot{x}_t(\theta) = \dot{x}(t+\theta),\ -\eta \le \theta \le 0;\ t \ge 0) \tag{1.1}$$

with the initial condition

$$x(t) = \Phi(t) \text{ for } t_0 - \eta \le t \le t_0\ (t_0 \ge 0), \tag{1.2}$$

where the function F maps $[0,\infty) \times (C([t_0-\eta, t_0], \mathbf{C}^n))^2$ into $\mathbf{C}^n$, and $\Phi(t) \in C^1([t_0-\eta, t_0];\ \mathbf{C}^n)$ is a given function.

A continuously differentiable function $x : [t_0 - \eta, T) \to \mathbf{C}^n$ $(t_0 < T \leq \infty)$ is said to be a solution of the initial problem (1.1), (1.2) on an interval $[t_0, T)$ if it satisfies (1.1) at each point $t \in [t_0, T)$, and (1.2) holds.

We will assume that the sewing condition

$$\dot{\Phi}(t_0) = F(t, \Phi_t, \dot{\Phi}_t)|_{t=t_0} \tag{1.3}$$

holds. Consider for instance the following system:

$$\dot{x} - \sum_{k=1}^{m_1} B_k \dot{x}(t - l_k) = \sum_{j=0}^{m} A_j x(t - h_j) + f(t, x(t), x(t - h_1), \dot{x}(t - l_1))$$

with constant matrices A_j, B_k, and constant delays $0 = h_0 < h_1 < ... < h_m \leq \eta$, and $0 < l_1 < ... < l_{m_1} \leq \eta$. This system can be written in the form (1.1). Besides, f maps $[0, \infty) \times \mathbf{C}^{3n}$ into $\mathbf{C}^n$. In this case the sewing condition is

$$\dot{\Phi}(t_0) - \sum_{k=1}^{m_1} B_k \dot{\Phi}(t_0 - l_k) = \sum_{j=0}^{m} A_j \Phi(t_0 - h_j) + f(t_0, \Phi(t_0), \Phi(t_0 - h_1), \dot{\Phi}(t_0 - l_1)).$$

An additional example of the nonlinearity gives the mapping

$$F(t, x_t, \dot{x}_t) \equiv f(t, \int_{-\eta}^{0} A(t, s)x(t + s)ds, \int_{-\eta}^{0} B(t, s)\dot{x}(t + s)ds), \tag{1.4}$$

where f maps $[0, \infty) \times \mathbf{C}^{2n}$ into $\mathbf{C}^n$, and $A(t, s), B(t, s)$ are matrix-valued functions defined on $[0, \infty) \times [-\eta, 0]$.

Let $||.||_n$ denotes an arbitrary norm in $\mathbf{C}^n$. The following theorem gives sufficient conditions for the existence of solutions.

Theorem 12.1.1 *Let*

$$F : E_\eta \equiv [0, \infty) \times (C([-\eta, 0], \mathbf{R}^n))^2 \to \mathbf{R}^n$$

satisfy in some neighborhood of each point of E_η for all $t \geq 0$, the condition

$$||F(t, \psi_1, \chi_1) - F(t, \psi_2, \chi_2)||_n \leq L||\psi_2 - \psi_1||_{C[-\eta,0]} + l||\chi_2 - \chi_1||_{C[-\eta,0]}$$

with constants $L \in (0, \infty)$, $l \in (0, 1)$. Suppose $\Phi \in C^1([-\eta, 0]; \mathbf{R}^n)$, and the sewing condition (1.3) holds with $t_0 = 0$. Then there is $T > 0$ such that

a) there is a solution x of (1.1), (1.2) with $t_0 = 0$ on $[0, T)$;

b) on each segment $[0, T_1] \subset [0, T)$ this solution is unique;

c) if $T < \infty$, then there is a finite limit of $\dot{x}(t)$ as $t \to T$;

d) the solution depends together with its derivative continuously on F, Φ.

For the proof of this theorem see (Kolmanovskii and Myshkis, p. 49, Theorem 3.1).

Throughout the present chapter the existence and uniqueness of considered solutions on $[0, \infty)$ are assumed.

Definition 12.1.2 *Let $F(t, 0, 0) \equiv 0$ for all $t \geq 0$. Then the zero solution of equation (1.1) is said to be stable if for any $\epsilon > 0$ and given t_0 there exists a $\delta(t_0) > 0$ such that the inequality*

$$\|\Phi\|_{C^1[t_0-\eta,t_0]} \leq \delta(t_0) \text{ implies } \|x(t)\|_n \leq \epsilon \ (t \geq t_0)$$

for a solution $x(t)$ with the initial function Φ. Otherwise the zero solution is called unstable.

The zero solution of (1.1) is said to be asymptotically stable if it is stable and there is an open set $B \subseteq C^1([t_0 - \eta, t_0], \mathbf{C}^n)$ such that $\Phi \in B$ implies $x(t) \to 0$ as $t \to \infty$. Besides, B is called the region of attraction of the zero solution.

If the zero solution of (1.1) is asymptotically stable and $B = C^1([t_0 - \eta, t_0], \mathbf{C}^n)$, then the zero solution is globally asymptotically stable.

We will say that the zero solution of (1.1) is exponentially stable if there are positive constants r_0, Λ_0 and l_0, such that for a given $t_0 \geq 0$, and an initial function $\Phi \in C^1([t_0 - \eta, t_0]$, the condition $\|\Phi\|_{C^1[t_0-\eta,t_0]} \leq r_0$ implies the relation

$$\|x(t)\|_n \leq l_0 \|\Phi\|_{C^1[t_0-\eta,t_0]} e^{-\Lambda_0 t} \quad (t \geq t_0).$$

The zero solution of (1.1) is globally exponentially stable if it is exponentially stable and $r_0 = \infty$.

Definition 12.1.3 *The zero solution of equation (1.1) is said to be uniformly stable if for any $\epsilon > 0$ there exists a $\delta > 0$ independent of t_0, such that the inequality*

$$\|\Phi\|_{C^1[t_0-\eta,t_0]} \leq \delta \text{ implies } \|x(t)\|_n \leq \epsilon \ (t \geq t_0).$$

The zero solution of (1.1) is said to be uniformly asymptotically stable if it is uniformly stable and there is an open set $B \subseteq C^1([-\eta, 0], \mathbf{C}^n)$, such that under (1.2) with $t_0 = 0$ the relation $\Phi \in B$ implies $x(t) \to 0$ as $t \to \infty$.

One can prove that in the linear case these definitions agree with the definitions from Sections 11.1 and 11.9.

12.2 Absolute stability

Again, let $R(\tau)$ and $W(\tau)$ be $n \times n$-matrix-valued functions defined on a finite segment $[0, \eta]$, having bounded variations and continuous on the left. Consider the system

$$\dot{x}(t) - \int_0^\eta dW(\tau)\dot{x}(t-\tau) = \int_0^\eta dR(\tau)x(t-\tau) + F(t, x_t, \dot{x}_t) \ (t \geq 0). \quad (2.1)$$

Besides, the function F maps $[0, \infty) \times (C([-\eta, 0), \mathbf{C}^n))^2$ into $\mathbf{C}^n$. Introduce the set

$$C^1 L_\eta \equiv \{u \in C^1([-\eta, \infty), \mathbf{C}^n) : u \in L^2([-\eta, \infty), \mathbf{C}^n), \dot{u} \in L^2([-\eta, \infty), \mathbf{C}^n)\}.$$

Let us assume that F has the following property: there are constants $q_0, q_1 \geq 0$ such that

$$||F(., u_t, \dot{u}_t)||_{L^2([0,\infty),\mathbf{C}^n)} \equiv [\int_0^\infty ||F(t, u_t, \dot{u}_t)||^2_{C^n} dt]^{1/2} \leq$$

$$q_0||u||_{L^2([-\eta,\infty),\mathbf{C}^n)} + q_1||\dot{u}||_{L^2([-\eta,\infty),\mathbf{C}^n)} \tag{2.2}$$

for all $u \in C^1 L_\eta$. For instance, let the inequality

$$||F(t, u_t, \dot{u}_t)||_{C^n} \leq \sum_{k=1}^{m_1} q_{1,k}||\dot{u}(t - l_k(t))||_{C^n} + \sum_{j=1}^{m} q_{0,j}||u(t - h_j(t))||_{C^n}$$

$$(t \geq 0) \tag{2.3}$$

hold for all $u \in C^1 L_\eta$. Here $h_j(t)$, $l_k(t)$ are differentiable functions with the properties

$$1 - \dot{h}_j(t) \geq b_{0,j}, \quad 1 - \dot{l}_k(t) \geq b_{1,k}, \text{ and } 0 \leq h_j(t), l_k(t) \leq \eta, \tag{2.4}$$

where $q_{0,j}$, $q_{1,k}$, $b_{0,j}$, $b_{1,k}$ are positive numbers; $t > 0$, $j = 1, ..., m$; $k = 1, ..., m_1$. Recall that $||.||_{C^n}$ is the Euclidean norm.

Lemma 12.2.1 *Let relations (2.3) and (2.4) be fulfilled. Then condition (2.2) holds with*

$$q_0 = \sum_{j=1}^{m} \frac{q_{0,j}}{\sqrt{b_{0,j}}}, \quad q_1 = \sum_{k=1}^{m_1} \frac{q_{1,k}}{\sqrt{b_{1,k}}}. \tag{2.5}$$

All the proofs are presented in the next section. An additional example of the nonlinearity satisfying condition (2.2) gives the function (1.4) if

$$||f(t, y, z)||_{C^n} \leq \tilde{q}_0||y||_{C^n} + \tilde{q}_1||z||_{C^n} \ (z, y \in \mathbf{C}^n, \ \tilde{q}_0, \tilde{q}_1 \equiv const),$$

and, in addition,

$$\sup_{t\geq 0} \int_0^\eta ||A(t, s)||_{C^n} ds < \infty, \sup_{t\geq 0} \int_0^\eta ||B(t, s)||_{C^n} ds < \infty.$$

Take the initial condition

$$x(t) = \Phi(t) \text{ for } -\eta \leq t \leq 0, \tag{2.6}$$

with a given $\Phi \in C^1([-\eta, 0], \mathbf{C}^n)$.

Definition 12.2.2 *We will say that the zero solution of equation (2.1) is absolutely stable in the class of nonlinearities (2.2), if under condition (2.2) that equation is globally asymptotically stable. Moreover, there is a constant χ which does not depend on a concrete F (but which depends on q_0 and q_1), such that the inequality*

$$||x(t)||_{C^n} \leq \chi||\Phi||_{C^1[-\eta,0]} \ (t \geq 0)$$

is valid for any solution $x(t)$ of (2.1) under the initial condition (2.6).

Recall that the matrix

$$\int_0^\eta |dW(s)|$$

is defined in Section 11.1. We shall assume that the spectral radius ρ_W of that matrix

$$\rho_W \equiv \rho_s\left(\int_0^\eta |dW(s)|\right) < 1. \tag{2.7}$$

Due to Lemma 11.1.1, there is a norm $\|.\|_{an}$ in $\mathbf{C}^n$ such that

$$V_a \equiv \left\| \int_0^\eta |dW(s)| \right\|_{an} < 1.$$

This condition according to Theorem 12.1.1 under the obvious restrictions provides the existence of solutions for equation (2.1).

To formulate the result recall that

$$K_W(p) = I - \int_0^\eta exp(-ps)dW(s), \quad K_R(p) = \int_0^\eta exp(-ps)dR(s),$$

$$K_0(p) = K_R(p) - pK_W(p),$$

$$\Gamma(K_0(p)) = \sum_{k=0}^{n-1} \frac{g^k(K(p))}{\sqrt{k!}d^{k+1}(K_0(p))} \text{ and } \Gamma_0(K_0) \equiv \sup_{\omega \in \mathbf{R}} \Gamma(K_0(\omega i)),$$

where $d(K_0(p))$ is the smallest absolute value of eigenvalues of matrix $K_0(p)$ with a fixed $p \in \mathbf{C}^n$ (see Section 11.4). We also use the quantities

$$V(R) = \int_0^\eta \|dR(s)\|_{C^n},$$

and

$$\tilde{\Gamma}_W = \sum_{k=0}^{n-1} \frac{g^k(\int_0^\eta |dW(s)|)}{\sqrt{k!}(1-\rho_W)^{k+1}}.$$

Theorem 12.2.3 *Let all the zeros of $\det K_0(p)$ belong to the open left half-plane. In addition, let the conditions (2.7),*

$$\tilde{\Gamma}_W q_1 < 1, \text{ and } \tilde{\Gamma}_W q_1 + \Gamma_0(K_0)[q_0(1-\tilde{\Gamma}_W q_1) + q_1(V(R)+q_0)] < 1 \tag{2.8}$$

hold. Then the zero solution of equation (2.1) is absolutely stable in the class of nonlinearities (2.2).

Example 12.2.4 *Consider the system*

$$\dot{x} = Ax + F(t, x_t, \dot{x}_t) \tag{2.9}$$

where A is a constant Hurwitz matrix, and F satisfies condition (2.2).

Under consideration we have $K_W(p) = I$, $K_0(p) = A - pI$, $W \equiv 0$, $\tilde{\Gamma}_W = 1$, $V(R) = ||A||_{C^n}$, and

$$\Gamma_0(K_0) \leq \Gamma(A) \equiv \sum_{k=0}^{n-1} \frac{g^k(A)}{\sqrt{k!}|\alpha(A)|^{k+1}}.$$

(See Example 11.9.5). If the inequalities $q_1 < 1$, and

$$q_1 + \Gamma(A)[q_0(1 - q_1) + q_1(||A||_{C^n} + q_0)] < 1,$$

are fulfilled, then equation (2.9) is absolutely stable in the class of nonlinearities (2.2) due to Theorem 12.2.3.

12.3 Proofs of Lemma 12.2.1 and Theorem 12.2.3

Proof of Lemma 12.2.1: Relation (2.3) gives

$$b_{0j}||u(t - h_j(t))||^2_{L[0,\infty)} \leq \int_0^\infty ||u(t - h_j(t))||^2_{C^n}(1 - \dot{h}_j(t))dt =$$

$$\int_{-h_j(0)}^\infty ||u(t)||^2_{C^n} dt$$

for $j = 1, ..., m$. Similarly,

$$b_{1k}||\dot{u}(t - l_k(t))||^2_{L[0,\infty)} \leq ||\dot{u}(t)||^2_{L[-l_k(0),\infty)}$$

for $k = 1, ..., m_1$. Exploiting these inequalities and taking into account that by (2.4),

$$||F(t, u_t, \dot{u}_t)||_L \leq \sum_{k=1}^{m_1} q_{1k}||\dot{u}(t - l_k(t))||_{L([0,\infty)} + \sum_{j=0}^{m} q_{0j}||u(t - h_j(t))||_{L([0,\infty)},$$

we arrive at the stated result. □

Recall that according to Lemmata 11.1.1 and 11.1.2

$$c_0 \equiv \sup_{\omega \in \mathbf{R}} ||K_0^{-1}(i\omega)||_{C^n} < \infty, \text{ and } c_W \equiv \sup_{\omega \in \mathbf{R}} ||K_W^{-1}(i\omega)||_{C^n} < \infty.$$

Lemma 12.3.1 *Let all the zeros of det $K_0(p)$ belong to the open left half-plane. In addition, let the conditions (2.7),*

$$c_W q_1 < 1, \text{ and } c_W q_1 + c_0[q_0(1 - c_W q_1) + q_1(V_R + q_0)] < 1 \tag{3.1}$$

be fulfilled. Then there are constants $\tilde{m}_0(q_0, q_1)$ and $\tilde{m}_1(q_0, q_1)$ independent of initial functions, such that the relations

$$\|x\|_{L[0,\infty)} \leq \tilde{m}_0(q_0, q_1)\|\Phi\|_{C^1[-\eta,0]} \text{ and } \|\dot{x}\|_{L[0,\infty)} \leq \tilde{m}_1(q_0, q_1)\|\Phi\|_{C^1[-\eta,0]}$$

are true for any solution $x(t)$ of equation (2.1) with the initial condition (2.6).

Proof: Take

$$x(t) = y(t) + \phi(t), \tag{3.2}$$

where ϕ is the solution of the linear homogeneous equation

$$\dot{\phi}(t) - \int_0^\eta dW(\tau)\dot{\phi}(t-\tau) = \int_0^\eta dR(\tau)\phi(t-\tau)$$

with a given initial function Φ. Substitute (3.2) in (2.1). Then y satisfies the equation

$$\dot{y}(t) - \int_0^\eta dW(\tau)\dot{y}(t-\tau) = \int_0^\eta dR(\tau)y(t-\tau) + f(t) \ (t \geq 0) \tag{3.3}$$

with

$$f(t) = F(t, \phi_t(t) + y_t(t), \dot{y}_t(t) + \dot{\phi}_t(t)).$$

Besides y has the zero initial conditions. Due to Lemmas 11.2.2, and 11.7.2 we have

$$\|\phi\|_{L[0,\infty)} \leq M_0\|\Phi(s)\|_{C^1[-\eta,0]}, \tag{3.4}$$

and

$$\|\dot{\phi}\|_{L[0,\infty)} \leq m_0\|\Phi(s)\|_{C^1[-\eta,0]} \ (M_0, \ m_0 = const). \tag{3.5}$$

So due to condition (2.2)

$$\|f(t)\|_{L[0,\infty)} = \|F(t, \phi_t + y_t, \dot{y}_t + \dot{\phi}_t)\|_{L[0,\infty)} \leq$$

$$q_1\|\dot{y}\|_{L[0,\infty)} + q_0\|y\|_{L[0,\infty)} + m_1(\Phi). \tag{3.6}$$

where

$$m_1(\Phi) = \tilde{m}_1\|\Phi(s)\|_{C^1[-\eta,0]} \ (\tilde{m}_1 = const).$$

Thanks to Lemma 11.7.1 it can be written

$$\|\dot{y}\|_{L[0,\infty)} \leq c_W[(V(R) + q_0)\|y\|_{L[0,\infty)} + q_1\|\dot{y}\|_{L[0,\infty)}] + m_2(\Phi),$$

where $m_2(\Phi) = c_W m_1(\Phi)$. Hence by inequality (3.1) we easily arrive at the relation

$$\|\dot{y}\|_{L[0,\infty)} \leq (1 - q_1 c_W)^{-1}(V(R) + q_0)\|y\|_{L[0,\infty)} + m_3(\Phi) \tag{3.7}$$

where $m_3(\Phi) = (1 - c_W q_1)^{-1} m_2(\Phi)$.

On the other hand, due to Lemma 11.2.1 and (3.6)

$$||y||_{L[0,\infty)} \le c_0||f||_{L[0,\infty)} \le c_0(q_1||\dot{y}||_{L[0,\infty)} + q_0||y||_{L[0,\infty)} + m_1(\Phi)).$$

Substituting relation (3.7) in this inequality, we get

$$||y||_{L[0,\infty)} \le c_0(q_0 + q_1(1-c_W q_1)^{-1}(V(R)+q_0))||y||_{L[0,\infty)} +$$

$$m_4||\Phi(s)||_{C^1[-\eta,0]} \ (m_4 = const).$$

Now condition (3.1) implies

$$\theta \equiv c_0(q_0 + q_1(1-c_W q_1)^{-1}(V(R)+q_0)) < 1.$$

Hence,

$$||y||_{L[0,\infty)} \le (1-\theta)^{-1} m_4 ||\Phi(s)||_{C^1[-\eta,0]}.$$

Moreover, due to (3.7)

$$||\dot{y}||_{L[0,\infty)} \le \frac{V(R)+q_0}{(1-c_W q_1)(1-\theta)} m_4||\Phi(s)||_{C^1[-\eta,0]} + m_3(\Phi).$$

Now relations (3.2), (3.4) and (3.5) imply the required result. □

Invoking Lemma 8.4.4 and the latter lemma we arrive at the following result.

Corollary 12.3.2 *Under the hypothesis of Lemma 12.3.1, the zero solution of equation (2.1) is absolutely stable in the class of nonlinearities (2.2).*

Proof of Theorem 12.2.3: Lemmas 11.4.1 and 11.1.1 yield the relations $c_0 \le \Gamma_0(K_0), c_W \le \tilde{\Gamma}_W$. Now the result is due to the previous corollary. □

12.4 Stability in the first approximation

12.4.1 Statement of the result

Let us consider in $\mathbf{C}^n$ the equation

$$\dot{x}(t) - \int_0^\eta dW(\tau)\dot{x}(t-\tau) - \int_0^\eta dR(\tau)x(t-\tau) =$$

$$F(t, x(t-h_1(t)), ..., x(t-h_m(t)), \dot{x}(t-l_1(t)), ..., \dot{x}(t-l_{m_1}(t))) \ (t \ge 0), \quad (4.1)$$

where W, R are the same as in Section 12.2, $h_j(t)$ $l_k(t) (j = 1, ..., m;\ k = 1, ..., m_1)$ are differentiable scalar-valued functions with the properties (2.4). Besides, F continuously maps $[0,\infty) \times \mathbf{C}^{n(m+m_1)}$ into $\mathbf{C}^n$. For a vector

$$(y_1, ..., y_m, z_1, ..., z_{m_1}) \in \mathbf{C}^{(m_1+m)n}$$

with $y_k, z_k \in \mathbf{C}^n$, put

$$\|(y_1, ..., y_m, z_1, ..., z_{m_1})\|_{C^{(m+m_1)n}} \equiv$$
$$(\|y_1\|^2_{C^n} + ... + \|y_m\|^2_{C^n} + \|z_1\|^2_{C^n} + ... + \|z_{m_1}\|^2_{C^n})^{1/2}.$$

We will say that (4.1) *is a quasilinear equation* if

$$\frac{\|F(t, y_1, ..., y_m, z_1, ..., z_{m_1})\|_{C^n}}{\|(y_1, ..., y_m, z_1, ..., z_{m_1})\|_{C^{(m+m_1)n}}} \to 0$$
$$\text{as } \|(y_1, ..., y_m, z_1, ..., z_{m_1})\|_{C^{(m+m_1)n}} \to 0 \tag{4.2}$$

uniformly in t.

Theorem 12.4.1 *Under conditions (2.4), (2.7), and (4.2), let all the zeros of* $\det K_0(p)$ *lie in the open left half-plane. Then the zero solution of equation (4.1) is asymptotically stable.*

12.4.2 Proof of Theorem 12.4.1

First, assume that there are positive constants q_{0j}; q_{1k} $(j = 1, ..., m;\ k = 1, ..., m_1)$ such that

$$\|F(t, y_1, ..., y_m, z_1, ..., z_{m_1})\|_{C^n} \le \sum_{j=1}^{m} q_{0j}\|y_j\|_{C^n} + \sum_{k=1}^{m_1} q_{1k}\|z_k\|_{C^n}$$
$$\text{for all } y_j, z_k \in \mathbf{C}^n \ (j = 1, ..., m;\ k = 1, ..., m_1;\ t \ge 0). \tag{4.3}$$

Lemma 12.4.2 *Let all the zeros of* $\det K_0(p)$ *belong to the open left half-plane. In addition, let conditions (2.4), (2.7) and (4.3) be fulfilled. Then there are sufficiently small* q_{0j} *and* q_{1k}*, such that the zero solution of (4.1) is globally asymptotically stable. Moreover, any solution of equation (4.1) under (2.6) satisfies the inequality*

$$\|x\|_{C^1[0,\infty)} \le N_1\|\Phi(s)\|_{C^1[-\eta,0)} \tag{4.4}$$

where the constant N_1 *is independent of* Φ.

Proof: Thanks to Lemma 12.2.1, conditions (2.4) and (4.3) imply condition (2.2). Invoking Theorem 12.2.3 we can assert that with sufficiently small q_{0j} and q_{1k}, the zero solution of (4.1) is absolutely stable in the class of nonlinearities (4.3). Therefore, the estimate

$$\|x\|_{C[0,\infty)} \le \chi\|\Phi\|_{C^1[-\eta,0)}. \tag{4.5}$$

holds. On the other hand, due to Lemma 11.1.1 there is a norm $\|.\|_{an}$, such that $V_a(W) < 1$. Now (4.1) yields

$$\|\dot{x}(t)\|_{an} \le \|\int_0^\eta W(\tau)\dot{x}(t-\tau)\|_{an} + \|\int_0^\eta dR(\tau)x(t-\tau)\|_{an} + \|F\|_{an}.$$

Hence, with the notation

$$||x||_{\hat{C}[a,b]} \equiv \sup_{t\in[a,b]} ||x(t)||_{an}$$

we have

$$||\dot{x}||_{\hat{C}[0,\infty)} \leq V_a(W)(||\dot{x}||_{\hat{C}[0,\infty)} + ||\dot{\Phi}||_{\hat{C}[-\eta,0]})+$$

$$V_a(R)(||x||_{\hat{C}[0,\infty)} + ||\Phi||_{\hat{C}[-\eta,0]}) + ||F||_{\hat{C}[0,\infty)}, \tag{4.6}$$

where

$$V_a(R) = \int_0^\eta ||dR||_{an}.$$

Since norms in $\mathbf{C}^n$ are equivalent, according (4.3) there are q_{0j}, q_{1k} such that

$$||F(t, y_1, ..., y_m, z_1, ..., z_{m_1})||_{an} \leq \sum_{j=1}^{m} q_{0j}||y_j||_{an} + \sum_{k=1}^{m_1} q_{1k}||z_k||_{an}$$

$$\text{for all } y_j, z_k \in \mathbf{C}^n \ (j = 1, ..., m;\ k = 1, ..., m_1;\ t \geq 0).$$

Consequently,

$$||F(t, x(t - h_1(t)), ..., x(t - h_m(t)), \dot{x}(t - l_1(t)), ..., \dot{x}(t - l_{m_1}(t)))||_{\hat{C}[0,\infty)}$$

$$\leq q_0||x||_{\hat{C}(-\eta,\infty)} + q_1||\dot{x}||_{\hat{C}(-\eta,\infty)}$$

with constants

$$q_1 = \sum_{k=1}^{m_1} q_{1k},\ q_0 = \sum_{k=1}^{m} q_{0k}.$$

Now (4.6) yields

$$||\dot{x}||_{\hat{C}[0,\infty)} \leq (V_a(W) + q_1)||\dot{x}||_{\hat{C}[0,\infty)} + c_1||\dot{\Phi}||_{\hat{C}[-\eta,0]}+$$

$$c_2||\Phi||_{\hat{C}[-\eta,0]} + c_3||x||_{\hat{C}[0,\infty)}\ (c_1, c_2, c_3 = const). \tag{4.7}$$

Clearly, if q_{1k} are sufficiently small, then q_1 is sufficiently small. Since $V_a(W) < 1$, we can take $\hat{q}_1$ in such a way that $V_a(W)+\hat{q}_1 < 1$. Now relation (4.7) implies the inequality

$$||\dot{x}||_{\hat{C}[0,\infty)} \leq (1 - V_a(W) - q_1)^{-1}(c_1||\dot{\Phi}||_{\hat{C}[-\eta,0]}+$$

$$c_2||\Phi||_{\hat{C}[-\eta,0]} + c_3||x||_{\hat{C}[0,\infty)}). \tag{4.8}$$

Taking into account inequality (4.5) and the equivalence of norms in $\mathbf{C}^n$, we arrive at the stated result. □

Lemma 12.4.3 *Let F continuously map $[0,\infty)\times \mathbf{C}^{n(m+m_1)}$ into $\mathbf{C}^n$ and satisfy the condition*

$$\|F(t,y_1,...,y_m,z_1,...,z_{m_1})\|_{C^n} \le \sum_{j=1}^{m} q_{0j}\|y_j\|_{C^n} + \sum_{k=1}^{m_1} q_{1k}\|z_k\|_{C^n}$$

$$\text{for all } y_j, z_k \in \Omega(r) \equiv \{h \in \mathbf{C}^n : \|h\|_{C^n} \le r\}$$
$$(j=1,...,m;\ k=1,...,m_1;\ t\ge 0) \tag{4.9}$$

with a $r<\infty$. Then for any $\epsilon>0$, there is a function

$$F_1 : [0,\infty)\times \mathbf{C}^{n(m+m_1)} \to \mathbf{C}^n$$

with the properties

$$\|F_1(t,y_1,...,y_m,z_1,...,z_{m_1})\|_{C^n} \le \sum_{j=1}^{m}(q_{0j}+\epsilon)\|y_j\|_{C^n} + \sum_{j=1}^{m_1}(q_{1j}+\epsilon)\|z_j\|_{C^n}$$

$$\text{for all } y_j, z_k \in \mathbf{C}^n, \tag{4.10}$$

and

$$F_1(t,y_1,...,y_m,z_1,...,z_{m_1}) = F(t,y_1,...,y_m,z_1,...,z_{m_1})$$
$$\text{for } y_j, z_k \in \Omega(r)\ \ (j=1,...,m;\ k=1,...,m_1;\ t\ge 0). \tag{4.11}$$

The proof of that lemma is the same as the proof of Lemma 10.4.3.

Lemma 12.4.4 *Let all the zeros of det $K_0(p)$ belong to the open left half-plane. Let conditions (2.4), (2.7), and (4.9) hold. Then the zero solution of equation (4.1) is asymptotically stable provided q_{0j}, q_{1k} are sufficiently small.*

Proof: Consider the equation

$$\dot{x} = \int_0^{\eta} dR(\tau)x(t-\tau)+$$
$$F_1(t, x(t-h_1(t)),...,x(t-h_m(t)),\dot{x}(t-l_1(t)),\dot{x}(t-l_{m_1}(t))), \tag{4.12}$$

where the vector-valued function F_1 defined on $[0,\infty)\times\mathbf{C}^{n(m+m_1)}$ satisfies conditions (4.11) and (4.10). The function F_1 can be constructed due to Lemma 12.4.3. According to Lemma 12.4.2, the zero solution of equation (4.12) is globally asymptotically stable, and estimate (4.4) hold. So taking $\|\Phi\|_{C^1[0,\infty)}$ sufficiently small, we can assert that the corresponding solution of (4.12) and its derivative lie in $\Omega(r)$. But according to (4.11) it is the solution of (4.1), simultaneously. This proves the result. □

Proof of Theorem 12.4.1: From condition (4.2) it follows that condition (4.9) holds for any positive $r<\infty$. Besides, $q_{0j}=q_{0j}(r)\to 0, q_{1j}=q_{1j}(r)\to 0$ as $r\to 0$. Take r sufficiently small, then due to Lemma 12.4.4 we arrive at the required result. □

12.5 Notes

Note 12.1 The sewing conditions are too restrictive for a number of problems. But solutions of the problem (1.1), (1.2) must be understood in some generalized sense (see (Kolmanovskii and Nosov, 1986, p. 22), and references therein). Introduce the space $L^\infty[a,b]$ of measurable and essentially bounded vector-valued functions defined on $[a,b]$ and equipped with the norm

$$||h||_\infty = ess\ sup_{t\in[a,b]}||h(t)||_{C^n}.$$

An absolutely continuous function $x : [-\eta, T) \to \mathbf{C}^n$ $(0 < T \leq \infty)$ with $\dot{x} \in L^\infty[-\eta, T)$ is said to be a solution on $[0,T)$ of the problem (1.1), (1.2) with $t_0 = 0$, if it satisfies (1.1) and (1.2) for almost all $t \in [-\eta, T)$. Another equivalent definition of solutions is given in (Achmerov et al. 1986).

Let $\Omega \subseteq \mathbf{R} \times C([-\eta, 0], \mathbf{C}^n)$ be open, $f : \Omega \to \mathbf{C}^n$, $D : \Omega \to \mathbf{C}^n$. In the book (Hale, 1977), instead of equation (1.1), the following one:

$$\frac{d}{dt}D(t, x_t) = f(t, x_t) \tag{5.1}$$

is considered. Under consideration one can avoid the sewing conditions. Namely, a function $x \in C([-\eta, T], \mathbf{C}^n)$ is said to be a solution of that equation on $[0,T)$ with an initial function Φ if $D(t, x_t)$ is differentiable, and under (1.2) with $t_0 = 0$, (5.1) hold for each $t \in [0,T)$.

Note 12.2 Section 12.2 is based on the paper (Gil', 1996). Few papers are devoted to the absolute stability of neutral systems. The paper (Achmerov et al., 1977) should be mentioned. It extends to neutral equations the well-known absolute stability criterion for ordinary differential equations by Krasnosel'skii and Pokrovskii (1977). Theorem 12.2.3 improves the absolute stability criterion from (Achmerov et al., 1977) in the case of systems with many nonlinearities.

Note 12.3 Various theorems on stability in the first approximations for neutral type systems are presented in (Achmerov et al., 1981, p. 97).

Note 12.4 Applications of the Lyapunov direct method to neutral systems are connected with nontrivial mathematical problems. Many relevant results can be found in the books (Hale and Lunel, 1993), (Achmerov et al, 1986), (Kolmanovskii and Myshkis, 1992), (Kolmanovskii and Nosov, 1986), (Burton, 1985) and the papers (Achmerov et al., 1981), (Naito, 1994), etc.

References for Chapter 12

Achmerov, R.R., Kamenskii, M.I., Potapov, A.C., Rodkina, A.E. and Sadovskii, B.N. (1981). Theory of equations of neutral type. *Matematicheskii Analiz., 19, Itogi Nauki i Techniki. VINITI*, 55-126. In Russian.

Achmerov, R.R., Kamenskii, M.I., Potapov, A.C., Rodkina, A.E. and Sadovskii, B.N. (1986). *Measures of Noncompactness and Condensing Operators.* Nauka,

Novosibirsk. In Russian.

Achmerov, R.R., Kazakova, N.G. and Pokrovskii, A. V. (1978). The absent bounded solution principle and absolute stability of neutral type equations, *Serdika Bulgaria Math. Letters 4. N1*, 61-69. In Russian.

Bellman, R. and K.L. Cooke (1963). *Differential-Difference Equations.* Academic Press, New York.

Burton, T.A. (1985). *Stability and and Periodic Solutions of Ordinary and Functional Differential Equations*, Academic Press, Orlando.

Gil', M.I. (1996). On absolute stability of neutral type functional differential systems, *Nonlinear Studies, 3, N1*, 59-72.

Hale, J.K. (1977). *Theory of Functional Differential Equations*, Springer-Verlag, New York.

Hale, J.K., Verduyn Lunel, S. M. (1993). *Introduction to Functional Differential Equations.* Springer-Verlag, New York.

Kolmanovskii, V.B. and A.D. Myshkis (1992). *Applied Theory of Functional Differential Equations.* Kluwer Academic Publishers, Dordrecht.

Kolmanovskii, V.B. and V.K. Nosov (1986). *Stability of Functional Differential Equations.* Ac. Press, New York.

Krasnosel'skii, M.A. and Pokrovskii, A.V. (1977). The absent bounded solution principle, *Soviet Math. Doklady, 233*, 293-296.

Naito, Y. (1994). Existence and asymptotic behavior of positive solutions of neutral differential equations. *J. of Math. Analysis and Appl., 188*, 227-244.

Chapter 13

STRONGLY CONTINUOUS SEMIGROUPS

This chapter is devoted to strongly continuous semigroups of operators. It is well-known that the mentioned semigroups are the basic instrument for investigation of differential equations with constant operators in Banach and Hilbert spaces. Wide classes of autonomous distributed parameter systems are governed by such equations.

In Sections 13.1 and 13.2 we recall some basic notions of unbounded linear operators in Banach and Hilbert spaces.

In Section 13.3 we consider dissipative operators in a Hilbert space.

Section 13.4 deals with spectral resolutions of selfadjoint and normal operators.

In Section 13.5 we investigate vector-valued functions of normal operators.

In Section 13.6 we review some of the standard facts from the theory of functions with values in a Banach space.

The notion of the strongly continuous semigroup is introduced in Section 13.7.

In Section 13.8 the well-known results about sectorial operators, fractional powers of operators and analytic semigroups are collected.

The solvability of the Cauchy problem for differential equations with constant operators in a Banach space is investigated in Section 13.9.

A norm estimate for a class of semigroups in a Hilbert space is established in Section 13.10.

Section 13.11 deals with stability of equations with constant operators.

In Sections 13.12 and 13.13 relevant examples are presented.

13.1 Unbounded linear operators

We shall now consider a linear operator A defined on some linear manifold $D(A)$ of a Banach space X, operating into X. We shall call the operator A *closed*, if from the fact that $x_n \to x$ $(x_n \in D(A))$, and $Ax_n \to y$ it follows that $x \in D(A)$, and $Ax = y$. A bounded operator is obviously closed. If A is not closed, then for it to have a closed extension it is necessary and sufficient that from $x_n \to 0$ $(x_n \in D(A))$, and $Ax_n \to y$ it follows that $y = 0$. The smallest closed extension of A is called its *closure* $\overline{A}$. If A admits closure, then from $x_n \to x$ $(x_n \in D(A))$, and $Ax_n \to y$ it follows that $\overline{A}x = y$. That may be written otherwise as

$$\overline{A} \lim_{n\to\infty} x_n = \lim_{n\to\infty} Ax_n,$$

if this limit exists.

Example 13.1.1 *In the space* $C[0,1]$ *of all continuous scalar-valued functions defined on* $[0,1]$, *the operator*

$$Ax = \frac{dx}{dt}$$

defined on continuously differentiable functions is closed.

But the same differentiation operator in $L^2[0,1]$ is not closed, but admits closure.

Example 13.1.2 *The operator (functional) acting from* $C[0,1]$ *into* **C** *defined by the formula* $Ax = x'(0)$ *does not admit closure.*

Indeed the sequence $x_n(t) = n^{-1}sin(nx)$ converges uniformly to 0, but $Ax_n = x_n'(0) = 1$.

The following fact is essential: a closed operator defined on the entire space is bounded.

If, for a linear operator A, the relation $Ax = 0$ implies $x = 0$, then it has on its range $Rang(A)$ an inverse operator A^{-1}. That is, an operator mapping $Rang(A)$ onto $D(A)$ with the properties $A^{-1}Ax = x$ $(x \in D(A))$ and $AA^{-1}y = y$ $(y \in Rang(A))$. If A is closed and has an inverse A^{-1}, then A^{-1} is closed. So if A has a bounded inverse, then it is closed.

Obviously, the linear operator $A - \lambda I$ with a domain $D(A)$ is closed or not closed along with the operator A. Therefore, if there exists a bounded inverse operator $R_\lambda(A) = (A - \lambda I)^{-1}$, the operator A is closed. In other words, if a linear operator has at least one regular point, then it is closed. If λ_0 is a regular point of A, then $AR_{\lambda_0}(A)$ is bounded, since

$$AR_{\lambda_0}(A) = I + \lambda_0 R_{\lambda_0}(A).$$

Let A be an unbounded closed linear operator whose domain is a dense set in X. Then its natural powers A^n defined by the relation $A^n = AA^{n-1}$ $(n = 1, 2, \ldots)$, generally speaking, have smaller domains than the domain of the operator itself.

Thus for example the domain $D(A^2)$ consists of those elements $x \in D(A)$, for which Ax also lies in $D(A)$.

Let A be densely defined. Let f be a linear bounded functional defined on X. Then on $D(A)$ the functional $g_0(x) = f(Ax)$ is defined. It can be happen that this functional is bounded on $D(A)$. Then in view of the denseness of $D(A)$ it admits a unique continuous extension to a linear functional $g(x)$ defined on the entire space X. In this case one says that *the dual (adjoint)* operator A^* is defined on f and we write $g = A^* f$. Thus the domain of A^* consists of all functionals $f \in X^*$ for which

$$|f(Ax)| \leq c\|x\|_X,$$

where c does not depend on x. Besides,

$$A^* f(x) = f(Ax) \; (x \in D(A)).$$

The dual operator is always closed.

13.2 Linear operators in a Hilbert space

Let H be a linear space over the field $\mathbf{C}$ of complex numbers. If for any pair $x, y \in H$ is defined a complex number $(x,y) = (x,y)_H$ with the properties

a) $(x,y) = \overline{(y,x)}$;

b) $(x_1 + x_2, y) = (x_1 + x_2, y)$ $(x_1, x_2 \in H)$;

c)$(\lambda x, y) = \lambda(x,x)$ $(\lambda \in \mathbf{C})$;

d) $(x,x) \geq 0$, besides, $(x,x) = 0$ iff $x = 0$.

then $(.,.)$ is called the scalar product. By the scalar product the norm

$$\|h\|_H = \sqrt{(h,h)} \; (h \in H) \tag{2.1}$$

is defined. In addition, the Schwarz inequality

$$|(x,y)| \leq \|x\|_H \, \|y\|_H$$

is valid. A Banach space with a scalar product and the norm defined by (2.1) is called a Hilbert space.

Example 13.2.1 *The Banach space $L^2(\omega, \mathbf{C}^n)$ of all vector-valued functions defined on a set $\omega \subseteq \mathbf{R}^m$ is the Hilbert space with the scalar product*

$$(u,v) = \int_\omega \sum_{k=1}^n u_k(x)\overline{v}_k(x)dx.$$

Here u_j, v_j are coordinate functions of $u, v \in L^2(\omega, \mathbf{C}^n)$.

The Riesz theorem on the general form of a linear functional holds in a Hilbert space: every linear bounded functional $f(x)$ on H may be uniquely represented in the form $f(x) = (x,u)$, where u is an element of H.

A bounded linear operator A^* acting in H is adjoint to A if

$$(Af, g) = (f, A^*g) \text{ for every } h, g \in H.$$

The relation $||A|| = ||A^*||$ is true. A bounded operator A *is a selfadjoint* one if $A = A^*$.

A is *a unitary operator* if $AA^* = A^*A = I$. A bounded linear operator satisfying the relation $AA^* = A^*A$ is called *a normal operator*. It is clear that unitary and selfadjoint operators are examples of normal ones.

The spectrum of a selfadjoint operator is real, the spectrum of a unitary operator lies on the unit circle.

Further, an operator P is called *a projector* if $P^2 = P$. If, in addition, $P^* = P$, then it is called *an orthogonal projector* (an orthoprojector).

If in a Hilbert space there is a countable set whose closure coincides with the entire space, then the space is said to be *separable* . Any separable Hilbert space H possesses an orthonormal basis. This means that there is a sequence $\{e_k \in H\}$ such that

$$(e_k, e_j) = 0 \text{ if } j \neq k \text{ and } (e_k, e_k) = 1$$

and any $h \in H$ can be represented as

$$h = \sum_{k=1}^{\infty} c_k e_k$$

with

$$c_k = (h, e_k), k = 1, 2, \ldots$$

Besides the series strongly converges.

Any nonzero eigenvalue of a completely continuous operator has the finite multiplicity. The spectrum is either finite or sequence $\{\lambda_j(A)\}$ or the eigenvalues of A converges to zero, only. Any completely continuous normal operator A can be represented in the form

$$A = \sum_{k=1}^{\infty} \lambda_k(A) E_k, \tag{2.2}$$

where $\lambda_k(A)$ are eigenvalues of A repeated according to their multiplicities, and E_k are eigenprojectors of A, i.e. the orthoprojectors defined by

$$E_k = (., e_k)e_k,$$

where e_k are the eigenvectors. The last equality means

$$E_k h = (h, e_k)e_k \text{ for any } h \in H.$$

Let A be unbounded and $D(A)$ be dense in H. Then the set of vectors g satisfying

$$|(Af, g)| \leq c||f|| \text{ for all } f \in D(A)$$

with a constant c is the domain of *the adjoint operator* A^*, and besides,

$$(Af, g) = (f, A^*g) \text{ for all } f \in D(A) \text{ and } g \in D(A^*).$$

An unbounded operator A is selfadjoint if

$$D(A) = D(A^*) \text{ and } Ah = A^*h$$

for each $h \in D(A)$. A selfadjoint operator A is positive definite if

$$(Ah, h) > 0 \text{ for every nonzero } h \in D(A).$$

A selfadjoint operator A is negative definite if $-A$ is positive definite. An unbounded selfadjoint operator possesses the unbounded real spectrum. An unbounded operator A is normal if $D(AA^*) = D(A^*A)$ and $AA^*h = A^*Ah$ for each $h \in D(A^*A)$. An unbounded normal operator has an unbounded spectrum.

13.3 Dissipative operators in a Hilbert space

The proofs of all the statements presented in this section can be found in (Tanabe, 1979, Section 2.1).

Definition 13.3.1 *Let A be a linear operator in a Hilbert space H and its domain is assumed to be dense. Then A is called a dissipative operator if $Re(Au, u) \le 0$ for all $u \in D(A)$. A dissipative operator which extends a dissipative operator A is called a dissipative extension of A. A is said to be a maximal dissipative operator if its only dissipative extension is A itself.*

Lemma 13.3.2 *Let A be a linear operator with a dense domain. Then the following three statements are equivalent:*

(i) A is a dissipative operator.

(ii) $\|(A-\lambda)u\| \ge Re\lambda\|u\|$ for all $u \in D(A)$ and all $\lambda \in \mathbf{C}$ satisfying $Re\lambda > 0$.

(iii) $\|(A-\lambda)u\| \ge \lambda\|u\|$ for all $u \in D(A)$ and all $\lambda > 0$.

Theorem 13.3.3 *Let A be a densely-defined operator. Then A is a maximal dissipative operator if and only if it is closed, its resolvent set contains the half-plane $\{\lambda \in \mathbf{C} : Re\lambda > 0\}$, and the inequality*

$$\|(A - I\lambda)^{-1}\| \le \frac{1}{Re\lambda}$$

holds here.

Theorem 13.3.4 *Let A be a closed dissipative operator. A is a maximal dissipative operator if and only if A^* is a dissipative operator. In this case A^* is also a maximal dissipative operator.*

13.4 Spectral resolutions of selfadjoint and normal operators

13.4.1 Selfadjoint operators

For the proofs of the assertions presented in this section see for instance (Ahiezer and Glazman, 1981), (Dunford and Schwartz, 1963).

A family of orthogonal projectors E_t $(-\infty < t < \infty)$ is said to be *an orthogonal resolution of the identity* if E_t is strongly left-continuous in t, and the following relations are satisfied:

$$E_t E_s = E_s E_t = E_s \text{ for any } s \le t,$$

and

$$E_{-\infty} \equiv \lim_{t \to -\infty} E_t = 0, \; E_\infty \equiv \lim_{t \to \infty} E_t = I,$$

where the limits are understood in the sense of the strong topology. For every bounded continuous scalar-valued function $F(t)$ defined on the entire real axis, one can define the Stieltjes operator integral

$$\int_a^b F(t) dE_t. \tag{4.1}$$

This integral is defined as the limit in the norm of H of integral sums of the form

$$\sum_{k=0}^{N} F(t_k)(E_{t_{k+1}} - E_{t_k}) \; (a = t_0 < t_1 ... < t_N = b),$$

if the segment is finite, and as an improper integral if $a = -\infty$, $b = \infty$. If the function $F(t)$ is bounded, then the inequality

$$\|\int_a^b F(t) dE_t\| = \sup_{a \le t \le b} |F(t)|$$

is valid. If the function $F(t)$ takes only real values, then the operator (4.1) is selfadjoint. If the function $F(t)$ is unbounded on the real axis, then the formula

$$\int_{-\infty}^{\infty} F(t) dE_t x$$

defines an unbounded operator whose domain consists of those and only those elements x for which integral

$$\int_{-\infty}^{\infty} |F(t)|^2 d(E_t x, x)$$

converges.

It turns out that to every selfadjoint operator A, there is some orthogonal resolution E_t of the identity, such that the equality

$$Ax = \int_{-\infty}^{\infty} t dE_t x \ (x \in D(A))$$

holds. The operator E_t commutes with any operators commuting with A.

13.4.2 Normal operators

A spectral measure in H is a homomorphic map of a Boolean algebra of sets of the complex plane $\mathbf{C}$ into a Boolean algebra of projection operators in H which has the additional property: it maps the unit in its domain into the identity operator I in its range (Dunford and Schwartz, 1963, p. 888). If, in addition, the projection operators are orthogonal, the spectral measure is called *the self-adjoint spectral measure.*

Theorem 13.4.1 *Let A be a bounded normal operator. There exists a unique self-adjoint spectral measure E such that*

$$A = \int_{\sigma(A)} t dE_t.$$

For the proof see (Dunford and Schwartz, 1963, p. 898). The uniquely defined spectral measure E associated with the normal operator A according to that equality is called the resolution of the identity for A.

For any scalar-valued continuous function f defined on $\sigma(A)$, a function of the operator A is defined by

$$f(A) = \int_{\sigma(A)} f(t) dE_t.$$

Moreover,

$$\|f(A)\| = \sup_{t \in \sigma(A)} |f(t)|. \tag{4.2}$$

Now let S be an unbounded normal operator having a bounded inverse one. Then $A = S^{-1}$ is a normal operator, and for any bounded scalar-valued continuous function h defined on $\sigma(S)$, we define the operator

$$h(S) = \int_{\sigma(A)} h(1/t) dE_t.$$

Hence, according to (4.2) and the spectrum mapping theorem (Dunford and Schwartz, 1963)

$$\|h(S)\| = \sup_{t \in \sigma(S)} |h(t)|.$$

In addition, define

$$D(S) = \{y \in H : \int_{\sigma(A)} 1/|t|^2 d(E_t y, y) < \infty\}$$

where E the resolution of the identity for A. Then with the notation $\tilde{E}_z = E_{1/z}$ we can write

$$Sy = \int_{\sigma(S)} t d\tilde{E}_t y \ (y \in D(S)).$$

Besides, $\tilde{E}$ is the resolution of the identity for S.

13.5 Vector-valued functions of normal operators

Let $H = E_1 \oplus ... \oplus E_n$ be an orthogonal sum of the same separable Hilbert spaces $E_1 = ... = E_n = E$ with a scalar product $(.,.)_E$. The scalar product in H is defined by the formula

$$(v, w)_H = \sum_{k=1}^{n} (v_k, w_k)_E, \text{ where } v_k, w_k \in E \text{ are coordinates of } v, w \in H.$$

Let S be a boundedly invertible normal operator in H commuting with any $n \times n$-matrix. More exactly, S commutes with operators in H defined by constant matrices.

Further let E_s be the resolution of the identity for S (see the previous section), and let $W(s)$ be a function defined on $\sigma(S)$ whose values are linear operators in $\mathbf{C}^n$. Suppose that $W(s)$ is continuous and bounded on $\sigma(S)$. Define an operator $W(S)$ by the formula

$$W(S) = \int_{\sigma(S)} W(s) dE_s. \tag{5.1}$$

This integral is the strong limit of the consequence of the sums

$$\sum_m W(s_m) E(\Delta_m) \ (\cup \Delta_m = \sigma(S), \ s_m \in \Delta_m)$$

as the measure of $\max \Delta_m$ tends to zero.

Lemma 13.5.1 *Let the condition*

$$m(W) \equiv \sup_{s \in \sigma(S)} ||W(s)||_{\mathbf{C}^n} < \infty,$$

be valid. Then the operator $W(S)$ defined by (5.1) is bounded, and the estimate $||W(S)||_H \leq m(W)$ is true.

Proof: For any $u \in H$ we have

$$(W(S)u, W(S)u)_H = (\int_{\sigma(S)} W(s)dE(s)u, \int_{\sigma(S)} W(z)dE(z)u)_H.$$

Take into account that $dE(s)dE(z) = 0$ if $s \neq z$. Since $W(s)$ commutes with $E(s)$, it can be written

$$(W(S)u, W(S)u)_H = (\int_{\sigma(S)} W^*(s)W(s)dE(s)u, u)_H$$

But

$$(W^*(s)W(s)dE(s)u, u) = (W^*(s)W(s)dE(s)u, dE(s)u) \leq m^2(W)(dE(s)u, u),$$

and

$$\int_{\sigma(S)} dE(s) = I_H.$$

We thus get the stated result. □

13.6 Functions with values in a Banach space

13.6.1 Differentiation

We will consider functions defined on a real segment $[0, T]$ with values in a Banach space X. A function $x : [0, T] \to X$ is continuous at a point t_0 if

$$||x(t) - x(t_0)|| \to 0 \text{ as } t \to t_0.$$

$x(t)$ is continuous if it is continuous at each point of $[0, T]$. The norm of a continuous function is a scalar continuous function.

Let $C([0, T];\ X)$ be the Banach space of all continuous functions acting from $[0, T]$ into X with the norm

$$||x||_{C([0,T];X)} = \max_{0 \leq t \leq T} ||x(t)||_X.$$

We will say that $x(t) \in C([0, T];\ X)$ has at the point t_0 a right (left) derivative if there exists $y \in X$ such that

$$||\frac{x(t_0 + \delta t) - x(t_0)}{\delta t} - y|| \to 0$$

as $\delta t \to +0$ ($\delta t \to -0$). One then writes

$$\frac{d_+ x(t_0)}{dt} = y \ (\frac{d_- x(t_0)}{dt} = y).$$

If the left and the right derivatives exist, x is continuous , and

$$\frac{d_-x(t_0)}{dt} = \frac{d_+x(t_0)}{dt},$$

then $x(t)$ is differentiable at the point t_0 and its derivative is

$$x'(t_0) = \frac{dx(t_0)}{dt} = y.$$

The function $x(t)$ is differentiable on a segment $[0,T]$ if it is differentiable at each point of $(0,T)$, has the right derivative at 0, and has the left derivative at T.

One introduces in the natural way the concepts of m-times differentiable and infinitely differentiable functions. The set of all continuously m-times differentiable functions acting from $[0,T]$ into X is denoted by $C^m([0,T];\, X)$, and the norm in $C^m([0,T];\, X)$ is

$$\|u\|_{C^m([0,T];\, X)} = \sum_{k=0}^{m} \|u^{(k)}\|_{C([0,T];\, X)}.$$

If A admits closure $\overline{A}$, then

$$\overline{A}\frac{dx}{dt} = \frac{dAx}{dt} \quad (x,\ \frac{dx}{dt} \in D(A)).$$

In particular, if A is closed, then

$$A\frac{dx}{dt} = \frac{dAx}{dt} \quad (x,\ \frac{dx}{dt} \in D(A)).$$

13.6.2 Integration

Let $x \in C([0,T];\, X)$, and

$$0 = t_0 < t_1 < \ldots < t_N = T$$

be a partition of segment $[0,T]$. Put

$$\int_0^T x(t)dt = \lim \sum_{k=1}^{N} x(t_k)\Delta t_k; \ (\Delta t_k = t_k - t_{k-1})$$

where the limit is understood in the sense of the strong topology as $\max \Delta t_k \to 0$. The estimate

$$\|\int_0^T x(t)dt\| \leq \int_0^T \|x(t)\|dt$$

is valid. Moreover

$$\int_0^T x(t)dt = T\overline{x},$$

where $\overline{x}$ is an element of the closed convex hull of the values of the functions on the segment $[0, T]$ (the mean-value theorem).

The function

$$y(t) = \int_0^t x(s)ds \ (x \in C([0, T];\ X))$$

is continuously differentiable, and $y'(t) = x(t)$. For any continuously differentiable function $x(t)$, $\int_0^T x'(t)dt = x(T) - x(0)$.

Lemma 13.6.1 *Let A be a closed linear operator in X and $u(t)$ be defined on $[0, T]$ such that $x(t) \in D(A)$ for each $t \in [0, T]$. If $x \in C([0, T];\ X)$ and $Ax \in C([0, T];\ X)$, then*

$$\int_0^T x(t)dt \in D(A), \ \text{and}\ A\int_0^T x(t)dt = \int_0^T Ax(t)dt.$$

Just as in the classical analysis, one may introduce the concept of an improper integral. For example, if $x(t)$ is continuous on $[a, \infty)$, then by its integral on $[a, \infty)$ we mean the limit

$$\int_a^\infty x(t)dt = \lim_{b\to\infty} \int_a^b x(t)dt.$$

13.7 Strongly continuous semigroups

The proofs of all statements presented in this section can be found in (Tanabe, 1979, Section 3.1).

Definition 13.7.1 *Let $T(t)$ for each $t \geq 0$ be a bounded linear operator in a Banach space X. Then $T(t)$ is called a C_0-semigroup (a strongly continuous semigroup) if*

(i) $T(0) = I$, $T(t)T(s) = T(t + s)$ for all $t, s \geq 0$,

(ii) $T(t) \to T(s)$ as $t \to s$ for each $s \geq 0$ in the sense of the strong topology.

Theorem 13.7.2 *For any C_0-semigroup $T(t)$ $(t \geq 0)$ there exist real numbers $M \geq 1$ and α such that*

$$\|T(t)\| \leq Me^{\alpha t} \tag{7.1}$$

for all $t \geq 0$.

Definition 13.7.3 *A linear operator A is called the infinitesimal generator of a C_0-semigroup $T(t)$ if it is defined by*

$$Au = \lim_{h \to +0} h^{-1}(T(h)u - u)$$

in the sense of the strong topology whenever the limit exists.

If A is the infinitesimal generator of a semigroup $T(t)$, we will sometimes write e^{At} instead of $T(t)$.

Theorem 13.7.4 *The generator A of a C_0-semigroup $T(t)$ is a closed operator. Let M and α be numbers such that (7.1) is satisfied. Then the half-plane $\{\lambda \in \mathbf{C} : Re\lambda > \alpha\}$ is contained in the resolvent set of A and therein*

$$(A - \lambda I)^{-1} = -\int_0^{\infty} e^{\lambda t} T(t) dt, \tag{7.2}$$

and the relation

$$\|(A - \lambda I)^{-n}\| \leq M(Re\lambda - \alpha)^{-n} \tag{7.3}$$

is satisfied for any $n = 1, 2, \ldots$.

Theorem 13.7.5 *Let A be a densely defined closed operator. Assume that the half-plane $\{\lambda \in \mathbf{C} : Re\lambda > \alpha\}$ is contained in the resolvent set of A and that relation (7.3) is satisfied for any $Re\lambda > \alpha$, and $n = 1, 2, \ldots$. Then A generates a C_0-semigroup $T(t)$. Moreover, the inequality (7.1) is valid.*

Definition 13.7.6 *A semigroup $T(t)$ satisfying $\|T(t)\| \leq 1$ $(t \geq 0)$ is called a contraction semigroup.*

Theorem 13.7.7 *Let H be a Hilbert space. An operator A is a generator of a contraction C_0-semigroup in H if and only if it is maximal dissipative.*

Definition 13.7.8 *A semigroup $T(t)$ is said to be exponentially stable if the inequality (7.1) holds with $\alpha < 0$.*

13.8 Sectorial operators, analytic semigroups and fractional powers of operators

For the convenience of the reader in this section we repeat the relevant material from the book (Henry, 1981, Sections 1.3 and 1.4) without proofs. Let X be a Banach space. We will call a linear operator A in X *a sectorial one* if A is a closed densely defined operator such that for some $\phi \in (0, \pi/2)$, and some $M_0 \geq 1$ and a real b the sector

$$S_{b,\phi} \equiv \{\lambda \in \mathbf{C} : \phi \leq |arg(\lambda - b)| \leq \pi, \ \lambda \neq b\}$$

is in the resolvent set of A, and

$$\|(I\lambda - A)^{-1}\| \leq M_0/|\lambda - b| \text{ for all } \lambda \in S_{b,\phi}.$$

Recall that I is the unit operator.

Examples:

1. If A is a bounded linear operator in X, then A is a sectorial operator.

2. If A is a selfadjoint densely defined operator in X and A is bounded from below, then A is a sectorial operator.

3. If $Au(x) = -\Delta u(x)$, $(x \in \Omega \subset \mathbf{R}^n)$, when A is the closure in $L^p(\Omega)$ $(1 \leq p \leq \infty)$ of $-\Delta$ defined on smooth functions defined on Ω. Then A is a sectorial operator (see (Henry, 1981, Section 1.2)).

4. Many other elliptic boundary-value problems define sectorial operators as shown by Friedman (1969).

A function $f(t) : (0,\infty) \to X$ is analytic if it is infinitely differentiable at each point $t > 0$, and there exists $\delta = \delta(t) > 0$ so that whenever $|h| \leq \delta$,

$$f(t+h) = \sum_{k=0}^{\infty} \frac{h^k}{k!} f^{(k)}(t)$$

the series converging in X-norm uniformly in $|h| \leq \delta$.

Definition 13.8.1 *An analytic semigroup $T(t)$ on a Banach space X is a strongly continuous semigroup with the property that for any $h \in X$, $t \to T(t)h$ is a real analytic function on $0 < t < \infty$.*

Lemma 13.8.2 *If A is a sectorial operator, then $-A$ generates an analytic semigroup e^{-At}.*

If A is a sectorial operator on X with $Re\sigma(A) > 0$, then for any $\nu > 0$, one can define

$$A^{-\nu} = \frac{1}{\Gamma(\nu)} \int_0^{\infty} t^{\nu-1} e^{-At} dt,$$

where $\Gamma(\nu)$ is the Euler Gamma function evaluated at ν. The operator $A^{-\nu}$ is a bounded linear operator on X which is one-to-one and satisfies

$$A^{-\nu} A^{-\mu} = A^{-(\nu+\mu)} \text{ for } \mu, \nu > 0.$$

One can now define A^{ν} for $\nu > 0$ as the inverse of $A^{-\nu}$. The operator A^{ν}, $\nu > 0$ is closed and densely defined. Define $A^0 = I$.

Lemma 13.8.3 *If A is sectorial with $Re\sigma(A) > \delta > 0$, then for any $\nu \geq 0$ there is a constant $C_\nu < \infty$ such that*

$$\|A^{\nu} e^{-At}\| \leq C_\nu t^{-\nu} e^{-\delta t} \ (t > 0).$$

If A is a sectorial operator in X, then there is a number a, such that $A_1 = A + aI$ has $Re\sigma(A_1) > 0$. If we define $X^\nu = D(A_1^\nu)$ with the graph norm

$$||x||_{X^\nu} = ||A_1^\nu x||, \; x \in X^\nu,$$

then X^ν is a Banach space with norm $||.||_{X^\nu}$. Furthermore if X^ν is defined by different a with the above property, then the norms are equivalent and so the dependence on a is suppressed. Besides, $X^0 = X$, and X^μ is a dense subspace of X^ν for any $\mu \geq \nu \geq 0$.

Theorem 13.8.4 *If*

$$0 \leq \nu \leq 1, x \in D(A,)$$

then

$$||A^\nu x|| \leq C||Ax||^\nu ||x||^{1-\nu}.$$

(Here C is a constant independent of ν).

When

$$\nu = \beta\theta + (1-\theta)\gamma, 0 \leq \theta \leq 1, \beta \geq 0, \gamma \geq 0,$$

then the latter theorem gives the inequality

$$||A^\nu x|| \leq C||A^\beta x||^\theta ||A^\gamma x||^{1-\theta}.$$

Corollary 13.8.5 *If A is a sectorial operator with $Re\sigma(A) > 0$ and if B is a linear operator such that $BA^{-\nu}$ is bounded for some $\nu \in [0, 1)$, then $A + B$ is sectorial.*

13.9 The Cauchy problem for linear equations with constant operators

Let X be a Banach space and A be a closely-defined linear operator in X. Consider the Cauchy (initial value) problem

$$du(t)/dt = Au(t) + f(t) \; (0 \leq t \leq T), \tag{9.1}$$

$$u(0) = u_0, \tag{9.2}$$

where u_0 and f are given elements of $D(A)$ and $C([0, T], X)$, respectively. A function u is called *a solution* of this initial value problem if, besides satisfying (9.1) and (9.2) it has the properties $u \in C^1([0, T], X)$, $u(t) \in D(A)$ for each $t \in [0, T]$, and $Au \in C([0, T], X)$. For the proof of the following two theorems see (Tanabe, 1979, p. 64-65).

Theorem 13.9.1 *Let A generate a C_0-semigroup $T(t), u_0 \in D(A)$ and $f \in C^1([0,T], X)$. Then the function*

$$u(t) = T(t)u_0 + \int_0^t T(t-s)f(s)ds \tag{9.3}$$

is the solution of the problem (9.1), (9.2).

In particular the function $v(t) = T(t)u_0$ is the solution of the homogeneous problem

$$dv(t)/dt = Av(t) \ (t \geq 0; \ v(0) = u_0). \tag{9.4}$$

Theorem 13.9.2 *Let A generate a C_0-semigroup $T(t)$, $u_0 \in D(A)$,*

$$f \in C([0,T], X), f(t) \in D(A) \ (t \in [0,T]), \text{ and } Af \in C([0,T], X).$$

Then the function u defined by (9. 3) is the solution of the problem (9.1), (9.2).

Now let us assume that $-A$ is a sectorial operator. In this case we can use stronger results. Let $C_\beta([a,b], X)$ with some $\beta \in (0,1]$ denote the space of Hölder-continuous functions defined on $[a,b]$ with values in X, and endowed with the norm

$$||f||_{C_\beta([a,b],X)} \equiv \sup_{t,s \in [a,b], t \neq s} \frac{||f(t) - f(s)||_X}{|t-s|^\beta}.$$

Theorem 13.9.3 *Suppose $-A$ is a sectorial operator in X, $u_0 \in D(A)$, and $f \in C_\beta([0,T], X) \cap \in L^1([0,T], X)$. Then the formula (9.3) gives a solution of the problem (9.1), (9.2).*

For the proof of that theorem see (Henry, 1981, p. 52).

13.10 Norm estimates for a class of semigroups in a Hilbert space

Estimates for norms of semigroups of general operators are unknown. Because of this we consider a class of operators which are pencils with matrix coefficients of a selfadjoint operator.

Let $H = E^n$ be an orthogonal sum of the same Hilbert spaces $E_1 = ... = E_n = E$. *Throughout this section S is a boundedly invertible selfadjoint operator in H commuting with any constant $n \times n$-matrix.*

For a positive number μ, define on the set $D(S^\mu)$ the graph norm $||h||_\mu = ||S^\mu h||_H$ $(h \in D(S^\mu))$. Denote the obtained Hilbert space by H^μ. Consider the operator

$$A = Q(S) \equiv c_0 S^2 + c_1 S + c_2, \tag{10.1}$$

where c_k $(k = 0, 1, 2)$ are constant $n \times n$-matrices. In addition,

$$c_0 = c_0^* \text{ is negative definite }. \tag{10.2}$$

Clearly, $D(Q(S)) = D(S^2)$, and $Q(S)$ commutes with S:

$$Q(S)S^\mu h = S^\mu Q(S)h \ (h \in D(S^{\mu+2}),\ \mu \geq 0).$$

Obviously,

$$(Q(S) - c_0S^2)(c_0S^2)^{-\mu} \ (1/2 \leq \mu < 1)$$

is a bounded operator. But $-c_0S^2$ is selfadjoint and, therefore, is sectorial with the positive real spectrum. Now Corollary 13.8.5, yields

Lemma 13.10.1 *The operator $-Q(S)$ is sectorial.*

Introduce the matrix pencil

$$Q(s) = s^2c_0 + sc_1 + c_2 \ (s \in \mathbf{C}).$$

Let us suppose that all the eigenvalues $\lambda_k(Q(s))$ lie in the left half-plane for any $s \in \sigma(S)$. Namely,

$$\alpha(Q(s)) \equiv \max_{k=1,\ldots,n} Re\lambda_k(Q(s)) \leq \alpha_0 < 0 \ (s \in \sigma(S)). \tag{10.3}$$

Hense, it follows that $\alpha(Q(S)) = \sup\ Re\sigma(Q(S)) \leq \alpha_0$. Further, assume that

$$v \equiv \sup_{s \in \sigma(S)} g(Q(s)) < \infty. \tag{10.4}$$

For example, in the scalar case $g(Q(s)) \equiv 0$. Moreover, if under (10.2) $c_1 = c_1^*$, then according to Proposition 2.1.1.

$$g(Q(s)) \leq \sqrt{1/2}N(Q(s) - Q^*(s)) = \sqrt{1/2}N(c_2 - c_2^*).$$

Recall that $N(.)$ is the Hilbert-Schmidt (Frobenius) norm. This gives $v \leq v(c_2) \equiv \sqrt{1/2}N(c_2 - c_2^*)$. Put

$$\chi(Q) = \sup_{t \geq 0} e^{t\alpha_0} \sum_{k=0}^{n-1} \frac{t^k v^k(c_2)}{(k!)^{3/2}},$$

and

$$\eta(\mu, t) = \sum_{k=0}^{n-1} \frac{t^k v^k}{(k!)^{3/2}} \sup_{s \in \sigma(S)} |s|^\mu e^{t\alpha(Q(s))} \ (0 \leq \mu < 2).$$

Clearly,

$$\chi(Q) = \sup_{t \geq 0} \eta(0, t) \text{ and } \eta(\mu, .) \in L^1[0, \infty).$$

Lemma 13.10.2 *Under conditions (10.3), (10.4), the analytic semigroup $T(t)$ generated in H by operator $Q(S)$, subordinates the inequalities $||T(t)||_{H^\mu} \leq \chi(Q)$ $(t \geq 0)$ and*

$$||S^\mu T(t)||_H \leq \eta(\mu, t) \ (t > 0;\ 0 \leq \mu < 2). \tag{10.5}$$

Proof: Let E_s be the resolution of identity for S. Consider the operator

$$e^{Q(S)t} = \int_{\sigma(S_0)} e^{t\,Q(s)} dE_s \tag{10.6}$$

According to Corollary 2.1.7 we have

$$\|s^\mu exp[Q(s)t]\|_{C^n} \leq \eta(\mu, t) \ (s \in \sigma(S), t > 0).$$

Hence, by virtue of Lemma 13.5.1, the operator $e^{Q(S)t}$ subordinates the inequality $\|S^\mu e^{tQ(S)}\|_H \leq \eta(\mu, t)$ $(t > 0)$. In particular,

$$\|e^{tQ(S)}u\|_H \leq \eta(0, t) \leq \chi(Q)$$

for every $t \geq 0$. But

$$\|S^\mu e^{tQ(S)} w\|_H = \|e^{tQ(S)} S^\mu w\|_H \leq \|e^{tQ(S)}\|_H \|S^\mu w\|_H.$$

So $\|e^{tQ(S)}\|_{H^\mu} \leq \chi(Q)$ $(t \geq 0)$. Moreover, according to (10.6),

$$Q(S)e^{Q(S)t}h = \int_{\sigma(S)} Q(s)e^{t\,Q(s)} dE_s h \ (h \in D(S^2)).$$

On the other hand,

$$\frac{de^{Q(S)t}h}{dt} = \int_{\sigma(S)} \frac{de^{t\,Q(s)}}{dt} dE_s h.$$

Thus, $d(e^{Q(S)t}h)/dt = Q(S)e^{Q(s)t}h$. That is, $e^{Q(S)t}$ is a strongly continuous semigroup generated by the operator $Q(S)$, really. Due to Lemma 13.8.2 it is analytic. □

13.11 Stability of linear equations with constant operators

Definition 13.11.1 *Let $Y \supseteq D(A)$ be a Banach subspace of X. Equation (9.4) is said to be Y-stable if for any initial vector $u_0 \in D(A)$ it has on $[0, \infty)$ a solution $v(t)$, which subordinates the inequality*

$$\|v(t)\|_Y \leq M\|u_0\|_Y \ (t \geq 0), \tag{11.1}$$

where M is a constant independent of u_0.

Equation (9.4) is exponentially Y-stable if for any initial vector $u_0 \in D(A)$ it has on $[0, \infty)$ a solution $v(t)$, which subordinates the inequality

$$\|v(t)\|_Y \leq M\|u_0\|_Y e^{-\alpha t} \ (t \geq 0),$$

where the positive constants M and α do not depend on u_0.

We will need the following trivial

Lemma 13.11.2 *Let A generate in X a strongly continuous semigroup, and let the equation*

$$dw(t)/dt = (A + \epsilon I)w(t) \tag{11.2}$$

be stable for a sufficiently small $\epsilon > 0$. Then equation (9.1) is exponentially X-stable.

Proof: Put in (9.1) $u(t) = w(t)e^{-\epsilon t}$. Then $w(t)$ satisfies (11.2). Since (11.2) is stable, relation (11.1) holds for w. Hence the result follows. □

Lemma 13.11.3 *Let $A = A_0 + B$, where $-A_0$ is sectorial and $Re\sigma(A_0) < 0$. Let the linear operator B have the following properties: $BA_0^{-\nu}$ is bounded for some $\nu \in (0, 1)$, and*

$$\xi \equiv \int_0^\infty \|e^{A_0 t}B\|_X \, dt < 1.$$

Then equation (9.1) is exponentially X-stable. Moreover, estimate (11.1) is true with

$$M = c_0(1 - \xi)^{-1} \text{ where } c_0 = \max_{t \geq 0} \|e^{A_0 t}\|.$$

Proof: Operator $-(A_0 + B)$ is sectorial due to Corollary 13.8.4. Furthermore, rewrite equation (9.1) in the form

$$u(t) = e^{A_0 t}u_0 + \int_0^t e^{A_0(t-s)}Bu(s)ds.$$

We have

$$\|u(t)\|_X \leq \|e^{A_0 t}u_0\|_X + \|u\|_{C([0,\infty),X)} \int_0^t \|e^{A_0(t-s)}B\|_X ds \leq$$

$$c_0\|u_0\|_X + \|u\|_{C([0,\infty),X)}\xi \ (t \geq 0).$$

Hence

$$\|u\|_{C([0,\infty),X)} \leq c_0\|u_0\|_X(1 - \xi)^{-1}.$$

This is the required estimate. The exponential stability can be proved by a small perturbation and the previous lemma. □

Let $H = E^n$ be an orthogonal sum of n exemplars of the same Hilbert space E. Consider in H the equation

$$\dot{u} = Q(S)u + Bu \tag{11.3}$$

where S is an invertible selfadjoint operator, $Q(S)$ is defined by (10.1), and B has the property

$$q_\mu = \|S^{-\mu}B\|_H < \infty \ (0 < \mu < 2). \tag{11.4}$$

Due to Lemma 13.10.2.

$$\|e^{Q(S)t}B\|_H \le \|e^{Q(S)t}S^{\mu}\| \, \|S^{-\mu}B\|_H \le q_\mu \eta(\mu, t).$$

Now the previous lemma yields

Lemma 13.11.4 *Let the conditions (10.3), (10.4), (11.4) and*

$$q_\mu \int_0^\infty \eta(\mu, t)dt < 1$$

hold. Then equation (11.3) is exponentially H-stable.

13.12 Parabolic systems with constant coefficients

Let us consider in the space $H = L^2([0, \pi], \mathbf{C}^n)$ the problem

$$\frac{\partial u}{\partial t} = \tilde{c}_0 u_{xx} + \tilde{c}_1 u_x + \tilde{c}_2 u \quad (t > 0, \ 0 < x < \pi) \tag{12.1}$$

with constant $n \times n$-matrices $\tilde{c}_0, \tilde{c}_1, \tilde{c}_2$. Besides, $\tilde{c}_0 = \tilde{c}_0^* > 0$. Take boundary condition

$$u(0, t) = u(\pi, t) = 0 \quad (t > 0) \tag{12.2}$$

and put

$$D(A_0) = \{u \in L^2 \equiv L^2([0, \pi], \mathbf{C}^n) : u_{xx} \in L^2; \ u(0) = u(\pi) = 0\}$$

and

$$A_0 u = \tilde{c}_0 u_{xx} + \tilde{c}_1 u_x + \tilde{c}_2 u \ (u \in D(A_0)).$$

Clearly, the set $D(A_0)$ consists of vector-valued functions defined on $[0, \pi]$ of the form

$$v(x) = \sum_{k=1}^{\infty} a_k sin(kx) \ (a_k \in \mathbf{C}^n) \text{ with the property } \sum_{k=1}^{\infty} \|a_k\|_{C^n}^2 k^4 < \infty.$$

We will complexify $D(A_0)$. Namely, introduce the set $D(A) \subset L^2([0, \pi], \mathbf{C}^n)$ of functions of the form

$$w(x) = \sum_{k=-\infty, k \neq 0}^{\infty} a_k e^{ikx} \ (a_k \in \mathbf{C}^n) \text{ with the property } \sum_{k=-\infty}^{\infty} \|a_k\|_{C^n}^2 k^4 < \infty.$$

Clearly, $D(A_0)$ is a subset of $D(A)$. By A we denote the extension of A_0 to $D(A)$. Thus, any solution $u(t)$ of problem (12.1), (12.2) is simultaneously a solution of equation (12.1) with the condition $u(t) \in D(A)$, $t \ge 0$. So we will analyze the stability of the equation (12.1) with a solution $u : [0,\infty) \to D(A)$. To this end, take

$$D(S) = \{w \in L^2([0,\pi]; \mathbf{C}^n) : w(x) = \sum_{k=-\infty, k\neq 0}^{\infty} a_k e^{ikx}; \sum_{k=-\infty}^{\infty} \|a_k\|^2_{C^n} k^2 < \infty\}.$$

Define S by the formula $Sy = -iy'$ for $y \in D(S)$. Then $A = Q(S)$ where $Q(S)$ is defined by (10.1) with

$$c_0 = -\tilde{c}_0, c_1 = -i\tilde{c}_1 \text{ and } c_2 = \tilde{c}_2.$$

It is clear that the normed eigenfunctions of S are $(\sqrt{\pi})^{-1}\, e^{ixk}$, and the eigenvalues are

$$\lambda_k(S) = k;\ k = \pm 1, \pm 2, \ldots. \tag{12.3}$$

Proposition 13.12.1 *With the notation*

$$Q_k = -\tilde{c}_0 k^2 + i\tilde{c}_1 k + \tilde{c}_0, \tag{12.4}$$

let the conditions

$$Re\lambda_l(Q_k) \leq \alpha_0 < 0\ (l = 1, ..., n;\ k = \pm 1, \pm 2, \ldots) \tag{12.5}$$

hold. Then problem (12.1), (12.2) is exponentially H-stable.

This result follows from Lemmata 13.8.3 and13.10.1.

In particular if (12.1) is a real scalar equation, then it is exponentially stable provided $-\tilde{c}_0 + \tilde{c}_2 < 0$.

13.13 Time-invariant parabolic systems with coefficients depending on spatial variables

Let us consider in the space $H = L^2([0,\pi], \mathbf{C}^n)$ the problem

$$\frac{\partial u}{\partial t} = \tilde{c}_0 u_{xx} + (\tilde{c}_1 + b_1(x))u_x + (\tilde{c}_2 + b_2(x))u\ \ (t > 0,\ 0 < x < \pi) \tag{13.1}$$

with constant $n \times n$-matrices $\tilde{c}_0, \tilde{c}_1, \tilde{c}_2$, and differentiable in x $n \times n$-matrices $b_j(x)$ $(j = 1, 2)$. Again take the boundary condition (12.2).

Define S as in the previous section, and put

$$Bu = b_1(x)\frac{\partial u}{\partial x} + b_2(x)u = \frac{\partial}{\partial x}(b_1(x)u) + (-b_1'(x) + b_2(x))u\ (u \in D(S)).$$

Then $B = iSb_1(x) + (-b_1'(x) + b_2(x))$. Hence,

$$\|S^{-1}B\|_H \leq q_1 \equiv \sup_x \|b_1(x)\|_{C^n} + \sup_x \|b_1'(x) - b_0(x)\|_{C^n} \|S^{-1}\|_H.$$

According to (12.3) $||S^{-1}||_H = 1$. So

$$q_1 = \sup_x ||b_1(x)||_{C^n} + \sup_x ||b_1'(x) - b_0(x)||_{C^n}.$$

Assume that $\tilde{c}_0 = \tilde{c}_0^* > 0$, and for simplicity of calculations suppose that

$$g(Q_k) \leq v < \infty, \ (k = \pm 1, \pm 2, \ldots). \tag{13.2}$$

Condition (13.2) holds, for example, if $\tilde{c}_1^* = -\tilde{c}_1$, or $n = 1$. Under consideration we have

$$\eta(1,t) = \sum_{k=0}^{n-1} \frac{t^k v^k}{(k!)^{3/2}} \zeta(t)$$

where

$$\zeta(t) = \sup_{k=\pm 1, \pm 2, \ldots} |k| e^{t\alpha(Q_k)},$$

and Q_k are defined by (12.4). Now Lemma 13.11.4 implies:

Proposition 13.13.1 *Let the conditions (13.2) and*

$$q_1 \int_0^\infty \eta(1,t)dt < 1 \tag{13.3}$$

hold. Then problem (13.1), (12.2) is exponentially H-stable.

In particular, if (13.1) is a real scalar equation, then

$$q_1 = \sup_x |b_1(x)| + \sup_x |b_1'(x) - b_2(x)|,$$

$$\eta(1,t) = \max_{k=1,2,\ldots} k \, exp[t(-\tilde{c}_0 k^2 + \tilde{c}_2)]$$

with real constants $\tilde{c}_0, \tilde{c}_2$. This maximum is simply calculated. Now condition (13.3) provides the result.

13.14 Notes

Note 13.1 The literature dealing with strongly continuous semigroups and differential equations with constant operators in a Banach space is extensive. Let us mention the following excellent books: (Hille and Phillips, 1957), (Tanabe, 1979), (Pazy. 1983), (Krein, 1971). Unlike finite dimensional systems, the spectrums of generators need not determine the growth bounds of the semigroups. This fact was demonstrated in an indirect way by an example in (Hille and Phillips, 1957, p. 665) and by a more straightforward example in (Zabczyk, 1975). Relevant counterexamples are presented in (Curtain and Zwart, 1995, Chapter 5).

Of course, it is useful to know which generators have the property that the spectrum of the generator determines the growth bound of the semigroup. Conditions for this are any of the following:

a. $-A$ is sectorial;

b. $T(t)z$ is differentiable for any $z \in X$;

c. $T(\tau)$ is compact for some $\tau > 0$;

d. A is a Riesz-spectral operator.

Conditions *a* and *b* were proved in (Henry, 1981) and (Triggiani, 1975), respectively. Condition *c* was proved in (Zabczyk, 1975), and *d* is proved in (Curtain and Zwart, 1995).

Note 13.2 The following concept of stability was investigated in (Benchimol, 1978): a C_0-semigroup $T(t)$ is weakly stable, if $\phi(T(t)h)$ converges to zero as t tends to ∞ for any $\phi \in X^*$ and $h \in X$. About some related works see (Huang, 1993), (Balakrishnan, 1981), (Curtain and Pritchard, 1978), (Kappel et al., 1989), and references therein.

Note 13.3 Stabilizability of linear distributed systems was considered in (Russell, 1978), (Sakawa, 1984), (Desch and Schappacher, 1985), (Rabah and Ionescu, 1987), (Slemrod, 1989), (Wang, 1990), (Zabczyk, 1992), (Nambu, 1994), etc.

Note 13.4 The materials of Sections 13.10-13.13 are adapted from (Gil', 1985) and (Gil', 1995, Chapter 10).

References for Chapter 13

Ahiezer, N. I. and Glazman, I. M. (1981). *Theory of Linear Operators in a Hilbert Space.* Pitman Advanced Publishing Program, Boston.

Balakrishnan, A.V. (1981). Strong stabilization and the steady state Riccati equation, *Applied Mathematics and Optimization*, *7*, 335-341.

Benchimol, C.D. (1978). A note on weak stabilizability of contraction semigroups, *J. Control and Optim.*, *16*, 373-379.

Curtain, R. F. and A. J. Pritchard. (1978). *Infinite Dimensional Linear Systems Theory.* Lectures Notes in Control and Information Sciences 8, Springer-Verlag, New York.

Curtain, R. F. and H.J. Zwart. (1995). *An Introduction to Infinite Dimensional Linear Systems Theory.* Springer-Verlag, New York.

Desch, W. and Schappacher, W. (1985). Spectral properties of finite-dimensional perturbed linear semigroups, *J. Diff. Eqs.*, *59*, 80-102.

Dunford, N and Schwartz, J. T. (1963). *Linear Operators, part II* , Interscience Publishers, New York, London.

Friedman, A. (1969). *Partial Differential Equations.* Holt, Rienart and Winston, New York.

Gil'. M. I. (1985). On absolute stability of nonlinear nonstationary distributed systems, *Automation and Remote Control, 6*, 12-19.

Gil', M. I. (1995). *Norm Estimations for Operator-valued Functions and Applications.* Marcel Dekker, Inc. New York.

Gohberg, I. C. and Krein, M. G. (1969). *Introduction to the Theory of Linear Nonselfadjoint Operators.* Trans. Mathem. Monographs, vol. 18, Amer. Math. Soc., R. I.

Henry, D. (1981). *Geometric Theory of Semilinear Parabolic Equations.* Lectures Notes in Mathematics, No 840. Springer-Verlag, New York.

Hille, E. and R. S. Phillips. (1957). *Functional Analysis and Semigroups.* AMS Colloquium Publications, Providence.

Huang, F. (1993). Strong asymptotic stability of linear dynamical systems in Banach spaces, *J. Diff. Equations, 104*, 307-324.

Kappel, F., Kunish, K. and W. Schappacher (Eds) (1989). *Control and Estimation of Distributed Parameter Systems.* Birkhäuser, Boston.

Krein, S.G. (1971). *Linear Differential Equations in Banach space.* AMS, Providence, R. I.

Nambu, T. (1994). Robustness of feedback control systems of parabolic type .... *J. Diff. Eqs., 110*, 256-253.

Pazy, A. (1983). *Semigroups of Linear Operators and Applications to Partial Differential Equations.* Springer-Verlag, Berlin.

Rabah, R. and Ionescu, D. (1987). Stabilization problem in Hilbert spaces, *Int. J. Control, 46, No 6*, 2035-2042.

Russell, D.L. (1978). Controllability and stabilizability theory for linear partial differential equations: recent progress and open problems, *SIAM Review, 20*, 639-739.

Sakawa, Y. (1984). Feedback control of second order evolution equations with damping, *SIAM Control and Optimization*, *22, No 3*, 343-353.

Slemrod, M. (1989). Feedback stabilization of a linear control system in a Hilbert space with an a priory bounded control. *Mathematics of Control Signals and Systems*, *2*, 265-285.

Tanabe, H. (1979). *Equations of Evolution*. Pitman, London - San-Francisco - Melburn.

Tanabe, H. (1997). *Functional Analytic Methods for Partial Differential Equations*. Marcel Dekker, Inc. New York.

Triggiani, R. (1975). On the stabilization problem in a Banach space, *J. Math. Anal. Applications*, *52*, 308-403.

Wang, P.K.C. (1990). Stabilization and control of distributed systems with time-dependent spatial domains, *Journal of Optimization Theory and Appl.*, *65, No 2*, 331-341.

Zabczyk, J. (1975). A note on C_0-semigroups, *Bull l' Acad. Pol. De Sc. Serie Math.*, *23*, 895-898.

Zabczyk, J. (1992). *Mathematical Control Theory. An Introduction*. Birkhäuser, Boston.

Chapter 14

LINEAR TIME-VARIANT EQUATIONS IN BANACH SPACES

This chapter is devoted to the stability of linear equations with variable operators in a Banach space. It is well-known that wide classes of nonautonomous linear distributed parameter systems are governed by such equations.

The solvability of the Cauchy problem for differential equations with variable operators in a Banach space is considered in Section 14.1.

The conception of the uniform stability for linear nonautonomous evolution equations is presented in Section 14.2.

The multiplicative representation for solutions of linear evolution equations is established in Section 14.3.

By virtue of the multiplicative representation of solutions in Sections 14.4 and 14.5, we extend the Lozinskii and Wazewski inequalities for ordinary differential equations (see Section 4.2) to linear evolution equations in Banach and Hilbert spaces.

In Section 14.6 we consider an equation whose operator is a sum of a constant operator and a variable one. Besides, the variable operator is bounded with respect to the constant operator.

In Sections 14.7 and 14.8 we extend the freezing method for ordinary differential equations (see Section 3.2) to linear evolution equations in a Banach space.

Sections 14.9, 14.10 and 14.11 deal with systems of linear evolution equations.

Throughout this chapter, $A(t)$ is a linear closed operator in a Banach space

X depending on $t \geq 0$ with a dense constant domain $D(A(t)) \equiv D$. In addition, $A(\tau)$, for each fixed $\tau \geq 0$, generates a strongly continuous semigroup $e^{A(\tau)t}$.

14.1 Evolution operators and solution existence

Let us consider in a Banach space X the initial problem (the Cauchy problem)

$$\dot{u} = A(t)u \quad (t \geq 0, \ \dot{u} \equiv du/dt), \tag{1.1}$$

$$u(0) = u_0 \in D. \tag{1.2}$$

A solution of problem (1.1), (1.2) is a strongly continuously differentiable function u defined on $[0, \infty)$ satisfying these relations and $u(t) \in D$ for each $t \geq 0$.

If the problem (1.1), (1.2) has a solution $u(t)$ for any $u_0 \in D$, then it is possible to introduce a linear operator $U(t, s)$ by the relation

$$U(t, s)u(s) = u(t) \ (0 \leq s \leq t < \infty). \tag{1.3}$$

That operator will be called *the evolution operator (the fundamental solution) for equation (1.1), and the operator $U(t) = U(t, 0)$ will be called the Cauchy operator*. It is simple to check that the evolution operator satisfies the relations

$$U(t,t) = I, \ U(t, s) = U(t, r)U(r, s) \ (t \geq s \geq r \geq 0). \tag{1.4}$$

Recall that $I = I_X$ is the unit operator in X.

Our assumptions in the present chapter are as follows:

Assumption 14.1A. *For all $t \geq 0$ and small enough $\delta > 0$, the operator $I - \delta A(t)$ is invertible.*

Assumption 14.1B. *$A(t)$ is continuously differentiable in $t \geq 0$ on D.*

That is, the function $A(t)v$ is strongly continuously differentiable for any $v \in D$.

In addition, assume that for any fixed $T < \infty$, there exist real numbers $M = M(T) \geq 1$ and $\beta = \beta(T)$, such that

$$\| \overleftarrow{\prod_{1 \leq k \leq n}} (A(t_k^{(n)}) - I\lambda)^{-1} \| \leq M(\lambda - \beta)^{-n} \tag{1.5}$$

for all $\lambda > \beta$, and

$$0 = t_0^{(n)} < t_1^{(n)} < ... < t_n^{(n)} = T.$$

Here the arrow over the symbol of the product means that the cofactor indices increase from right to left:

$$\overleftarrow{\prod_{1 \leq k \leq n}} (A(t_k^{(n)}) - I\lambda)^{-1} \equiv (A(t_n^{(n)}) - I\lambda)^{-1} ... (A(t_1^{(n)}) - I\lambda)^{-1}.$$

We will use the following result (Tanabe, 1979, p.102):

Lemma 14.1.1 *Under Assumptions 14.1A and 14.1B suppose that (1.5) holds. Then there exists a bounded operator-valued function $U(t,s): D \to D$ satisfying the following conditions: relations (1.4) hold, and $U(t,s)$ is strongly continuous in s and t. Furthermore, $U(t,s)$ is strongly continuously differentiable in t and s, and*

$$(\partial/\partial t)U(t,s)v = A(t)U(t,s)v, \tag{1.6}$$

$$(\partial/\partial s)U(t,s)v = -U(t,s)A(s)v \ (v \in D;\ t,s \geq 0). \tag{1.7}$$

The function

$$u(t) = U(t,0)u_0$$

gives a solution of the problem (1.1), (1.2). So the operator $U(t,s)$ in Lemma 14.1.1 is the evolution operator. Let us consider the initial value problem for the inhomogeneous equation

$$\dot{u} = A(t)u + f(t) \quad (t \geq 0). \tag{1.8}$$

The solution is defined as in the homogeneous case.

Lemma 14.1.2 *Suppose the assumptions of Lemma 14.1.1 are fulfilled and $A(t)f(t)$ is defined and continuous. Then for any $u_0 \in D$, the function*

$$u(t) = U(t,0)u_0 + \int_0^t U(t,s)f(s)ds \tag{1.9}$$

is a solution on $[0,T]$ for the problem (1.8), (1.2).

For the proof see (Krein, 1971, p. 196).

We will also need the following result.

Theorem 14.1.3 *For all $t \geq 0$, let $-A(t)$ be sectorial with a constant domain, and $Re\sigma(A(t)) < 0$. In addition, let there be constants $L > 0$, $\nu \in (0,1]$, such that*

$$\|(A(t) - A(s))A^{-1}(0)\| \leq L|t-s|^{\nu} \ (t,s \geq 0).$$

Then the problem (1.1), (1.2) has a solution.

For the proof see (Tanabe, 1979, p. 127).

14.2 Stability definitions for linear evolution equations

Let $Y \supseteq D$ be a Banach subspace of X. As in Section 13.11 *we will say that equation (1.1) is Y-stable*, if it has a solution $u(t)$ for any initial vector $u_0 \in D$, and there is a constant N_0 such that $\|u(t)\|_Y \leq N_0 \ \|u(0)\|_Y$ for all $t \geq 0$. Equation (1.1) is *asymptotically Y-stable* if it is Y-stable, and any of its solution has the property $\|u(t)\|_Y \to 0$ as $t \to \infty$.

Lemma 14.2.1 *A necessary and sufficient condition for Y-stability of equation (1.1) is the uniform boundedness of its Cauchy operator:*

$$N_0 = \sup_{t \geq 0} \|U(t)\|_Y < \infty. \tag{2.1}$$

Proof: The sufficiency follows at once from the formula $u(t) = U(t)u_0$. The necessity is due to the uniform boundedness principle (see e.g. (Daleckii and Krein, 1974, p. 11 and p. 112)). □

Clearly, if (2.1) holds and $\|U(t)\|_Y \to 0$ as $t \to \infty$, then equation (1.1) is asymptotically Y-stable.

It is easily seen that the choice of the interval $[0, \infty)$ is not essential for the definition of stability. It can be replaced by any fixed interval $[t_0, \infty)$. The role of the Cauchy operator $U(t)$ in the above formulas will be played by the operator $U(t, t_0)$.

Equation (1.1) is said to be uniformly Y-stable if it has a solution $u(t)$ for any initial condition $u(0) = u_0 \in D$, and there is a constant $N > 0$ such that

$$\|u(t)\|_Y \leq N \, \|u(s)\|_Y \quad (t \geq s \geq 0).$$

It can be shown that uniform stability is equivalent to the relation

$$\sup_{t \geq s \geq 0} \|U(t, s)\|_Y < \infty.$$

Suppose $A(t) \equiv A$ is a (constant) generator of a strongly continuous semigroup e^{At}. In this case, $U(t, s) = e^{A(t-s)}$ and, as is easily seen, the stability condition

$$\sup_{t \geq 0} \|e^{At}\|_Y < \infty$$

coincides with the uniform stability condition.

The exponential Y-stability for equation (1.1) is defined as in Section 13.11.

14.3 The multiplicative representation of solutions

For the partitioning of a segment $[0, t]$: $t_k^{(n)} = k\delta$ with $\delta = kn^{-1}$ $(k = 0, ..., n)$, let us denote

$$U_{n,k} = (I - A(t_n^{(n)})\delta)^{-1}(I - A(t_{n-1}^{(n)})\delta)^{-1} ... (I - A(t_{k+1}^{(n)})\delta)^{-1},$$

for $k < n$, and $U_{n,n} = I$. Assume that for any finite $t > 0$ there is a constant $N = N(t)$ independent of n and k such that

$$\|U_{n0}\|_X \leq N. \tag{3.1}$$

Theorem 14.3.1 *Under Assumptions 14.1A and 14.1B, suppose (3.1) holds. Then for any $u_0 \in D$ there exists a unique solution $u(t)$ of the problem (1.1), (1.2). Moreover, the relation*

$$u(t) = \lim U_{n0}u_0 \text{ as } n \to \infty \tag{3.2}$$

is valid in the sense of the norm of space X.

Proof: Condition (3.1) implies

$$\| \overleftarrow{\prod_{1 \le k \le n}} (A(t_k^{(n)}) - I\delta^{-1})^{-1} \| \le M\delta^n.$$

That is, condition (1.5) holds with $\lambda = \delta^{-1}$ and $\beta = 0$. So Lemma 14.1.1 implies the existence of solutions.

Now we prove relation (3.2). Indeed, from (1.1) the equality

$$\delta^{-1}[u(t_j) - u(t_{j-1})] = A(t_j)u(t_j) + h_{n,j} \quad (t_j = t_j^{(n)}),$$

follows, where $h_{n,j} = u'(t_j) - \delta^{-1}[u(t_j) - u(t_{j-1})]$ $(j = 1, 2, ...)$. That is,

$$u(t_j) = (I - A(t_j)\delta)^{-1}[u(t_{j-1}) + \delta h_{n,j}] \; (j = 1, 2, ...).$$

By this relation, omitting the simple calculations, we get

$$u(t_n) = U_{n,0}u_0 + \sum_{k=0}^{n-1} U_{n,k}h_{n,k+1}\delta. \tag{3.3}$$

Due to Lemma 14.1.1, solutions of (1.1) have continuous derivatives, and therefore the expression

$$u'(s) - \delta^{-1}[u(s+\delta) - u(s)]$$

is uniformly bounded with respect to $\delta > 0$ and $s \le t < \infty$. Thus, $h_{n,k}$ are uniformly bounded with respect to $k \le n$ and $n = 2, ...$ Taking into account that $h_{n,k} \to 0$ as $\delta \to 0$, and employing the Lebesgue theorem on passing to the limit under the integral sign, we get

$$\sum_{k=0}^{n-1} \|h_{n,k+1}\|_X \delta \to 0, \; n \to \infty.$$

Now (3.1) and (3.3) yield the required relation (3.2). □

14.4 The Lozinskii and Wazewski inequalities for evolution equations

Again, let $A(t)$ be a linear operator in a Banach space X with a dense constant domain D. Let us suppose that for small enough $\delta > 0$, there is a continuous scalar-valued function $a(t)$ defined on $[0, \infty)$ such that

$$||(I - \delta A(t))^{-1}||_X < 1 + a(t)\delta \ (t \geq 0). \tag{4.1}$$

Theorem 14.4.1 *Suppose Assumptions 14.1A, 14.2B, and condition (4.1) hold. Then for any initial vector $u_0 \in D$, problem (1.1), (1.2) has a solution $u(t) \in C^1([0,\infty), X)$. Moreover, it subordinates the inequality*

$$||u(t)||_X \leq ||u(0)||_X exp\,[\int_0^t a(\tau)d\tau] \ (t \geq 0; \tag{4.2}$$

Proof: First, we check the solution existence. According to (4.1) and the definition of $U_{n,k}$ (see the previous section) we easily get

$$||U_{n,j}||_X \leq \prod_{1\leq k\leq n} (1 + a(t_k)\delta_k). \tag{4.3}$$

Now with the notation $m = max_{0\leq s\leq t}||a(s)||_B$ it can be written

$$||U_{n,j}||_X \leq e^{tm} \ (j = 1, ..., n).$$

Thus, due to Theorem 14.3.1 the problem (1.1), (1.2) has the solution. Furthermore, obviously

$$\prod_{1\leq k\leq n} (1 + a(t_k)\delta) = exp[\sum_{k=1}^{n} a(t_k)\delta + 0(\delta)] \to exp\,[\int_0^t a(\tau)d\tau]$$

as $\delta \to 0$. Now the desired assertion follows from inequality (4.3) and Theorem 14.3.1. □

Corollary 14.4.2 *Under Assumptions 14.1A, 14.1B, let $A(t)$ satisfy the condition*

$$\overline{\lim}_{h\to+0} h^{-1}[||(I - hA(t))^{-1}||_X - 1] = a(t), \tag{4.4}$$

where $a(t)$ is a continuous function. Then for any initial vector $u_0 \in D$, the problem (1.1), (1.2) has a solution $u(t)$. Moreover, it subordinates inequality (4.2).

Indeed, from (4.4) it follows the relation

$$||(I_X - \delta A(t))^{-1}||_X \leq 1 + a(t)\delta + o(\delta) \ (\delta \downarrow 0).$$

Now the result is due to the previous theorem.

Corollary 14.4.3 *Let $X = H$ be a Hilbert space with a scalar product $(.,.)_H$. In addition, let the conditions*

$$Re(A^*(t)g, g)_H < a(t)\|g\|_H^2,$$

and

$$Re(A(t)h, h)_H \leq a(t)\|h\|_H^2$$

$$(g \in D(A^*(t)),\ h \in D;\ t \geq 0)$$

be fulfilled. Then for any initial vector $u_0 \in D$, problem (1.1), (1.2) has a solution $u(t)$. Moreover, it subordinates inequality (4.2).

Indeed, we have

$$\|(I_H - \delta A(t))h\|_H^2 =$$

$$1 - 2Re(A(t)h, h)_H\delta + (A(t)h, A(t)h)_H\delta^2 \geq$$

$$1 - 2a(t)\delta \ (h \in D, \|h\|_H = 1).$$

Indeed, the operator $I_H - \delta A(t)$ is clearly invertible. Omitting simple calculations, we get inequality (4.1). The result is due to Theorem 14.4.1.

Corollary 14.4.3 extends the Wazewski inequality to equation (1.1). Furthermore, if $A(t)$ is bounded, then for small enough $\delta > 0$, and for all $t \geq 0$, one can write

$$(I_X - \delta A(t))^{-1} = \sum_{k=0}^{\infty} \delta^k A^k(t).$$

So

$$(I_X - \delta A(t))^{-1} = I_X + \delta A(t) + o(\delta) \ (\delta \downarrow 0).$$

Hence,

$$\overline{\lim}_{h \to +0} \frac{1}{h}[\|(I_X - hA(t))^{-1}\|_X - 1] =$$

$$\overline{\lim}_{h \to +0} \frac{1}{h}[\|I_X + hA(t)\|_X - 1].$$

Therefore, Corollary 14.4.2 generalizes the Lozinskii inequality.

14.5 Lower estimates for solutions

Now we are going to establish lower solution estimates for equation (1.1).

Theorem 14.5.1 *For all $t \geq 0$, and small enough $\delta > 0$, let the operator $I + \delta A(t)$ be invertible and the relation*

$$\|(I + \delta A(t))^{-1}\|_X \leq 1 + b_-(t)\delta \ (t \geq 0;\ h \in X) \tag{5.1}$$

hold, where $b_-(t)$ is a scalar-valued continuous function. Then under Assumption 14.1B any solution $u(t)$ of (1.1) (if it exists) subordinates the estimate

$$\|u(t)\|_X \geq \|u(0)\|_X exp\,[-\int_0^t b_-(\tau)d\tau] \ \ (t \geq 0).$$

Proof: For a fixed t_0 and every $t \in [0, t_0]$, put in (1.1)

$$\tau = t_0 - t,\ A_1(\tau) = -A(t_0 - \tau),\ \text{and } y(\tau) = u(t_0 - \tau). \tag{5.2}$$

Then equation (1.1) takes the form

$$dy/d\tau = A_1(\tau)y(\tau) \ \ (0 \leq \tau \leq t_0). \tag{5.3}$$

Condition (5.1) gives

$$\|(I_X - \delta A_1(\tau))^{-1}\| \leq 1 + b_-(t_0 - \tau)\delta.$$

According to Theorem 14.4.1 the following estimate for a solution $y(\tau)$ of equation (5.3) holds:

$$\|y(\tau)\| \leq exp\,[\int_0^\tau b_-(t_0 - s)ds]\|y(0)\| =$$

$$exp\,[-\int_{t_0}^{t_0-\tau} b_-(z)dz]\|y(0)\| \ \ (0 \leq \tau \leq t_0).$$

So

$$\|y(t_0)\| \leq exp\,[\int_0^{t_0} b_-(z)dz]\|y(0)\|.$$

But according to (5.2), $y(0) = u(t_0)$, $y(t_0) = u(0)$. Now taking into account that t_0 is arbitrary, we arrive at the stated result. □

Furthermore, let for instance, $X = H$ be a Hilbert space with a scalar product $(.,.)_H$, and

$$A(t) = iS(t) + K(t) \ (t \geq 0),$$

where $S(t)$ is a selfadjoint operator with a constant domain D, and $K(t)$ is a continuous in t bounded operator. Then

$$Re(A(t)h, h)_H = Re(K(t)h, h)_H \geq -\|K(t)\|_H$$

$$(h \in D, \|h\|_H = 1),$$

and

$$\|(I_H + \delta A(t))\, h\|_H^2 =$$

$$1 + 2Re(A(t)h, h)_H\delta + (A(t)h, A(t)h)_H\delta^2 \geq$$

$$1 - 2\|K(t)\|_H\delta \ (h \in D, \|h\|_H = 1).$$

Thus

$$\|(I_H + \delta A(t))^{-1}\|_H \leq 1 + \|K(t)\|_H\delta + o(\delta) \ \ (\delta \downarrow 0).$$

Now thanks to Theorem 14.5.1,

$$\|u(t)\|_H \geq \|u(0)\|_H exp \,[-\int_0^t \|K(\tau)\|_H d\tau] \ (t \geq 0).$$

14.6 Equations with relatively bounded operators

Consider in X the initial problem

$$\dot{u}(t) = A_0 u(t) + B(t)u(t) \ (t \geq 0) \tag{6.1}$$

$$u(0) = u_0 \in D(A_0). \tag{6.2}$$

It is assumed that $-A_0$ is a sectorial operator, and the linear operator $B(t)$ has the following property: for some positive $\nu < 1$, $A_0^{-\nu}B(t)$ is a bounded Hölder continuous operator-valued function. Then due to Theorem 14.1.3 the problem (6.1), (6.2) has a solution.

Lemma 14.6.1 *Under the above assumptions, let $\alpha(A_0) < 0$ and the condition*

$$\xi \equiv \sup_{t\geq 0} \int_0^t \|e^{A_0(t-s)}B(s)\|ds < 1 \tag{6.3}$$

hold. Then equation (6.1) is X-stable. Moreover, the constant N_0 in (2.1) can be chosen as

$$N_0 = (1-\xi)^{-1}c_0 \ \ where \ c_0 = \sup_{t\geq 0}\|e^{A_0 t}\|.$$

Proof: Rewrite the equation in the form

$$u(t) = e^{A_0 t}u_0 + \int_0^t e^{A_0(t-s)}B(s)u(s)ds.$$

Hence,

$$\sup_{t\geq 0}\|u(t)\| \leq \|e^{A_0 t}u_0\| + \sup_{t\geq 0}\|u(t)\| \int_0^t \|e^{A_0(t-s)}B(s)\|ds \leq$$

$$c_0\|u_0\| + \xi \sup_{t\geq 0}\|u(t)\|.$$

Now condition (6.3) implies the required result. □

Now let $X = H = E^n$ be an orthogonal sum of the same Hilbert spaces $E_1 = ... = E_n = E$. Again let S be a boundedly invertible selfadjoint operator in H commuting with any constant matrix, and $Q(S) \equiv c_0 S^2 + c_1 S + c_2$ where c_k $(k = 0, 1, 2)$ are constant $n \times n$-matrices. In addition, $c_0 = c_0^*$ is negative definite. Consider in H the equation

$$\dot{u} = Q(S)u + B(t)u, \tag{6.4}$$

where the linear operator $B(t)$ has the following property: for some $\mu \in (0, 2)$ the function

$$q_\mu(t) = ||S^{-\mu} B(t)||_H$$

is Hölder continuous. To formulate the result put $Q(s) = s^2 c_0 + s c_1 + c_2$ for $s \in \mathbf{C}$, and assume that

$$\alpha(Q(s)) \equiv \max_{k=1,\dots,n} Re\lambda_k(Q(s)) \leq \alpha_0 < 0 \ (s \in \sigma(S)) \tag{6.5}$$

and

$$v \equiv \sup_{s \in \sigma(S)} g(Q(s)) < \infty. \tag{6.6}$$

For example, this condition holds in the scalar case, or under the restriction $c_1 = c_1^*$ (see Section 13.10). Recall that

$$\chi(Q) = \sup_{t \geq 0} e^{t\alpha_0} \sum_{k=0}^{n-1} \frac{t^k v^k}{(k!)^{3/2}} \text{ and } \eta(\mu, t) = \sum_{k=0}^{n-1} \frac{t^k v^k}{(k!)^{3/2}} \sup_{s \in \sigma(S)} |s|^\mu e^{t\alpha(Q(s))}.$$

Due to Lemma 13.10.2,

$$||e^{Q(S)t} B(t)||_H \leq ||e^{Q(S)t} S^\mu|| \, ||S^{-\mu} B(t)||_H \leq q_\mu \eta(\mu, t).$$

Now the previous lemma yields:

Lemma 14.6.2 *Let the conditions (6.5), (6.6) and*

$$\sup_{t \geq 0} \int_0^t q_\mu(s) \eta(\mu, t-s) ds < 1$$

hold. Then equation (6.4) is H-stable.

Example 14.6.3 *Let us consider in the space $H = L^2([0, \pi], \mathbf{C}^n)$ the problem*

$$\frac{\partial u}{\partial t} = \tilde{c}_0 u_{xx} + (\tilde{c}_1 + b_1(t, x)) u_x + (\tilde{c}_2 + b_2(t, x)) u \ \ (t \geq 0; \ 0 < x < \pi),$$

$$u(0, t) = u(\pi, t) = 0 \ (t > 0) \tag{6.7}$$

with variable in t and differentiable in x $n \times n$-matrices $b_j(t, x)$ $(j = 1, 2)$.

With the notation

$$Q_k = -\tilde{c}_0 k^2 + i\tilde{c}_1 k + \tilde{c}_0,$$

assume that the conditions

$$g(Q_k) \leq v < \infty,\ Re\lambda_l(Q_k) \leq \alpha_0 < 0\ (l = 1, ..., n;\ k = \pm 1, \pm 2, \ldots) \tag{6.8}$$

are fulfilled. Define $D(S)$ as in Section 13.12, and put $Sy = -iy'$ $(y \in D(S))$. In addition, take

$$B(t)u = b_1(t,x)\frac{\partial u}{\partial x} + b_2(t,x)u = \frac{\partial}{\partial x}(b_1(t,x)u) + (-b_{1x}(t,x) + b_2(t,x))u$$

$$(u \in D(S)).$$

Then

$$B(t) = iSb_{1x}(t,x) + (-b'_{1x}(t,x) + b_2(t,x)).$$

Hence, taking into account that $||S^{-1}||_H = 1$ (see Section 13.13), we get

$$||S^{-1}B(t)||_H \leq q_1(t) \equiv \sup_x ||b_1(t,x)||_{C^n} + \sup_x ||b'_{1x}(t,x) - b_2(t,x)||_{C^n}.$$

As shown in Section 13.13, in the considered case we have

$$\eta(1,t) = \sum_{k=0}^{n-1} \frac{t^k v^k}{(k!)^{3/2}} \zeta(t) \text{ where } \zeta(t) = \sup_{k=\pm 1, \pm 2, \ldots} |k| e^{t\alpha(Q_k)}.$$

Now the previous lemma implies:

Proposition 14.6.4 *Let the conditions (6.8) and*

$$\sup_{t \geq 0} \int_0^t q_1(s)\eta(1, t-s)ds < 1 \tag{6.9}$$

hold. Then problem (6.7) is H-stable.

In particular, if (6.7) is a real scalar equation, then

$$q_1(t) = \sup_x |b_1(t,x)| + \sup_x |b'_1(t,x) - b_2(t,x)|,$$

$$\eta(1,t) = \max_{k=1,2,\ldots} k\ exp[t(-\tilde{c}_0 k^2 + \tilde{c}_0)].$$

This maximum is simply calculated. Condition (6.9) provides the result.

14.7 The freezing method for evolution equations

14.7.1 Statements of the results

In the present section we extend the freezing method for ordinary differential equations (see Section 3.2) to a class of linear evolution equations.

Let X be a Banach space, and Y be a dense imbedded subspace of X with norms $\|.\|_X$ and $\|.\|_Y$, respectively. Consider in X the equation

$$\dot{u} = A(t)u \ (t \geq 0, \ \dot{u} = du/dt, \ u = u(t)), \tag{7.1}$$

where $-A(\tau)$ for each fixed $\tau \geq 0$ is a sectorial operator in X with a constant dense domain $D(A(\tau)) \equiv D \subset Y$. Moreover,

$$\|(A(t) - A(s))h\|_X \leq \tilde{q}|t - s|\|h\|_Y \quad (s, t \geq 0; \ h \in D) \tag{7.2}$$

with a constant $\tilde{q}$ independent of t, s, h. In addition, the analytic semigroup $e^{A(\tau)t}$ generated by operator $A(\tau)$ has the properties

$$\chi \equiv \sup_{t \geq 0} \sup_{\tau \geq 0} \|e^{A(\tau)t}\|_Y < \infty, \tag{7.3}$$

and

$$\theta \equiv \tilde{q} \int_0^\infty t \sup_{\tau \geq 0} \|e^{A(\tau)t}\|_{X \to Y} dt < 1. \tag{7.4}$$

Theorem 14.7.1 *Let conditions (7.2)-(7.4) hold. Then equation (7.1) is Y-stable. Moreover the constant N_0 in (2.1) can be taken as $N_0 = \chi(1 - \theta)^{-1}$.*

All the proofs are presented in the next section.

Now we are going to specialize Theorem 14.7.1 in the case of a Hilbert space. Let ω be a set with a positive measure ρ, and $H \equiv L^2_\rho(\omega, \mathbf{C}^n)$ be the space of vector-valued functions defined on ω and endowed with the norm

$$\|h\|_H \equiv \|h\|_{L^2_\rho(\omega, \mathbf{C}^n)} = [\int_\omega \|h(x)\|^2_{C^n} d\rho(x)]^{1/2}.$$

Clearly, $L^2_\rho(\omega, \mathbf{C}^n)$ is the orthogonal sum of n exemplars of space $E = L^2_\rho(\omega)$ of scalar-valued functions defined on ω.

As the previous theorem shows, the extension of the freezing method to evolution equations is based on norm estimates for relevant semigroups. But obtaining of these estimates is usually not easy. Because of this we restrict ourselves by equations in a Hilbert space with operators having the form of pencils of selfadjoint operators, besides, coefficients of the pencils are variable matrices.

Throughout the present section S is an invertible selfadjoint operator in space $L^2_\rho(\omega, \mathbf{C}^n)$ commuting with all matrices independent of x.

Consider in $L^2_\rho(\omega, \mathbf{C}^n)$ the equation

$$\frac{\partial}{\partial t}u(t,x) = (-S^2 + b_1(t,x)S + b_2(t,x))u(t,x) \ (x \in \omega, \ t \geq 0), \tag{7.5}$$

where $b_k(t,x)$ $(k = 1,2)$ are measurable $n \times n$-matrix-valued functions bounded for all $x \in \omega$ and finite $t \geq 0$.

Our main assumptions in this section are as follows:

1) The coefficients are presented in the form

$$b_k(t,x) = c_k(t) + a_k(t,x) \ (k = 1,2), \tag{7.6}$$

where matrix-valued functions $c_k(t)$ are independent of $x \in \omega$, and matrix-valued functions $a_k(t,x)$ $(k = 1,2)$ are sufficiently small in the sense pointed out below.

2) There are constants q_k such that

$$||c_k(t) - c_k(\tau)||_{C^n} \leq q_k|t - \tau| \ (t, \tau \geq 0; \ k = 1,2). \tag{7.7}$$

Note that if $b_k(t,x)$ are real, one can take

$$c_k(t) = 1/2(\sup_{x \in \omega} b_k(t,x) + \inf_{x \in \omega} b_k(t,x)) \ \text{ and } a_k(t,x) = b_k(x,t) - c_k(t).$$

Furthermore, for the matrix pencil

$$Q(\tau, s) = -s^2 I_{C^n} + c_1(\tau)s + c_2(\tau) \ \ (\tau \geq 0; \ s \in \sigma(S))$$

put

$$\alpha(Q(\tau,s)) \equiv \max_{k=1,\dots,n} Re\lambda_k(Q(\tau,s)) = -s^2 + \max_{k=1,\dots,n} Re\lambda_k(c_1(\tau)s + c_2(\tau)).$$

It is assumed that all the eigenvalues of $Q(s)$ lie in the left half-plane:

$$\alpha_0 \equiv \sup_{\tau \geq 0, s \in \sigma(S)} \alpha(Q(\tau,s)) < 0.$$

In addition,

$$v(\tau) \equiv \sup_{s \in \sigma(S)} g(c_1(\tau)s + c_2(\tau)) \leq v_0 < \infty \ (\tau \geq 0). \tag{7.8}$$

Recall that the quantity $g(.)$ is defined in Section 2.1. For instance, due to Proposition 2.1.1, if

$$c_1(\tau) = c_1^*(\tau) \text{ then } v(\tau) \leq \sqrt{1/2}N(c_2(\tau) - c_2^*(\tau)). \tag{7.9}$$

If $c_1(\tau) = -c_1^*(\tau)$, then Proposition 2.1.1 gives

$$v(\tau) \leq \sqrt{1/2}N(c_2(\tau) + c_2^*(\tau)).$$

Let us introduce the following notation:

$$\eta(\tau, t) \equiv \sup_{s \in \sigma(S)} |s| e^{t\alpha(Q(\tau, s))} \sum_{k=0}^{n-1} \frac{t^k v^k(\tau)}{(k!)^{3/2}}.$$

$$\gamma(\tau) = \sup_{x \in \omega} ||a_1(\tau, x)||_{C^n} + \sup_{x \in \omega} ||a_2(\tau, x)||_{C^n} ||S^{-1}||_H,$$

and

$$I_1(\tau) \equiv \int_0^\infty \eta(\tau, s) ds.$$

It is simple to show that $\eta(\tau, .) \in L^1(0, \infty)$ for any fixed $\tau \geq 0$. Recall that H^1 is the subspace of H with the norm

$$||w||_{H^1} = ||Sw||_H \ (w \in D(S)).$$

We shall prove the following result.

Theorem 14.7.2 *Let conditions (7.6)- (7.8) be fulfilled. In addition, with the notation $q = q_1 + q_2||S^{-1}||_H$, let the inequalities*

$$\gamma(\tau) I_1(\tau) < 1, \tag{7.10}$$

and

$$q \int_0^\infty t\eta(\tau, t) dt < (1 - \gamma(\tau) I_1(\tau))^2 \tag{7.11}$$

hold for all $\tau \geq 0$. Then equation (7.5) is H^1- stable.

Consider the case

$$b_1(t, x) \equiv 0. \tag{7.12}$$

Then $v(\tau) = g(c_2(\tau))$, $q = q_2||S^{-1}||_H$, and

$$\gamma(\tau) = \sup_{x \in \omega} ||a_2(\tau, x)||_{C^n} ||S^{-1}||_H.$$

Omitting simple calculations, we get $\max_{s \in \sigma(S)} e^{-s^2 t}|s| = \zeta(t)$, where

$$\zeta(t) = \begin{cases} \exp[t\alpha(-S^2)]\sqrt{|\alpha(S^2)|} & \text{if } t \geq \frac{1}{2|\alpha(-S^2)|}, \\ \frac{exp[-1/2]}{\sqrt{2t}} & \text{if } t \leq \frac{1}{2|\alpha(-S^2)|}. \end{cases}$$

Consequently, under (7.12)

$$\eta(\tau, t) = \zeta(t) e^{\alpha(c_2(\tau))t} \sum_{k=0}^{n-1} \frac{t^k v^k(\tau)}{(k!)^{3/2}}.$$

The integrals in (7.1), (7.11) are now simply calculated.

14.7.2 Example

Consider in $H = L^2([0,\pi], \mathbf{C}^n)$ the system

$$\frac{\partial u}{\partial t} = \frac{\partial^2 u}{\partial x^2} + \tilde{b}_1(t,x)\frac{\partial u}{\partial x} + \tilde{b}_2(t,x)u \quad (t \geq 0;\ 0 < x < \pi;\ u = u(t,x)) \tag{7.13}$$

with bounded for all finite t real $n \times n$-matrices $\tilde{b}_1(t,x), \tilde{b}_2(t,x)$, along with the periodic boundary conditions

$$u(0,t) = u(\pi,t),\ u'_x(0,t) = u'_x(\pi,t)\ (t \geq 0). \tag{7.14}$$

In addition, it is assumed that

$$\tilde{b}_1(x,t) = \tilde{b}_1^*(x,t)\ (x \in [0,\pi],\ t \geq 0). \tag{7.15}$$

Take

$$D(S) = \{w \in L^2 = L^2([0,\pi],\mathbf{C}^n),\ w' \in L^2, w(0) = w(\pi)\}.$$

Define S by the formula $Sy = -iy' - y$, for $y \in D(S)$. Then (7.13) takes the form

$$\frac{\partial u}{\partial t} = [-(S+I)^2 + i\tilde{b}_1(t,x)(S+I) + \tilde{b}_2(t,x)]u = [-S^2 + b_1(t,x)S + b_2(t,x)]u$$

with $b_1(t,x) = -2I + i\tilde{b}_1(t,x)$ and $b_2(t,x) = -I + i\tilde{b}_1(t,x) + \tilde{b}_2(t,x)$. That is, problem (7.13), (7.14) can be written in the form (7.5). It is clear that the eigenvalues of S are

$$\lambda_k(S) = 2k - 1\ (k = 0, \pm 1, \pm 2, \ldots).$$

Hence, $\alpha(-S^2) = -1$ and $\|S^{-1}\|_H = 1$. Take

$$\tilde{c}_1(t) = \pi^{-1}\int_0^\pi \tilde{b}_1(x,t)dx,\ c_1(t) = -2I + i\tilde{c}_1(t),$$

and

$$c_2(t) = \pi^{-1}\int_0^\pi b_2(x,t)dx, \text{ and } a_k(x,t) = b_k(x,t) - c_k(t)\ (k = 1,2).$$

We have by Proposition 2.1.1,

$$\begin{aligned} g(c_1(t)s + c_2(t)) &= g(-2Is + is\tilde{c}_1(t) + c_2(t)) = \\ g(is\tilde{c}_1(t) + c_2(t)) &= g(s\tilde{c}_1(t) - ic_2(t)) \leq \\ &\sqrt{1/2}N(s\tilde{c}_1(t) - ic_2(t) - s\tilde{c}_1^*(t) - ic_2^*(t))). \end{aligned}$$

According to (7.15) this implies

$$g(c_1(t)s + c_2(t)) \leq \sqrt{1/2}N(-ic_2(t) - ic_2^*(t))) = \sqrt{1/2}N(c_2(t) + c_2^*(t)).$$

So $v(\tau) \leq v_1(\tau) \equiv \sqrt{1/2}N(c_2(\tau) + c_2^*(\tau))$. We thus get

$$\eta(\tau,t) \leq \eta_1(\tau,t) \equiv \max_{k=0,\pm1,\pm2,\ldots} |2k-1|e^{t\alpha(Q(\tau,2k-1))}\sum_{j=0}^{n-1}\frac{t^j v^j(\tau)}{(j!)^{3/2}}.$$

Now Theorem 14.7.2 can be immediately applied.

14.8 Proofs of Theorems 14.7.1 and 14.7.2

Proof of Theorem 14.7.1: The solution existence is due to Theorem 14.1.3. Further, rewrite equation (7.1) in the form

$$du/dt - A(\tau)u = [A(t) - A(\tau)]u \tag{8.1}$$

with a fixed $\tau \geq 0$. This equation is equivalent to the following one:

$$u(t) = e^{tA(\tau)}u_0 + \int_0^t e^{(t-s)A(\tau)}[(A(s) - A(\tau))]u(s)\, ds. \tag{8.2}$$

Hence,

$$||u(t)||_Y \leq \chi||u_0||_Y + \int_0^t ||e^{(t-s)A(\tau)}[A(\tau) - A(s)]u(s)||_Y ds \leq$$

$$\chi||u_0||_Y + \int_0^t ||e^{(t-s)A(\tau)}||_{X\to Y}||[A(\tau) - A(s)]u(s)||_X ds.$$

Due to (7.2), taking $t = \tau$, we get

$$||u(t)||_Y \leq \chi||u_0||_Y + \int_0^t \tilde{\eta}(t-s)\tilde{q}(t-s)||u(s)||_Y ds.$$

where

$$\tilde{\eta}(t) = \sup_{\tau\geq 0} ||e^{tA(\tau)}||_{X\to Y}.$$

Consequently, for a fixed t_0,

$$\sup_{t\leq t_0} ||u(t)||_Y \leq \chi||u_0||_Y + \tilde{q}\sup_{t\leq t_0} ||u(t)||_Y \int_0^{t_0} \tilde{\eta}(t_0 - s)(t_0 - s)ds \leq$$

$$\chi||u_0||_Y + \tilde{q}\sup_{t\leq t_0} ||u(t)||_Y \int_0^{\infty} \tilde{\eta}(t_1)t_1 dt_1 \leq \chi||u_0||_Y + \sup_{t\leq t_0} ||u(t)||_Y \tilde{q}\theta.$$

According to (7.4) the latter inequality implies

$$\sup_{t\leq t_0} ||u(t)||_Y \leq \chi||u_0||_Y (1 - \tilde{q}\theta)^{-1}.$$

Since t_0 is arbitrary, the later inequality proves the required result. □

Now we will prove Theorem 14.7.2. In the space $H = L^2_\rho(\omega, \mathbf{C}^n)$ for a positive number μ, define on the set $D(S^\mu)$ the graph norm by the equality $||h||_{H^\mu} = ||S^\mu h||_H$ $(h \in D(S^\mu))$. Denote the obtained Hilbert space by H^μ. Consider in $L^2_\rho(\omega, \mathbf{C}^n)$, the operator A defined by

$$(Ah)(x) = (-S^2 + b_1(x)S + b_2(x))h(x) \ (h \in D(S^2),\ x \in \omega), \tag{8.3}$$

where $b_k(x)$ are $n \times n$-matrix-valued functions defined and bounded on ω. Let us assume that

$$b_k(x) = c_k + a_k(x) \ (k = 1, 2), \tag{8.4}$$

where c_k are constant $n \times n$-matrices with the property

$$v \equiv \sup_{s \in \sigma(S)} g(c_1 s + c_2) < \infty. \tag{8.5}$$

Furthermore, for the matrix pencil

$$Q(s) = -s^2 I_{C^n} + sc_1 + c_2 \ (s \in \mathbf{C}),$$

assume that

$$\alpha(Q(s)) \equiv \max_{k=1,\dots,n} Re\lambda_k(Q(s)) < 0 \ (s \in \sigma(S)).$$

Due to Lemma 13.10.1, under condition (8.5), the operator $Q(S)$ defined by the formula

$$(Q(S)h)(x) = (-S^2 + c_1 S + c_2)h(x) \ (h \in D(S^2);\ x \in \omega),$$

generates an analytic semigroup $e^{Q(S)t}$ in H, which subordinates the inequalities

$$\|e^{Q(S)t}\|_{H^\mu} \le \chi(Q) \ (t \ge 0),\ \|S^\mu e^{Q(S)t}\|_H \le \eta_\mu(t) \ (0 \le \mu < 2;\ t > 0),$$

where

$$\eta_\mu(t) \equiv \sup_{s \in \sigma(S)} |s|^\mu e^{t\alpha(Q(s))} \sum_{k=0}^{n-1} \frac{t^k v^k}{(k!)^{3/2}},$$

and $\chi(Q) \equiv \sup_{t \ge 0} \eta_0(t)$. Moreover, under (8.4) put

$$m_k = \sup_{x \in \omega} \|a_k(x)\|_{C^n} \ (k = 1, 2), \text{ and } \gamma_A = m_1 + m_2 \|S^{-1}\|_H.$$

Lemma 14.8.1 *Let the inequality*

$$I_\eta \equiv \int_0^\infty \eta_1(s) ds < \gamma_A^{-1} \tag{8.6}$$

be fulfilled. Then the operator A defined by (8.3) generates in H an analytic semigroup e^{At} which subordinates the inequalities

$$\|e^{At}\|_{H^1} \le \chi(Q)(1 - \gamma_A I_\eta)^{-1} \ (t \ge 0) \tag{8.7}$$

and

$$\int_0^\infty t \|Se^{At}\|_H dt \le J_\eta (1 - \gamma_A I_\eta)^{-2}, \tag{8.8}$$

where

$$J_\eta = \int_0^\infty t\eta_1(t) dt.$$

Proof: The operator $(A-S^2)S^{-1}$ is clearly bounded. Since S^2 is selfadjoint, operator $-A$ is sectorial due to Corollary 13.8.4. So it generates in H an analytic semigroup. Consider in H the equation

$$u = Au = Q(S)u + Cu, \tag{8.9}$$

where $C = A - Q(S)$. Omitting simple calculations, we arrive at the inequality

$$||Cu||_H \le m_1||Su||_H + m_2||u||_H \le \gamma_A||Su||_H. \tag{8.10}$$

Furthermore, equation (8.9) is equivalent to the following one:

$$u(t) = e^{Q(S)t}u_0 + \int_0^t e^{Q(S)(t-s)}Cu(s)ds \; (u(0) = u_0).$$

Hence,

$$||Su(t)||_H \le ||e^{Q(S)t}Su_0||_H + \int_0^t ||Se^{Q(S)(t-s)}||_H||Cu(s)||_H ds. \tag{8.11}$$

Consequently, with the notation $||u(t)||_{H^1} = ||u(t)||_1$ we have

$$||u(t)||_1 \le \chi(Q)||u_0||_1 + \int_0^t \eta_1(t-s)||Cu(s)||_H ds$$

$$\le \chi(Q)||u_0||_1 + \gamma_A \int_0^t \eta_1(t-s)||u(s)||_1 ds.$$

Thus,

$$\sup_{t\ge 0}||u(t)||_1 \le \chi(Q)||u_0||_1 + \gamma_A \sup_{t\ge 0}||u(t)||_1 \int_0^\infty \eta_1(s)ds.$$

Condition (8.6) ensures the inequality

$$\sup_{t\ge 0}||u(t)||_1 \le \chi(Q)||u_0||_1(1-\gamma_A I_\eta)^{-1}.$$

That inequality is equivalent to (8.7). Furthermore, due to (8.10) and (8.11)

$$||u(t)||_1 \le \eta_1(t)||u_0||_H + \int_0^t \eta_1(t-s)||Cu(s)||_H ds \le$$

$$\eta_1(t)||u_0||_H + \gamma_A \int_0^t \eta_1(t-s)||u(s)||_1 ds.$$

Put for a fixed $T > 0$

$$\psi(t) = \frac{||u(t)||_1}{||u_0||_H}, \; J_T = \int_0^T t\psi(t)dt \text{ and } I_T = \int_0^T \psi(\tau)d\tau.$$

Then

$$\psi(t) \leq \eta_1(t) + \gamma_A \int_0^t \eta_1(t-s)\psi(s)ds. \tag{8.12}$$

Hence,

$$J_T \leq \int_0^T t\eta_1(t)dt + \gamma_A \int_0^T t \int_0^t \eta_1(t-s)\psi(s)dsdt. \tag{8.13}$$

But

$$\int_0^T t \int_0^t \eta_1(t-s)\psi(s)dsdt = \int_0^T \int_s^T t\eta_1(t-s)\psi(s)dtds \leq$$

$$\int_0^T \int_s^\infty t\eta_1(t-s)\psi(s)dtds =$$

$$\int_0^T \psi(s) \int_0^\infty (\tau + s)\eta_1(\tau)d\tau ds = \int_0^T \psi(s)ds \int_0^\infty \tau\eta_1(\tau)d\tau +$$

$$\int_0^T \psi(s)sds \int_0^\infty \eta_1(\tau)d\tau = I_T J_\eta + J_T I_\eta.$$

Therefore (8.13) yields

$$J_T \leq J_\eta + \gamma_A(I_T J_\eta + J_T I_\eta).$$

Under (8.6) this gives

$$J_T \leq J_\eta(1 + \gamma_A I_T)(1 - \gamma_A I_\eta)^{-1}. \tag{8.14}$$

But according to (8.12)

$$I_T \leq I_\eta + \gamma_A \int_0^T \int_0^t \eta_1(t-s)\psi(s)dsdt = I_\eta + \gamma_A \int_0^T \int_s^T \eta_1(t-s)\psi(s)dtds \leq$$

$$I_\eta + \gamma_A I_T \, I_\eta.$$

That is, under (8.6)

$$I_T \leq I_\eta(1 - \gamma_A I_\eta)^{-1}.$$

Now (8.14) implies

$$J_T \leq J_\eta(1 + \gamma_A I_\eta(1 - \gamma_A I_\eta)^{-1})(1 - \gamma_A I_\eta)^{-1} = J_\eta(1 - \gamma_A I_\eta)^{-2}.$$

Passing to the limit as $T \to \infty$, we get to the required estimate (8.8). □

Proof of Theorem 14.7.2: Let $A(\tau)$ be an operator defined in H by the formula

$$A(\tau)h(x) = (-S^2 + b_1(\tau, x)S + b_2(\tau, x))h(x) \quad (h \in D(S^2),\ x \in \omega;\ \tau \geq 0).$$

Due to (7.6)

$$\|(A(t) - A(\tau))h\|_H \leq q_1|t-\tau|\|Sh\|_H + q_2|t-\tau|\|h\|_H.$$

$$\leq |t-\tau|(q_1 + q_2\|S^{-1}\|_H)\|Sh\|_H = q|t-\tau|\|Sh\|_H.$$

Thus, condition (7.2) holds with $\tilde{q} = q$, $X = H$ and $Y = H^1$.

Put

$$\eta_\mu(\tau, t) \equiv \sup_{s\in\sigma(S)} |s|^\mu e^{t\alpha(Q(\tau,s))} \sum_{k=0}^{n-1} \frac{t^k v^k(\tau)}{(k!)^{3/2}} \ (0 \leq \mu < 2).$$

That is, $\eta_1(\tau,t) = \eta(\tau,t)$.

By the previous lemma under conditions (7.6)-(7.8), operator $A(\tau)$ with a fixed $\tau \geq 0$ generates in H a strongly continuous semigroup $e^{A(\tau)t}$ which subordinates the inequalities

$$\|e^{A(\tau)t}\|_{H^1} \leq \chi_\tau(1 - \gamma(\tau)I_1(\tau))^{-1}$$

and

$$\int_0^\infty t\|Se^{A(\tau)t}\|_H dt \leq J_1(\tau)(1-\gamma(\tau)I_1(\tau))^{-2} < q^{-1} \ (\tau \geq 0),$$

where $\chi_\tau = max_{t\geq 0}\eta_0(\tau,t)$, and

$$J_1(\tau) = \int_0^\infty t\eta_1(\tau,t)dt.$$

Thanks to condition (7.8),

$$\sup_{\tau\geq 0}\sup_{t\geq 0} e^{t\alpha_0} \sum_{k=0}^{n-1} \frac{t^k v^k(\tau)}{(k!)^{3/2}} < \infty.$$

Consequently,

$$\sup_{\tau,t\geq 0} \|e^{A(\tau)t}\|_{H^1} < \infty \text{ and } \int_0^\infty t\|e^{A(\tau)t}\|_{H\to H^1} dt < q^{-1} \ (\tau \geq 0).$$

Now the result follows from Theorem 14.7.1. □

14.9 Systems of linear equations with majorants

14.9.1 Solution estimates

Let $X = E_1 + ... + E_n$ be a direct sum of Banach spaces E_k ($k = 1, ..., n < \infty$) with norms $\|.\|_{E_k}$. For an element $h = (h_k \in E_k)_{k=1}^n$ of X, the norm is defined by $\|h\|_X = [\|h_1\|_{E_1}^2 + ... + \|h_n\|_{E_n}^2]^{1/2}$.

Furthermore, let $A_{jk}(t)$ be linear operators acting from E_k into E_j $(j,k = 1,...,n)$. Consider the system

$$\dot{u}_j = \sum_{k=1}^{n} A_{jk}(t)u_k \quad (t \geq 0). \tag{9.1}$$

That system can be rewritten in space X as equation (1.1) with the operator matrix $A(t) = (A_{jk}(t))_{j,k=1}^n$.

It is assumed that the operators $A_{kk}(t)$ have dense constant domains $D_k \subset E_k$ and are continuously differentiable on D_k. For $j \neq k$ operators $A_{jk}(t)$ are assumed bounded and continuously differentiable. Moreover,

$$\|A_{jk}(t)\|_{E_k \to E_j} < a_{jk}(t) \ (t \geq 0). \tag{9.2}$$

In addition, for all small enough $\delta > 0$ the operators $I_{E_k} - \delta A_{kk}(t)$ are invertible, and the inequalities

$$\|(I_{E_k} - \delta A_{kk}(t))h_k\|_{E_k} \geq (1 - a_{kk}(t)\delta)\|h_k\|_{E_k} \ (h_k \in D_k) \tag{9.3}$$

hold, where $a_{jk}(t)$ $(j,k = 1,...,n)$ are continuous scalar-valued functions. Clearly, the domain D of $A(t)$ is the direct sum of D_j $(j = 1,...,n)$.

Theorem 14.9.1 *Let conditions (9.2) and (9.3) be fulfilled. Then for any initial vector $u_0 \in D$, system (9.1) has a solution $u(t) = (u_j(t)) : [0,\infty) \to D$. Moreover, it subordinates the inequalities*

$$\|u_j(t)\|_{E_j} \leq z_j(t) \ (t \geq 0), \tag{9.4}$$

where $z_j(t)$ are the solutions of the ordinary differential system

$$\dot{z}_j(t) = \sum_{k=1}^{n} a_{jk}(t)z_k(t) \ (j = 1,...,n, \ t \geq 0) \tag{9.5}$$

with the initial conditions $z_j(0) = \|u_j(0)\|_E$, $(j = 1,...,n)$.

The proof of that theorem is given in Subsection 14.9.4 below.

Now we are going to specialize assumption (9.3) in the case of Hilbert spaces. Namely, let $X = H \equiv E_1 \oplus E_2 \oplus ... \oplus E_n$ be an orthogonal sum of Hilbert spaces E_k.

Corollary 14.9.2 *Let conditions (9.2) and*

$$Re(A_{kk}(t)h,h)_{E_k} < a_{kk}(t)\|h\|_{E_k}^2; \ Re(A_{kk}^*(t)g,g)_{E_k} \leq a_{kk}(t)\|g\|_{E_k}^2$$
$$(h \in D_k, g \in D(A_{kk}^*(t)); \ k = 1,...,n) \tag{9.6}$$

be fulfilled. Then for any initial vector $u_0 \in D$, system (9.1) has a solution $u(t)$. Moreover, it subordinates inequalities (9.4).

Indeed, considering that

$$\|(I_{E_k} - \delta A_{kk}(t)h\|_{E_k}^2 = 1 - 2\Re(A_{kk}(t)h,h)_{E_k}\delta + (A_{kk}(t)h, A_{kk}(t)h)_{E_k}\delta^2 \geq$$
$$1 - 2a_{kk}(t)\delta \ (h \in D_k, \|h\|_{E_k} = 1),$$

we easily get inequality (9.3). Now the result is due to Theorem 14.9.1

14.9.2 Stability conditions

The following result immediately follows from the latter theorem

Corollary 14.9.3 *Let conditions (9.2) and (9.3) be fulfilled with*

$$a_{jk}(t) \equiv a_{jk} = const \ (j, k = 1, ..., n),$$

and let the constant matrix (a_{jk}) be a Hurwitz one. Then system (9.1) is exponentially X-stable.

Furthermore, consider system (9.5) assuming that matrix $a(t) = (a_{jk}(t))$ satisfies the conditions

$$v = \sup_{t \geq 0} g(a(t)) < \infty, \tag{9.7}$$

and

$$\|a(t) - a(s)\|_{C^n} \leq q_0 |t - s| \text{ for } t, s \geq 0, \tag{9.8}$$

where q_0 is a constant. Denote by $z(q_0, v)$ the extreme right-hand root of the algebraic equation

$$z^{n+1} = q_0 \sum_{k=0}^{n-1} \frac{(k+1)\, v^k}{\sqrt{k!}} z^{n-k-1}. \tag{9.9}$$

Due to Theorem 3.2.2 above, under conditions (9.7), (9.8) equation (9.5) is stable provided the matrix $a(t) + z(q_0, v) I_{C^n}$ is a Hurwitz one for all $t \geq 0$. Combining this result with Theorem 14.9.1, we obtain

Theorem 14.9.4 *Let conditions (9.2) and (9.3) be fulfilled. Let matrix $a(t) = (a_{jk}(t))$ satisfy conditions (9.7) and (9.8). Then system (9.1) is X-stable provided that $a(t) + z(q_0, v) I_{C^n}$ is a Hurwitz matrix for all $t \geq 0$.*

Put

$$w_n = \sum_{k=0}^{n-1} \frac{k+1}{\sqrt{k!}}. \tag{9.10}$$

Combining Corollary 3.2.3 with Theorem 14.9.1, we get

Corollary 14.9.5 *Let conditions (9.2) and (9.3) be fulfilled. Let matrix $a(t) = (a_{jk}(t))$ satisfy conditions (9.7) and (9.8). In addition, let $q_0 w_n \leq v$. Then system (9.1) is X-stable provided that*

$$a(t) + v^{1-1/n} [q_0 w_n]^{1/n} I_{C^n}$$

is a Hurwitz matrix for all $t \geq 0$.

Remark 14.9.6 *Due to Corollary 14.9.2, in the case of a Hilbert space condition (9.3) in Theorem 14.9.4 and Corollary 14.9.5 can be replaced by condition (9.6).*

14.9.3 Example

Consider the problem

$$\frac{\partial u_j}{\partial t} = \frac{\partial}{\partial x}\phi_j(x,t)\frac{\partial u_j}{\partial x} + \sum_{k=1}^{n}\psi_{jk}(x,t)u_k,\ (j=1,...,n;\ 0<x<1)$$

$$u(t,0) = u(t,1) = 0\ (t \geq 0). \tag{9.11}$$

Here $u = u(t,x) = (u_k(t,x)) \in \mathbf{C}^n$; ϕ_j and ψ_{jk} $(j,k=1,...,n)$ are continuously differentiable in t real scalar-valued functions defined on $[0,1]\times[0,\infty)$.

Take $X = L^2([0,1],\mathbf{C}^n)$. Clearly, $L^2([0,1],\mathbf{C}^n)$ is an orthogonal sum of the same spaces $E_k \equiv E = L^2[0,1]$ $(k=1,...,n)$ of scalar-valued functions. Furthermore, put

$$D_1 = \{v \in L^2[0,1] : v'' \in L^2[0,1],\ v(0) = v(1) = 0\},$$

and $D = (D_1)^n$. According to (9.1) define operators $A_{jk}(t)$ by the formulae

$$(A_{jk}(t)v_k)(x) = \psi_{jk}(x,t)v_k(x)\ (k \neq j),$$

and

$$(A_{jj}(t)v_j)(x) = \frac{\partial}{\partial x}\phi_j(x,t)\frac{\partial v_j}{\partial x} + \psi_{jj}(x,t)v_j\ (v_j \in D_1;\ j,k=1,...,n).$$

Assume that

$$a_{jk}(t) \equiv \sup_{x\in[0,1]} |\psi_{jk}(x,t)| < \infty\ (k \neq j,\ t \geq 0), \tag{9.12}$$

and

$$b_j(t) \equiv \sup_{x\in[0,1]} \psi_{jj}(x,t) < \infty, \text{ and } m_j(t) \equiv \inf_{x\in[0,1]} \phi_j(x,t) > 0\ (t \geq 0). \tag{9.13}$$

Hence,

$$\|A_{jk}(t)\|_E \leq a_{jk}(t)\ (j \neq k, t \geq 0).$$

Further, integration by parts and the condition $v_j(0) = v_j(1) = 0$ give

$$(\frac{d}{dx}\phi_j(.,t)v_j', v_j)_E = -(\phi_j(.,t)v_j', v_j')_E \leq -m_j(t)(v_j', v_j')_E \leq$$

$$-m_j(t)\pi^2\|v_j\|_E^2,$$

since $\beta(S) = \pi^2$ is the smallest eigenvalue of the operator S defined on D_1 by the formula

$$(Sv_j)(x) = -v_j''(x)\ (v_j \in D_1,\ j=1,...,n).$$

Moreover, relations (9.12) and (9.13) imply

$$(A_{jj}(t)v_j, v_j)_E \leq a_{jj}(t)(v_j, v_j)_E$$

with

$$a_{jj}(t) = -m_j(t)\pi^2 + b_j(t)\ (j=1,...,n). \tag{9.14}$$

Now according to Reamark 14.9.6, Theorem 14.9.4 implies

Proposition 14.9.7 *Let conditions (9.12), (9.13) be fulfilled. In addition, let the matrix $a(t) = (a_{jk}(t))$ defined by (9.12), (9.14) satisfy conditions (9.7) and (9.8). Then system (9.11) is X-stable provided that $a(t) + z(q_0, v)I_{C^n}$ is a Hurwitz matrix for all $t \geq 0$.*

14.9.4 Proof of Theorem 14.9.1

Let us define in X a vector-valued function $M(.)$ by the formula

$$M(h) = (\|h_k\|_{E_k})_{k=1}^n.$$

That is, $M(h)$ is the vector whose coordinates are $\|h_k\|_{E_k}$ $(k = 1, ..., n)$. We have

$$M_j((I_X - \delta A(t))h) = \|(I_{E_j} - A_{jj}(t)\delta)h_j + \delta \sum_{k \neq j} A_{jk}(t)h_k\|_{E_j}.$$

Here $M_j((I_X - \delta A(t))h)$ $(j = 1, ..., n)$ are the coordinates of $M((I_X - \delta A(t))h) \in \mathbf{C}^n$. Therefore,

$$M_j((I_X - \delta A(t))h) \geq \|(I_{E_j} - A_{jj}(t)\delta)h_j\|_{E_j} - \delta \sum_{k \neq j} \|A_{jk}(t)h_k\|_{E_j} \geq$$

$$(1 - a_{jj}(t)\delta)\|h_j\|_{E_j} - \delta \sum_{k \neq j} a_{jk}(t)\|h_k\|_{E_k}.$$

That inequality can be written in the vector form

$$M((I_X - \delta A(t))h) > (I_B - a(t)\delta)M(h) \ (h \in D)$$

with the matrix $a(t) = (a_{jk}(t))$. It is simple to check that the invertibility of the operators $I_{E_k} - \delta A_{kk}(t)$ $(k = 1, ..., n)$ implies the invertibility of the operator $I_X - \delta A(t)$ for small enough $\delta > 0$, because $A_{jk}(t)$ $(j \neq k)$ are bounded. We thus easily get

$$M((I_X - \delta A(t))^{-1}v) < (I_{C^n} - a(t)\delta)^{-1}M(h) \ (v \in X).$$

But

$$(I_{C^n} - a(t)\delta)^{-1} = I_{C^n} + a(t)\delta + o(\delta) \ (\delta \downarrow 0).$$

Consequently,

$$M((I_X - \delta A(t))^{-1}v) \leq (I_{C^n} + a(t)\delta)M(h) \ (v \in X) \tag{9.15}$$

for sufficiently small $\delta > 0$. Furthermore, for some partitioning of a segment $[0, t]$: $0 = t_0^{(n)} < t_1^{(n)} < ... < t_n^{(n)} = t$, let us define operators $U_{n,0}$ as in Section 14.3. First, we check the solution existence. According to (9.15) we easily get

$$M(U_{n,0}u_0) \leq \overleftarrow{\prod_{1 \leq k \leq n}} (I_{C^n} + a(t_k)\delta_k)M(u_0). \tag{9.16}$$

With the notation $m = max_{0 \leq s \leq t} \|a(s)\|_B$ it can be written

$$\|I_B + \delta a(t)\|_B \leq 1 + m\delta \leq e^{m\delta} \ (t \geq 0).$$

The latter inequality implies

$$\|U_{k,0}u_0\|_X = \|M(U_{n,0}u_0)\|_B \leq e^{tm} \|M(u_0)\|_B \ \ (k = 1, ..., n).$$

Thus condition (3.1) holds, and due to Theorem 14.3.1 equation (9.1) has a solution $u(t)$. Moreover, $u(t) = \lim U_{n0}u_0$ as $\max_k \delta_k^{(n)} \to 0$, in the sense of the norm of space X.

On the other hand, due to Lemma 4.1.1, the limit in the right part of (9.16) is the solution of equation (9.5). Now the desired assertion follows from inequality (9.16). □

14.10 Systems of linear equations with unbounded off-diagonal operators

Let $H \equiv E_1 \oplus E_2 \oplus ... \oplus E_n$ be an orthogonal sum of Hilbert spaces E_k with scalar products $(.,.)_k$, and norms $\|.\|_{E_k} = \sqrt{(.,.)_k}$, $(k = 1, ..., n)$. For an element $h = (h_k \in E_k, k = 1, ..., n)$ from H, the norm is defined by $\|h\|_H = [\|h_1\|^2_{E_1} + ... + \|h_n\|^2_{E_n}]^{1/2}$.

Let $A_{jk}(t)$ be linear operators acting from E_j into E_k with constant dense domains $D_k \subseteq E_k$ $(j, k = 1, ..., n)$. Consider the system

$$\dot{u}_j = \sum_{k=1}^{n} A_{jk}(t)u_k \quad (t \geq 0;\ u_j = u_j(t),\ j = 1, ..., n). \tag{10.1}$$

That system can be rewritten in space H as

$$\dot{u} = A(t)u, \quad (t \geq 0;\ u = u(t)) \tag{10.2}$$

with the operator matrix $A(t) = (A_{jk}(t))_{j,k=1}^n$. Let D be the (constant) domain of $A(t)$. We assume the existence of a solution of (10.2) for any initial vector from D.

Furthermore, let us suppose that there are invertible constant selfadjoint operators S_k acting in E_k with domains $D(S_k) \supseteq D_k$, and real scalar-valued functions $d_{jk}(t)$, $e_{jk}(t)$ $(j, k = 1, ..., n)$, such that

$$Re(A_{kk}(t)h, h)_k \leq -d_{kk}(t)\|S_k h\|^2_{E_k} + c_{kk}(t)\|h\|^2_{E_k} \ (h \in D_k). \tag{10.3}$$

In addition, the domain $D(A^*_{jk}(t))$ of $A^*_{jk}(t)$ has the property $D(A^*_{jk}(t)) \supseteq D_j$, and

$$\|A^*_{jk}(t)h\|_{E_k} \leq d_{jk}(t)\|S_j h\|_{E_j} + e_{jk}(t)\|h\|_{E_j} \ \ (j \neq k,\ h \in D_j,\ t \geq 0). \tag{10.4}$$

Besides, it is assumed that

$$d_{jj}(t),\ e_{jk}(t),\ d_{jk}(t) \geq 0,\quad (j \neq k;\ j,k = 1, ..., n\ , t \geq 0).$$

Functions $e_{jj}(t)$ $(j = 1, ..., n)$ are not assumed to be positive.

Since S_k are selfadjoint operators, it can be written

$$||S_k h||_{E_k} \geq \beta(S_k)||h||_{E_k}\ (h \in D(S_k),\ k = 1, ..., n), \tag{10.5}$$

where $\beta(S_k) = \inf_{\lambda \in \sigma(S_k)} |\lambda|$. Put

$$a_{jj}(t) = -2(\beta^2(S_j)d_{jj}(t) - e_{jj}(t)) + \sum_{k \neq j}[\beta^2(S_j)d_{jk}(t) + e_{jk}(t)]$$

$$\text{and } a_{jk}(t) = d_{jk}(t) + e_{jk}(t) \tag{10.6}$$

with $j \neq k$.

Theorem 14.10.1 *Let conditions (10.3) and (10.4) be fulfilled, and let*

$$\sum_{k=1,\ k \neq j}^{n} d_{jk}(t) < 2d_{jj}(t) \text{ for all } t \geq 0,\ j = 1, ..., n. \tag{10.7}$$

Then a solution $u(t) = (u_j(t))_{j=1}^n$ of (10.1) (if it exists) is subject to the inequalities

$$||u_j(t)||^2_{E_j} \leq w_j(t)\ \ (j = 1, ..., n,\ t \geq 0),$$

where $w(t) = (w_j(t))_{j=1}^n$ is the solution of the ordinary differential system

$$\dot{w}_j(t) = \sum_{k=1}^{n} a_{jk}(t)w_k(t) \tag{10.8}$$

with the initial conditions $w_j(0) = ||u_j(0)||^2_{E_j}$ $(j = 1, ..., n)$.

The proofs are presented in the next section.

Corollary 14.10.2 *Under the hypothesis of Theorem 14.10.1, let the ordinary differential system (10.8) be (asymptotically) stable. Then system (10.1) is (asymptotically) H-stable. In particular, let $a(t) \equiv (a_{jk})$ be a constant Hurwitz matrix. Then (10.1) is exponentially H-stable.*

Let the matrix $a(t)$ with the entries $a_{jk}(t)$ $(j, k = 1, ..., n)$ satisfy the conditions (9.7) and (9.8). Again denote by $z(q_0, v)$ the extreme right root of the algebraic equation (9.9). Now, combining Theorem 3.2.2 with Theorem 14.10.1, we obtain

Corollary 14.10.3 *Let conditions (10.3), (10.4) and (10.7) be fulfilled. Let matrix $a(t)$, defined by (10.6), satisfy conditions (9.7) and (9.8). Then system (10.1) is H-stable, provided that $a(t) + z(q_0, v)I_{C^n}$ is a Hurwitz matrix for all $t \geq 0$.*

Let w_n be defined by the formula (9.10). Then combining Corollary 3.2.3 with Theorem 14.10.1 we get

Corollary 14.10.4 *Under the hypothesis of the previous corollary, let $q_0 w_n \leq v$. Then system (10.1) is H-stable provided that*

$$a(t) + v^{1-1/n}[q_0 w_n]^{1/n} I_{C^n}$$

is a Hurwitz matrix for all $t \geq 0$.

14.11 Proof of Theorem 14.10.1

Doing the scalar product in (10.1) by u_j and considering that

$$\frac{d}{dt}(u_j(t), u_j(t))_j = 2\ Re(\dot{u}_j(t), u_j(t))_j\ (j = 1, 2, ..., n),$$

we obtain

$$\frac{d}{dt}(u_j, u_j)_j = 2Re\sum_{k=1}^{n}(A_{jk}(t)u_k, u_j)_j. \qquad (11.1)$$

But due to the Schwartz inequality,

$$Re(A_{jk}(t)u_k, u_j)_j = Re(u_k, A^*_{jk}(t)u_j)_k \leq ||A^*_{jk}(t)u_j||_{E_k}||u_k||_{E_k}.$$

So, with the notation $x_k = ||u_k(t)||_{E_k}$ and $z_k = ||S_k u_k||_{E_k}$ we get by virtue of (10.3) and (10.4) the relations

$$Re(A_{jk}(t)u_k, u_j)_j \leq d_{jk}(t)z_j x_k + e_{jk}(t)x_j x_k,\ (j \neq k) \qquad (11.2)$$

and

$$Re(A_{jj}(t)u_j, u_j)_j \leq -d_{jj}(t)z_j^2 + e_{jj}(t)x_j^2\ (u_j \in D_j). \qquad (11.3)$$

Now equation (10.1) gives

$$\frac{d}{dt}x_j^2 \leq -2d_{jj}(t)z_j^2 + 2e_{jj}(t)x_j^2 + 2\sum_{k\neq j} d_{jk}(t)z_j x_k + e_{jk}(t)x_j x_k.$$

Inequality $2ab \leq a^2 + b^2$ $(a, b \geq 0)$ implies

$$2z_j x_k \leq x_k^2 + z_j^2 \text{ and } 2x_j x_k \leq x_k^2 + x_j^2.$$

Hence,

$$\frac{d}{dt}x_j^2 \leq z_j^2\ (-2d_{jj}(t) + \sum_{k\neq j} d_{jk}(t)) + x_j^2\ (2e_{jj}(t) + \sum_{k\neq j} e_{jk}(t)) +$$

$$\sum_{k \neq j} x_k^2 \, (d_{jk}(t) + e_{jk}(t)). \tag{11.4}$$

But due to (10.5), $\beta^2(S_j)x_j^2 \leq z_j^2$. Thus, (10.7) yields

$$\frac{d}{dt}x_j^2 \leq x_j^2\beta^2(S_j)(-2d_{jj}(t) + \sum_{k \neq j} d_{jk}(t)) + x_j^2(2e_{jj}(t)+$$

$$\sum_{k \neq j} e_{jk}(t)) + \sum_{k \neq j} x_k^2(d_{jk}(t) + e_{jk}(t)).$$

Or, according to (10.6),

$$\frac{d}{dt}x_j^2 \leq \sum_{k=1}^{m} a_{jk}(t)x_k^2.$$

With the notation $y(t) = (x_1^2(t), ..., x_n^2(t))$ rewrite the latter inequality in the vector form:

$$\frac{d}{dt}y(t) \leq a(t)y(t).$$

Hence,

$$y(t) \leq y(0) + \int_0^t a(s)y(s)ds.$$

The result now follows from Lemma 1.7.1. □

14.12 Notes

Note 14.1 A lot of papers and books are devoted to differential linear equations with a variable operator in a Banach space. Mainly solvability conditions were investigated. We refer the reader to the famous books (Krein, 1971), (Tanabe, 1979 and 1997), (Pazy, 1983).

Note 14.2 Stability conditions for linear nonautonomous evolution equations with slowly and rapidly varying coefficients were established in (Henry, 1981).

A fundamental approach to the stability of diffusion parabolic equations is the method of upper and lower solutions. A systematical treatment of that approach is given in (Pao, 1992). We also mention the existence of Lyapunov stability theory for proving the stability of linear equations via a Lyapunov functional (see (Walker, 1976)).

Note 14.3 Sections 14.3-14.5 are based on the papers (Gil', 1989b) and (Gil', 1996). Theorem 14.6.1 is adapted from (Gil', 1989a) and (Gil', 1990, Chapter 6). The contents of Sections 14.7 and 14.8 are based on the paper (Gil', 1997). The material of Section 14.9 is taken from (Gil', 1996). The material of Section 14.10 is new.

About some other related works see Notes to Chapter 13.

References for Chapter 14

Daleckii, Yu L. and Krein, M. G. (1974). *Stability of Solutions of Differential Equations in Banach Space*. Amer. Math. Soc., Providence, R. I.

Gil', M.I. (1989a). Bounds for solutions of quasilinear parabolic equations, *Differential Eqs. 25*, 723-726.

Gil', M.I. (1989b). Estimates for solutions of nonlinear diffusion systems, *Differential Eqs. 25*, 2082-2090.

Gil', M.I. (1990). *Operator Functions Method in Theory of Differential Equations*. Nauka, Moscow. In Russian.

Gil', M.I. (1996). Stability of linear evolution equations in lattice normed spaces, *Zeitschrift fur Analysis, 15, No 4*, 949-959.

Gil', M.I. (1997). Freezing method for evolution equations, *Communications in Applied Analysis, 1, No 2*, 245-256.

Henry, D. (1981). *Geometric Theory of Semilinear Parabolic Equations*, Lectures Notes in Mathematics, No 840. Springer-Verlag, New York.

Krein, S.G. (1971). *Linear Differential Equations in Banach space*. AMS, Providence, R. I.

Pao, C.V. (1992). *Nonlinear Parabolic and Elliptic Equations*. Plenum Press, New York.

Pazy, A. (1983). *Semigroups of Linear Operators and Applications to Partial Differential Equations*. Springer-Verlag, Berlin.

Tanabe, H. (1979). *Equations of Evolution*. Pitman, London - San-Francisco - Melburn.

Tanabe, H. (1997). *Functional Analytic Methods for Partial Differential Equations*. Marcel Dekker, Inc. New York.

Walker, J.A. (1976). On the application of Lyapunov's direct method to linear distributed systems, *J. Math. Analysis and Appl., 53*, 187-220.

Chapter 15

SEMILINEAR EQUATIONS IN BANACH SPACES WITH CONSTANT LINEAR PARTS

This chapter is devoted to nonlinear evolution equations in a Banach space with separated constant linear operators and continuous nonlinearities. These nonlinearities are subject to the linear operators in a certain sense. Such equations are called semilinear equations. Wide classes of nonlinear distributed parameter systems are governed by semilinear equations in a Banach space.

Stability definitions for nonlinear equations in a Banach space are introduced in Section 15.1.

Stability conditions for semilinear evolution equations in a Banach space are considered in Sections 15.2 and 15.3.

Sections 15.4 and 15.5 are devoted to systems of semilinear equations in a Hilbert space. In particular, Section 15.4 absolute stability conditions are established. In Section 15.5 explicit stability conditions and bounds for the regions of attraction of stationary solutions are derived.

Section 15.6 deals with the generalized Aizerman hypothesis for differential equations in a Hilbert space.

Throughout the present chapter. A is a closed linear operator in a Banach space with a dense domain.

15.1 Definitions

Let X be a Banach space, and $Y \subseteq X$ be a densely imbedded Banach subspace. Consider in X the problem

$$du/dt = Au + F(u,t) \ (t \geq 0), \tag{1.1}$$

$$u(0) = u_0. \tag{1.2}$$

It is assumed that $D(A) \subseteq Y$. For a $r \in (0,\infty)$, put

$$\Omega(Y,r) = \{h \in Y : \|h\|_Y \leq r\}.$$

Our main assumptions in the present chapter are as follows.

Assumption 15.1A. The function F has the Lipschitz property

$$\|F(h,t) - F(v,t)\|_X \leq w\|h - v\|_Y \ (w = const;\ h,v \in \Omega(Y,r);\ t \geq 0). \tag{1.3}$$

Assumption 15.1B. The operator A generates a strongly continuous semigroup e^{At} with the properties

$$\chi \equiv \sup_{t \geq 0} \|e^{At}\|_Y < \infty \tag{1.4}$$

and

$$\xi \equiv \int_0^\infty \|e^{At}\|_{X \to Y} dt < \infty. \tag{1.5}$$

Following Browder's terminology cf. (Vrabie, 1987), (Henry, 1981, p. 55), we will call a continuous function $u(t) : [0,\infty) \to Y$ satisfying the equation

$$u(t) = e^{tA}u_0 + \int_0^t e^{(t-s)A}F(u(s),s)ds \ (t \geq 0) \tag{1.6}$$

a (global) *mild solution* of equation (1.1).

In the following definitions, it is assumed that there is a positive $r \leq \infty$, such that for any $u_0 \in \Omega(Y,r)$, problem (1.1), (1.2) has a unique mild solution $u(t)$.

Definition 15.1.1 *Let $F(t,0) \equiv 0$ for all $t \geq 0$.*

We will say that the zero solution of (1.1) is Y-stable if for given $t_0 \geq 0$ and any sufficiently small $\epsilon > 0$, there is a $\delta(t_0) > 0$, such that condition $u(t_0) \in \Omega(Y,\delta(t_0))$ implies the relation $\|u(t)\|_Y \leq \epsilon$ $(t \geq t_0)$.

The zero solution of (1.1) is asymptotically Y-stable, if it is Y-stable and there is an open set $B \subseteq Y$, such that for given t_0, the relation $u(t_0) \in B$ implies $u(t) \to 0$ as $t \to \infty$. Besides, B is called the region of attraction of the zero solution. If $B = Y$, then the zero solution of (1.1) is globally asymptotically stable.

We will say that the zero solution of (1.1) is uniformly Y-stable, if for any sufficiently small $\epsilon > 0$, there is a constant $\delta > 0$ independent of $t_0 \geq 0$, *such that for any mild solution* $u(t)$, *the condition* $u(t_0) \in \Omega(Y,\delta)$ *implies the relation* $\|u(t)\|_Y \le \epsilon$ $(t \ge t_0)$.

The zero solution of (1.1) is exponentially Y-stable if there are positive constants r_0, Λ_0 *and* l_0, *such that, for a given* $u_0 \in \Omega(Y, r_0)$, *the relation*

$$\|u(t)\|_Y \le l_0 \|u_0\|_Y \, \epsilon^{-\Lambda_0 t} \quad (t \ge 0)$$

holds. The zero solution of (1.1) is globally exponentially stable if it is exponentially Y-stable and the region of attraction coincides with Y.

If (1.1) has strong (continuously differentiable) solutions. then it can show that these definitions in the linear case agree with the stability definitions introduced in the previous chapter.

Definition 15.1.2 *We will say that the zero solution of equation (1.1) is absolutely Y-stable in the class of nonlinearities satisfying the inequality*

$$\|F(h,t)\|_X \le l\|h\|_Y \quad (l = const;\ h \in Y;\ t \ge 0). \tag{1.7}$$

if, for any $u_0 \in Y$, *problem (1.1), (1.2) has a unique mild solution* $u(t)$ *and the zero solution of (1.1) is globally asymptotically stable. and there is a positive constant* N, *which does not depend on the specific form of the function* F *(but which depends on* l*), such that the inequality*

$$\|u(t)\|_Y \le N\|u_0\|_Y \quad (t \ge 0) \tag{1.8}$$

holds.

15.2 Stability of equations in a Banach space

Theorem 15.2.1 *Under Assumptions 15.1A and 15.1B, let the conditions*

$$\|F(h,t)\|_X \le l\|h\|_Y \quad (l = const;\ h \in \Omega(Y,r);\ t \ge 0). \tag{2.1}$$

and

$$l\xi < 1 \tag{2.2}$$

be fulfilled. Then the zero solution of (1.1) is exponentially Y-stable. Moreover, any initial vector satisfying the inequality

$$\chi\|u_0\|_Y \le r(1 - l\xi), \tag{2.3}$$

belongs to the region of attraction of the zero solution, and the corresponding solution satisfies the estimate.

$$\|u(t)\|_Y \le \chi\|u_0\|_Y(1 - l\xi)^{-1} \le r \quad (t \ge 0). \tag{2.4}$$

The proof of this theorem is presented in the next section.

Let us indicate a few corollaries of that theorem.

The following result can be considered as the stability in the first approximation.

Corollary 15.2.2 *For some $r > 0$, let Assumptions 15.1A and 15B hold. In addition, let*

$$\lim_{\|h\|_Y \to 0} \frac{\|F(h,t)\|_X}{\|h\|_Y} = 0 \tag{2.5}$$

uniformly in t. Then the zero solution of equation (1.1) is asymptotically Y-stable.

In fact, for a sufficiently small r condition (2.5) implies relations (2.1) and (2.2). Now Theorem 15.2.1 yields the required result.

Recall that X^ν is defined in Section 13.8. The following result is due to Corollary 15.2.2 and Lemma 13.8.3.

Corollary 15.2.3 *Let $-A$ be a sectorial operator with $Re\sigma(A) < 0$. Let conditions (1.3), (2.5) hold with $Y = X^\nu$ ($\nu \in [0,1)$). Then under (2.5) the zero solution of equation (1.1) is asymptotically X^ν-stable.*

Note that the inequality

$$\|F(h,t)\|_X \le l_p \|h\|_Y^p \ (l_p = const;\ h \in Y;\ t \ge 0).$$

with $p > 1$ ensures condition (2.5) with $r < \infty, l = l_p r^{p-1}$.

Corollary 15.2.4 *Let Assumptions 15.1A, 15.1B hold with $r = \infty$. Let condition (2.2) be fulfilled. Then the zero solution of equation (1.1) is absolutely Y-stable in the class of nonlinearities (1.7). Moreover, in (1.8) one can take the constant $N = \chi(1 - l\xi)^{-1}$.*

15.3 Proof of Theorem 15.2.1

We need the following trivial

Lemma 15.3.1 *Let Ψ be a continuous mapping of a closed subset $C_1 \subseteq C([0,T], X)$ into itself, satisfying the inequality*

$$\|(\Psi x)(t) - (\Psi y)(t)\|_X \le (M\|x - y\|_X)(t) \ (t \in [0,T];\ x, y \in C_1).$$

where M is a bounded linear positive operator acting in the space of real continuous scalar-valued functions $C(0,T)$. If, in addition, the spectral radius $r_s(M)$ of M is less than one: $r_s(M) < 1$, then Ψ has a unique fixed point $\bar{x} \in C_1$. Moreover, that point can be found by the method of successive approximations.

Proof: Take an arbitrary $x_0 \in C_1$ and define the successive approximations

$$x_k = \Psi\, x_{k-1} \quad (k = 1, 2, \ldots).$$

Hence,

$$\|x_{k+1}(t) - x_k(t)\|_X = \|(\Psi x_k)(t) - (\Psi x_{k-1})(t)\|_X \le$$

$$(M\|x_k - x_{k-1}\|_X)(t) \le \ldots \le (M^k\|x_1 - x_0\|_X)(t) \ (0 \le t \le T).$$

Inasmuch as $r_s(M) < 1$, $\|M^k\|_{C(0,T)} \to 0 \ (k \to \infty)$. Thus some iteration of Ψ is a contraction in $C([0,T], X)$. This proves the result. □

Lemma 15.3.2 *Under Assumptions 15.1A and 15.1B, let conditions (2.1)-(2.3) be fulfilled. Then problem (1.1), (1.2) has a unique mild solution.*

Proof: With a fixed $\tau \ge 0$ define in $C([0,T], Y)$ the mapping

$$(\Psi z)(t) = e^{tA} u_0 + \int_0^t e^{(t-s)A} F(z(s), s) ds$$

$$(t \in [0,T];\ z \in C([0,T], Y)).$$

Clearly. (1.6) can be written as $(\Psi z)(t) = z(t)$. According to (1.3),

$$\|(\Psi z_1)(t) - (\Psi z_2)(t)\|_Y \le \int_0^t \|e^{(t-s)A}\|_{X\to Y} \|F(z_1(s), s) -$$

$$F(z_2(s), s)\|_X ds \le w \int_0^t \eta(t-s)\|z_1(s) - z_2(s)\|_Y ds$$

$$(t \in [0,T];\ z_1, z_2 \in C([0,T], Y)), \tag{3.1}$$

where

$$\eta(t) = \|e^{tA}\|_{X\to Y}.$$

By virtue of relations, (1.4), (1.5) and (2.3) we arrive at the inequality

$$\|(\Psi z)(t)\|_Y \le \chi\|u_0\|_Y + \int_0^t \|e^{(t-s)A}\|_{X\to Y}\|F(z(s), s)\|_X ds \le$$

$$\chi\|u_0\|_Y + l\int_0^t \eta(t-s)\|z(s)\|_Y ds \le \chi\|u_0\|_Y + \xi l r \tag{3.2}$$

for any

$$z \in C_r \equiv \{f \in C([0,T]; Y) : \|f\|_{C([0,T];Y)} \le r\}$$

So Ψ maps the ball C_r into itself. Further. denote by V a Volterra operator in $C(0,T)$ with the kernel

$$K(t, s) = w\eta(t-s).$$

Then relation (3.1) can be written in the form

$$||(\Psi z_1)(t) - (\Psi z_2)(t))||_Y \le (V||z_1 - z_2||_Y)(t).$$

Simple calculations show that, according to (1.8), the spectral radius of V is equal to zero, since $T < \infty$. The existence of solutions is now due to Lemma 15.3.1.

Proof of Theorem 15.2.1: By virtue of relations (1.6), (3.2) and (2.2) we easily have

$$\sup_{0\le t\le T} ||u(t)||_Y \le \chi ||u_0||_Y (1 - l\xi)^{-1}$$

Since T is arbitrary, this proves estimate (2.4). This estimate means the Y-stability. The substitution $v(t) = exp(t\epsilon)u(t)$ with a sufficiently small $\epsilon > 0$ in equation (1.1) leads to an estimate of the type (2.4) for $v(t)$ by the previous reasonings. Thus, $||u(t)||_Y \le exp(-t\epsilon)||v(t)||_Y \to 0$ as $t \to \infty$. This proves the exponential stability. The proof is complete. □

15.4 Absolute stability of equations with constant operators in a Hilbert space

Let $H = E^n$ be an orthogonal sum of the same Hilbert spaces $E_1 = ... = E_n = E$. That is, H is the n copies of E. In addition, S is a boundedly invertible selfadjoint operator in H commuting with all constant matrices. Recall that H^μ ($\mu \ge 0$) is the set $D(S^\mu)$ equipped with the graph norm $||h||_{H^\mu} = ||S^\mu h||_H$ ($h \in D(S^\mu)$). Again, put

$$Q(S) \equiv c_0 S^2 + c_1 S^1 + c_2.$$

Here c_k ($k = 0, 1, 2$) are constant $n \times n$-matrices. Now consider in H the equation

$$du/dt = Q(S)u + F(u,t) \ (t \ge 0), \tag{4.1}$$

where F maps $H^\mu \times [0, \infty)$ with some $\mu \in [0, 2)$ and has the Lipschitz property (1.3) with $X = H, r = \infty$ and $Y = H^\mu$. In addition, it satisfies the inequality

$$||F(h,t)||_H \le l||h||_{H^\mu} \quad (h \in H^\mu, \ t \ge 0). \tag{4.2}$$

Again, introduce the characteristic matrix pencil

$$Q(s) = s^2 c_0 + s c_1 + c_2 \ (s \in \mathbf{C}).$$

It is assumed that, $c_0 = c_0^*$ is negative definite and the spectrum of $Q(S)$ lie in the open half-plane:

$$\alpha(Q(s)) \le \alpha_0 < 0 \ (s \in \sigma(S)), \text{ and } v \equiv \sup_{s\in\sigma(S)} g(Q(s)) < \infty. \tag{4.3}$$

Recall that $g(.)$ was introduced in Section 2.1. According to Proposition 2.1.1, the restriction $v < \infty$ holds, for example, if $c_1 = c_1^*$, or in the scalar case. See also Section 13.10. Again put

$$\chi(Q) = \sup_{t \geq 0} e^{t\alpha_0} \sum_{k=0}^{n-1} \frac{t^k v^k}{(k!)^{3/2}},$$

and

$$\eta(\mu, t) = \sum_{k=0}^{n-1} \frac{t^k v^k}{(k!)^{3/2}} \sup_{s \in \sigma(S)} |s|^\mu e^{t\alpha(Q(s))} \ (0 \leq \mu < 2).$$

Theorem 15.4.1 *For some $\mu \in [0, 2)$, let the conditions (4.3) and*

$$\xi_\mu(Q) \equiv l \int_0^\infty \eta(\mu, s) ds < 1 \tag{4.4}$$

be fulfilled. Then the zero solution of (4.1) is absolutely H^μ-stable in the class of nonlinearities (4.2). Moreover, the constant in relation (1.8) with $Y = H^\mu$ can be taken as $N = \chi(Q)(1 - \xi_\mu)^{-1}$.

Proof: From Lemma 13.10.2 it follows that $Q(S)$ generates an analytic semigroup satisfying

$$||S^\mu e^{Q(S)t}||_H \leq \eta(\mu, t), \text{ and } ||e^{Q(S)t}||_{H^\mu} \leq \chi(Q).$$

Now the result is due to Corollary 15.2.4. □

Let ω be a set with a positive measure ρ. Let us specialize the previous theorem in the case $H = L^2_\rho(\omega, \mathbf{C}^n)$. Consider in $L^2_\rho(\omega, \mathbf{C}^n)$ equation (4.1) with

$$F(u, t) = F_0(x, u, Su, t) \ (u \in D(S);\ x \in \omega),$$

where the function F_0 maps $\omega \times \mathbf{C}^{2n} \times [0, \infty)$ into $\mathbf{C}^n$. First, assume that $F_0(x, h_0, h_1, t)$ has the Lipschitz property in h_0, h_1, and

$$||F_0(x, h_0, h_1, t)||_{C^n} \leq l_0||h_0||_{C^n} + l_1||h_1||_{C^n} \ (l_k = const;\ h_k \in \mathbf{C}^n,\ k = 0, 1). \tag{4.5}$$

Since

$$||F_0(x, w(x), (Sw)(x), t)||_{C^n} \leq l_0||w(x)||_{C^n} + l_1||(Sw)(x)||_{C^n},$$

we get

$$||F_0(x, w, Sw, t)||_H \leq l||Sw(x)||_H$$

with $l = l_0||S^{-1}||_H + l_1$. So condition (4.4) with $\mu = 1$ takes the form

$$(l_0||S^{-1}||_H + l_1) \sum_{k=0}^{n-1} \int_0^\infty \frac{t^k v^k}{(k!)^{3/2}} \sup_{s \in \sigma(S)} |s| e^{t\alpha(Q(s))} dt < 1. \tag{4.6}$$

Now Theorem 15.4.1 yields

Corollary 15.4.2 *Let conditions (4.3) and (4.6) be fulfilled. Then the equation*

$$du/dt = Q(S)u + F_0(x, u, Su, t) \quad (u = u(x,t),\ x \in \omega, t \geq 0) \tag{4.7}$$

is H^1-absolutely stable in the class of nonlinearities (4.5).

The case of $l_1 = 0$ has considerably more simple calculations. Namely, let

$$||F_0(x, h_0, h_1, t)||_{C^n} \leq l_0||h_0||_{C^n} (h_0 \in \mathbf{C}^n) \tag{4.8}$$

For example, F does not depend on h_1. Clearly,

$$\eta(0,t) = e^{\alpha_0 t} \sum_{k=0}^{n-1} \frac{t^k v^k}{(k!)^{3/2}}, \quad \int_0^\infty \eta(0,t)dt = \sum_{k=0}^{n-1} \frac{t^k v^k}{|\alpha_0|^{k+1}\sqrt{k!}}.$$

So (4.4) takes the form

$$l_0 \sum_{k=0}^{n-1} \frac{t^k v^k}{|\alpha_0|^{k+1}\sqrt{k!}} < 1. \tag{4.9}$$

We have derived

Corollary 15.4.3 *Let conditions (4.3) and (4.9) be fulfilled. Then the equation (4.7) is H-absolutely stable in the class of nonlinearities (4.8).*

Example 15.4.4 *In space $H = L^2([0,\pi], \mathbf{C}^n)$, let us consider the problem*

$$\frac{\partial u}{\partial t} = -c_0 u_{xx} + c_2 u = f(t, x, u, u_x) \quad (t \geq 0;\ 0 < x < \pi), \tag{4.10}$$

$$u(0,t) = u(\pi, t) = 0 \quad (t > 0), \tag{4.11}$$

where $f(t, x, h, w)$ continuously maps $[0,\infty) \times [0,\pi] \times (\mathbf{C}^n)^2$ into $\mathbf{C}^n$ and has the Lipschitz property with respect to h and w. Assume that

$$||f(t, x, h, w)||_{C^n} \leq l_0||h||_{C^n} + l_1||w||_{C^n} \quad (t \geq 0,\ 0 \leq x \leq \pi) \tag{4.12}$$

with every vectors h, w. Define S by the formula $Sy = -iy'$ for $y \in D(S)$ with

$$D(S) = \{w \in L^2([0,\pi]; \mathbf{C}^n) : w(x) = \sum_{k=-\infty, k\neq 0}^{\infty} a_k e^{ikx};\ a_k \in \mathbf{C}^n;$$

$$\sum_{k=-\infty}^{\infty} ||a_k||^2_{C^n} k^2 < \infty\}.$$

As it was shown in Section 13.12, any solution $u(t)$ of the problem (4.10), (4.11) is simulteneously a solution of (4.10) with the condition $u(t) \in D(S^2)$, $t \geq 0$. Furthermore, it can be written $f(t, x, u, u_x) = f(t, x, u, iSu)$ $(u \in D(S))$, and

$$||f(t, x, u, u_x)||_H \leq l_0||u||_H + l_1||Su||_H.$$

Thus, equation (4.10) takes the form (4.7). Since the eigenvalues $\lambda_k(S) = k$ $(k = \pm 1, \pm 2, ...)$, we have $||S^{-1}||_H = 1$ and

$$\eta(1, t) = \zeta(t) \sum_{k=0}^{n-1} \frac{t^k v^k}{(k!)^{3/2}},$$

where

$$\zeta(t) = \sup_{k=\pm 1, \pm 2, \ldots} |k| e^{t\alpha(Q_k)} \text{ with } Q_k = c_0 k^2 + c_2.$$

In particular, if $c_0 = -I_{C^n}$, then $\alpha(Q_k) = -k^2 + \alpha(c_2)$. In the considered case H^1 is the space $D(S)$ with the norm

$$||w||_{H^1} = ||w'||_{L^2([0,\pi], \mathbf{C}^n)} \ (w \in D(S)).$$

Now Corollary 15.4.3 implies:

Proposition 15.4.5 *Let the conditions (4.3) and*

$$(l_0 + l_1) \int_0^\infty \eta(1, s) ds < 1$$

hold. Then the zero solution of problem (4.10), (4.11) is absolutely H^1-stable in the class of nonlinearities (4.12).

Note that in the real scalar case $n = 1$, $v = 0$, and $\eta(1, t) = \zeta(t)$ with

$$\zeta(t) = \sup_{k=1,2,\ldots} k e^{t\alpha(Q_k)}, \ \alpha(Q_k) = c_0 k^2 + c_2 \ (c_0 < 0).$$

15.5 The region of attraction for equations in Hilbert spaces

Let H, S and $Q(S)$ be the same as in the previous section. In addition,

$$\Omega(H^\mu, r) = \{h \in H^\mu : ||h||_{H^\mu} \le r\}$$

for positive numbers μ, $r \le \infty$. Again consider equation (4.1) where F maps $\Omega(H^\mu, r) \times [0, \infty)$ into H with some $\mu \in [0, 2)$, and satisfies the inequality

$$||F(h, t)||_H \le l||h||_{H^\mu} \quad (l \equiv const;\ h \in \Omega(H^\mu, r)\ , t \ge 0). \tag{5.1}$$

Moreover, the Lipschitz property (1.3) holds with $X = H$, and $Y = H^\mu$.

Theorem 15.5.1 *For some $r < \infty$ and $\mu \in [0,2)$, let conditions (4.3), (4.4) and (5.1) be fulfilled. Then the zero solution of (4.1) is asymptotically H^μ-stable. Any initial vector $u_0 \in H^\mu$ satisfying the inequality*

$$\chi(Q)\|u_0\|_{H^\mu} \leq (1-\xi_\mu(Q))r$$

belongs to the region of attraction of the zero solution. Moreover, the corresponding mild solution $u(t)$ is subject to the estimate

$$\|u(t)\|_{H^\mu} \leq \chi(Q)\|u_0\|_{H^\mu}(1-\xi_\mu(Q))^{-1} \ (t \geq 0).$$

Proof: From Lemma 13.10.2 it follows that $Q(S)$ generates an analytic semigroup satisfying

$$\|S^\mu e^{Q(S)t}\|_H \leq \eta(\mu, t), \text{ and } \|e^{Q(S)t}\|_{H^\mu} \leq \chi(Q).$$

Now the result is due to Theorem 15.2.1. □

Note that this theorem can be directly use for calculations of a bound for the region of attraction of the zero solutions of parabolic systems of the type (4.10).

15.6 The Aizerman hypothesis for equations in a Hilbert space

15.6.1 Statement of the result

Let H and E be Hilbert spaces, and let A be a closed operator in H that generates a strongly continuous exponentially stable semigroup. In addition, let B and C be linear bounded operators that act from E to H, and from H to E, respectively. Consider the equation

$$du/dt = Au + B\phi(Cu, t) \ (t \geq 0) \tag{6.1}$$

where $\phi : E\times[0, \infty) \to E$ is a function that ensures the existence and uniqueness of the mild solutions for any initial vector belonging to $D(A)$, and that satisfies the inequality

$$\|\phi(w, t)\|_E \leq q\|w\|_E \ (w \in E, \ t \geq 0) \tag{6.2}$$

with a constant q independent of w.

Following Definition 15.1.2 we will say that the zero solution of (6.1) is absolutely E-stable in the class of nonlinearities (6.2) if the zero solution of (6.1) is globally asymptotically H-stable, and there is a constant N independent of the specific form of ϕ, such that $\|Cu(t)\|_E \leq N\|u(0)\|_H$ for all $t \geq 0$.

Let $W(\lambda)$ ($\lambda \in \mathbf{C}$) be the transfer function of the linear part of the equation (6.1): $W(\lambda) = C(\lambda I - A)^{-1}B$. We shall consider the case

$$W(\lambda) = P^{-1}(\lambda, S)Q(\lambda, S), \tag{6.3}$$

where

$$P(\lambda, S) = \lambda^n + \sum_{k=1}^{n}\sum_{j=0}^{m} \lambda^{n-k} p_{kj} S^{m-j} \text{ and } Q(\lambda, S) = \sum_{k=0}^{n_1}\sum_{j=0}^{m_1} \lambda^{n_1-k} q_{kj} S^{m_1-j}$$

$(n > n_1, m > m_1)$ are polynomials with real coefficients of $\lambda \in \mathbf{C}$ and of a selfadjoint operator S which is semibounded from below. That is

$$\inf_{h\in D(S)} (Sh, h)(h, h)^{-1} > -\infty.$$

Let us write

$$K(t, s) = (2\pi)^{-1} \int_{-\infty}^{\infty} e^{ti\omega} Q(i\omega, s) P^{-1}(i\omega, s) d\omega$$

assuming that all the roots of the polynomial

$$P(\lambda, s) = \lambda^n + \sum_{k=1}^{n}\sum_{j=0}^{m} \lambda^{n-k} p_{kj} s^{m-j}$$

lie in a half-plane $Rez \leq \alpha_0 < 0$ for each $s \in \sigma(S)$.

Theorem 15.6.1 *Let (6.3) be satisfied and let*

$$max_{s\in\sigma(S)} |K(t, s)| = K(t, s_0) \geq 0 \tag{6.4}$$

where s_0 does not depend on t. If, in addition, the polynomial $P(\lambda, s_0) - qQ(\lambda, s_0)$ is Hurwitzian, then the zero solution of equation (6.1) will be absolutely E-stable in the class of nonlinearities (6.2).

The proof is presented in the next subsection.

Corollary 15.6.2 *Let conditions (6.3) and (6.4) be satisfied. Then for the absolute E-stability of the zero solution of equation (6.1) in the class of nonlinearities (6.2) it is necessary and sufficient that this equation be asymptotically stable for*

$$\phi(z, t) = qz \ (z \in E).$$

Thus Theorem 15.6.1 specifies a class of equations in a Hilbert space that satisfy the generalized (in the evident sense) Aizerman hypothesis.

Consider in E the equation

$$u_{tt} + 2Su_t + cS^2 u = \phi(u, t) \ (0 < c = const < 1) \tag{6.5}$$

with a positive definite selfadjoint operator S.

Corollary 15.6.3 *Let the condition $c\beta^2(S) > q$ hold, where $\beta(S) \equiv \inf \sigma(S) > 0$. Then the zero solution of equation (6.5) is absolutely E-stable in the class of nonlinearities (6.2).*

In fact, under consideration, the roots of $P(\lambda, s) = \lambda^2 + 2s\lambda + cs^2$, are $r_{1,2}(s) = -s(1 \pm b)$, with $b = \sqrt{1-c}$, and

$$K(t, s) = (2bs)^{-1}[e^{-st(1-b)} - e^{-st(1+b)}] \geq 0.$$

Hence the derivative in s is

$$K_s(t, s) = -(2bs^2)^{-1}[((1-b)ts + 1)e^{-st(1-b)} - ((1+b)ts + 1)e^{-st(1+b)}].$$

The function $(1+z)e^{-z}$ $(z \geq 0)$ decreases. Therefore,

$$((1-b)ts + 1)e^{-st(1-b)} - ((1+b)ts + 1)e^{-st(1+b)} \geq 0 \ (t, s > 0).$$

Thus

$$K_s(t, s) \leq 0 \ (s > 0) \text{ and } \max_{s \geq \beta(S)} K(t, s) = K(t, \beta(S)) \geq 0.$$

So due to Theorem 15.6.1 if the polynomial $\lambda^2 + 2\lambda\beta(S) + c\beta^2(S) - q$ is Hurwitzian, then the zero solution of (6.5) is absolutely E-stable. As claimed.

Example 15.6.4 *Consider in $E = L^2[0, 1]$ the problem*

$$u_{tt} - 2u_{txx} + cu_{xxxx} = \phi(u, x, t) \ (t > 0; \ 0 < x < 1; \ c \in (0, 1))$$
$$u^{(k)}(0, t) = u^{(k)}(1, t) = 0 \ (k = 0, 2; \ t \geq 0), \tag{6.6}$$

where the scalar-valued function ϕ defined on $\mathbf{R} \times [0, 1] \times [0, \infty)$ satisfies the condition

$$|\phi(z, x, t)| \leq q|z| \ (z \in \mathbf{R}; \ t \geq 0; \ 0 \leq x \leq 1). \tag{6.7}$$

On the set

$$D(S) = \{w \in L^2[0, 1] : w'' \in L^2[0, 1] : w(0) = w(1) = 0\}$$

let us define the operator S by the equation $(Sw)(x) = -w''(x)$. Then (6.6) takes the form (6.5). Moreover, $\beta(S) = \pi^2$. Further, clearly, condition (6.7) implies (6.2). Thanks to Corollary 15.6.3 under the condition $c\pi^4 > q$, the zero solution of the problem (6.6) is absolutely E-stable in the class of nonlinearities (6.7).

15.6.2 Proof of Theorem 15.6.1

Let $U(t)$ be a semigroup generated by A. Rewrite (6.1) in the form

$$u(t) = U(t)u_0 + \int_0^t U(t-s)\phi(Cu(s), s)ds.$$

Hence,

$$Cu(t) = CU(t)u_0 + \int_0^t K(t-\tau)\phi(Cu(\tau), \tau)d\tau. \tag{6.8}$$

where $K(t) = CU(t)B$. Let E_s be the resolution of the identity for S. Simple calculations show that under condition (6.3),

$$K(t) = \int_{-\infty}^{\infty} K(t,s)dE_s.$$

Therefore, by virtue of condition (6.4),

$$\|K(t)\|_E = max_{s\in\sigma(S)}|K(t,s)| = K(t,s_0) \geq 0.$$

It follows from (6.8) that

$$\|Cu(t)\|_E \leq \|CU(t)u_0\|_E + \int_0^t K(t-\tau,s_0)\|\phi(Cu(\tau),\tau)\|_E d\tau.$$

Now (6.2) yields

$$\|Cu(t)\|_E \leq \|CU(t)u_0\|_E + q\int_0^t K(t-\tau,s_0)\|Cu(\tau)\|_E d\tau.$$

Since $U(t)$ is exponentially stable,

$$\|CU(t)u_0\|_E \leq c_0 e^{-\epsilon_0 t}\|u_0\|_H \ (t \geq 0,\ c_0, \epsilon = const > 0)$$

and

$$\|Cu(t)\|_E \leq c_0 e^{-\epsilon t}\|u_0\|_H + q\int_0^t K(t-\tau,s_0)\|Cu(\tau)\|_E d\tau.$$

Due to Lemma 1.7.1, $\|Cu(t)\|_E \leq \theta(t)$, where θ is a solution of the equation

$$\theta(t) = c_0 e^{-\epsilon t}\|u_0\|_H + q\int_0^t K(t-\tau,s_0)\theta(\tau)d\tau.$$

We shall solve this equation with the aid of the Laplace transformation. By virtue of the fact that a transform of a convolution is equal to a product of the transforms, and that the transforms of $K(t,s_0)$ and $e^{-\epsilon t}$ are equal to $P^{-1}(\lambda,s_0)Q(\lambda,s_0)$ and $(\lambda+\epsilon)^{-1}$, respectively, we obtain

$$\tilde{\theta}(\lambda) = c_0(\lambda+\epsilon)\|u_0\|_H + q\tilde{\theta}(\lambda)P^{-1}(\lambda,s_0)Q(\lambda,s_0)$$

where $\tilde{\theta}(\lambda)$ is the transform of $\theta(t)$. Consequently,

$$\tilde{\theta}(\lambda) = c_0\|u_0\|_H(\lambda+\epsilon)(1-qP^{-1}(\lambda,s_0)Q(\lambda,s_0))^{-1} =$$

$$c_0(\lambda+\epsilon)\|u_0\|_H P(\lambda,s_0)(P(\lambda,s_0)-qQ(\lambda,s_0))^{-1}.$$

By using the inverse Laplace transformation and the residue theorem we arrive at the inequality

$$\theta(t) \leq Me^{-\epsilon_1 t}\|u_0\|_H \ (M = const,\ \epsilon_1 < \min\{\epsilon,\alpha_1\}),$$

where α_1 is the largest real part of the roots of $P(\lambda,s_0) - qQ(\lambda,s_0)$. This proves the result. □

15.7 Passage to an imbedded norm

By virtue of results presented in this section we can reduce stability conditions in a Hilbert norm to stability ones in a stronger norm.

Let H be a Hilbert space, and Z be a continuously imbedded Banach subspace of H. That is, there is a constant $c(H,Z)$ such that

$$\|h\|_H \le c(H,Z)\|h\|_Z \ (h \in Z).$$

In addition, S is a positive definite selfadjoint operator in H: $\beta(S) = \inf \sigma(S) > 0$. Further, for a non-negative number $\mu < 1$, let us introduce the function

$$\zeta(S,\mu,t) = \begin{cases} (\frac{\mu}{te})^\mu & \text{if } 0 \le t \le \mu\beta^{-1}(S), \\ \beta^\mu(S)e^{-t\beta(S)} & \text{if } \mu\beta^{-1}(S) < t < \infty \end{cases}$$

and for a positive $\Lambda < \beta(S)$, put

$$J(S,\mu,\Lambda) = \int_0^\infty \zeta(S,\mu,t)e^{\Lambda t}dt.$$

Clearly, $J(S,\mu,\Lambda) < \infty$. Consider the equation

$$\dot{u} = Su + F(u,t), \tag{7.1}$$

where the function $F : \Omega(Z,r) \times [0,\infty) \to H$ satisfies the condition

$$\|F(h,t)\|_H \le \gamma_r\|h\|_H \ (h \in \Omega(Z,r);\ \gamma_r = const). \tag{7.2}$$

Lemma 15.7.1 *For non-negative numbers $\mu < 1$ and $r \le \infty$, let the relations (7.2), and*

$$\theta_\mu \equiv \|S^{-\mu}\|_{H\to Z} < \infty \tag{7.3}$$

hold. In addition, let a solution $u(t)$ of equation (7.1) satisfy the relations $\|u(t)\|_Z \le r$ $(t \ge 0)$, $u(0) \in D(S^\mu)$, and

$$\|u(0)\|_H \le \|u(0)\|_H C_\Lambda e^{-\Lambda t} \ (t \ge 0,\ C_\Lambda,\ \Lambda = const > 0) \tag{7.4}$$

with $0 < \Lambda < \beta(S)$. Then that solution is subject to the estimate

$$\|u(t)\|_Z \le \theta_\mu e^{-t\Lambda}\ [\|S^\mu u(0)\|_H + \|u(0)\|_H c_\Lambda \gamma_r J(S,\mu,\Lambda)] \ (t \ge 0). \tag{7.5}$$

Proof: It is directly checked that (7.1) is equivalent to the following equation:

$$u(t) = e^{-tS}u(0) + \int_0^t e^{-(t-s)S}F(u(s),s)ds.$$

Hence,

$$S^\mu u(t) = e^{-tS}S^\mu u(0) + \int_0^t S^\mu e^{-(t-s)S}F(u(s),s)ds.$$

Since S is selfadjoint, $\|e^{-tS}\|_H \leq e^{-t\beta(S)}$ $(t \geq 0)$, and

$$\|S^\mu e^{-tS}\|_H = \sup_{s\in\sigma(S)} |s^\mu e^{-ts}|.$$

Simple calculations show that $\|S^\mu e^{-tS}\|_H = \zeta(S,\mu,t)$. Due to (7.2) we get the relation

$$\|S^\mu u(t)\|_H \leq e^{-t\beta(S)}\|S^\mu u(0)\|_H + \int_0^t \zeta(S,\mu,t-s)\gamma_r\|u(s)\|_H ds.$$

So according to (7.4)

$$\|S^\mu u(t)\|_H \leq e^{-t\beta(S)}\|S^\mu u(0)\|_H + c_\Lambda\gamma_r\|u(0)\|_H \int_0^t \zeta(S,\mu,t-s)\, e^{-\Lambda s}ds. \quad (7.6)$$

Take into account that

$$\int_0^t \zeta(S,\mu,t-s)\, e^{-\Lambda s}ds = e^{-\Lambda t}\int_0^t \zeta(S,\mu,z)\, e^{\Lambda z}dz \leq e^{-\Lambda t}J(S,\mu,\Lambda).$$

Now (7.6) and the relations

$$\|u(t)\|_Z = \|S^{-\mu}S^\mu u(t)\|_Z \leq \theta_\mu\|S^\mu u(t)\|_H$$

yield the result. □

Put $k(\Lambda,\mu) = \theta_\mu(1 + \beta^{-\mu}(S)c_\Lambda\gamma_r J(S,\mu,\Lambda))$.

Theorem 15.7.2 *For non-negative $\mu < 1$ and $r \leq \infty$, let relations (7.2) and (7.3) hold. In addition, let a solution $u(t)$ of equation (7.1) satisfy inequality (7.4) with $0 < \Lambda < \beta(S)$, and let*

$$k(\Lambda,\mu)\|S^\mu u(0)\|_H < r. \quad (7.7)$$

Then for that solution, the following estimate is valid:

$$\|u(t)\|_Z \leq k(\Lambda,\mu)\|S^\mu u(0)\|_H e^{-\Lambda t} \ (t \geq 0). \quad (7.8)$$

Proof: Take into account that

$$\|h\|_H = \|S^{-\mu}S^\mu h\|_H \leq \|S^{-\mu}\|_H\|S^\mu h\|_H = \beta^{-\mu}(S)\|S^\mu h\|_H \ (h \in D(S)).$$

Thus (7.5) implies (7.8). Due to (7.7) and (7.3) we have $\|u(0)\|_Z < r$. Since the solution is a continuous function, there is $t_0 > 0$, such that $\|u(t)\|_Z \leq r$ $(t \leq t_0)$. So estimate (7.8) holds for $t \leq t_0$. According to that estimate and (7.7), the solution remains in $\Omega(Z,r)$. So we can extend the estimate to the whole half-line. □

15.8 Notes

Note 15.1 One of the basic methods in the stability theory of nonlinear evolution equations is the Lyapunov functionals method (see e.g. (Pazy, 1972), (Lakshmikantham et al., 1989), (Lakshmikantham et al., 1991), and references therein). By this method, many very strong results are obtained. But finding the Lyapunov functionals is usually a difficult task.

As was mentioned, a fundamental approach to the stability of scalar diffusion equations and coupled systems with monotone and quasimonotone nonlinearities is the method of upper and lower solutions. A systematical treatment of that approach is given in (Pao, 1992). About some other stability results for diffusion equations, see the excellent survey (Wollkind et al, 1994).

Our approach in this chapter is connected with the method of linearization, which develops the classical Lyapunov exponent method (Daleckii and Krein, 1971), (Henry, 1981), (Lunardi, 1995), etc. It is based on estimations for semigroups generated by linear parts of equations or on estimations for corresponding resolvents.

Note 15.2 Stabilizability of nonlinear distributed systems was considered in (Chan et al., 1992), (Kunimatsu and Sano, 1994) and in the papers cited therein.

Note 15.3 Contents of Sections 15.2-15.5 are based on the papers (Gil', 1985a, 1989, 1996). Our absolute stability conditions supplement the main results from the papers (Krasnoselskii et al, 1983), (Yakubovich, 1983). The results presented in Section 15.6 are adapted from (Gil', 1985b).

References for Chapter 15

Chan, W.L., Xiao, M.Q. and Y. Zhao (1992). Feedback stabilization of a class of multivalued nonlinear distributed parameter systems, *Nonlinear Analysis. Theory, Methods & Applications, 19, No 10*, 911-921

Daleckii, Yu L. and Krein, M. G. (1974). *Stability of Solutions of Differential Equations in Banach Space*. Amer. Math. Soc., Providence, R. I.

Gil', M.I. (1985a). On absolute stability of nonlinear nonstationary distributed systems, *Automation and Remote Control, No 6*, 12-19.

Gil', M.I. (1985b). On a class of absolutely stable distributed systems, *Automation and Remote Control, No 12*, 54-59.

Gil', M.I. (1989). Bounds for solutions of quasilinear parabolic equations, *Differential Eqs. 25*, 723-726.

Gil', M.I. (1996). On solvability of nonlinear equations in a lattice normed

Banach space. *Acta Sci.*, *62*, 201-215.

Henry, D. (1981). *Geometric Theory of Semilinear Parabolic Equations.* Lectures Notes in Mathematics, No 840. Springer-Verlag, New York.

Krasnosel'skii, M.A., Pokrovskii, A., and Zarubin A. (1983). On a principle bounded regimes absence in absolute stability of distributed parameters systems, *Automation and Remote Control, No 3*, 22-29.

Kunimatsu, N. and Sano, H. (1994). Compensator designs of semilinear parabolic systems. *Int. J. Control, 60, No 2*, 243-263.

Lakshmikantham, V., Leela, S., and Martynyuk, A.A. (1989). *Stability Analysis of Nonlinear Systems.* Marcel Dekker, New York.

Lakshmikantham V., Matrosov V.M., and Sivasundaram, S. (1991). *Vector Lyapunov Functions and Stability Analysis of Nonlinear Systems.* Kluwer Academic Publishers, Dordrecht, Boston, London.

Lunardi, A. (1995). *Analytic Semigroups and Optimal Regularity in Parabolic Problems.* Birkhäuser, Basel.

Pao, C.V. (1992). *Nonlinear Parabolic and Elliptic Equations.* Plenum Press, New York.

Pazy, A. (1972). On the applicability of Lyapunov's theorem in Hilbert spaces. *SIAM J. Math. Anal.*, *3*, 291-295.

Vrabie, I.I. (1987). *Compactness Methods for Nonlinear Evolutions.* Pitman, New York.

Wollkind D.J., Manoranjan V.S., and Zhang L. (1994). Weakly nonlinear stability of prototype reaction-diffusion model equation, *SIAM Review, 36:* 176-214.

Yakubovich V.A. (1983). Absolute stability of nonlinear distributed parameters systems, *Automation and Remote Control, N6*, 53-61.

Chapter 16

SEMILINEAR EQUATIONS IN BANACH SPACES WITH TIME-VARIANT LINEAR PARTS

This chapter is devoted to nonlinear evolution equations with separated linear variable operators in a Banach space and continuous nonlinearities.

Stability conditions for semilinear equations with time-variant linear operators in a Banach space are considered in Sections 16.1 and 16.2.

In Section 16.3 global stability conditions for systems in a Hilbert space with time-variant linear parts are established.

Throughout this chapter, $A(t)$ *for each* $t \geq 0$ *is a linear closed operator in a Banach space* X *with a dense constant domain* $D(A(t)) \equiv D$.

16.1 General equations in a Banach space

Let X be a Banach space and $Y \supseteq D$ be a dense imbedded Banach subspace of X. Consider in X the equation

$$\dot{u} = A(t)u + F(u,t) \ (t \geq 0, \ \dot{u} = du/dt, \ u = u(t)), \tag{1.1}$$

where the function F maps $Y \times [0, \infty)$ into X. Furthermore, suppose that the Cauchy problem for the "shorten" equation

$$\dot{u} = A(t)u \ (t \geq 0) \tag{1.2}$$

is well posed in X, and $U(t, s)$ is its evolution operator. We will call a continuous function $u(t) : [0, \infty) \to Y$ satisfying the equation

$$u(t) = U(t, 0)u_0 + \int_0^t U(t, s)F(u(s), s)ds \ (t \geq 0) \tag{1.3}$$

the mild solution of equation (1.1). Clearly, if $A(t) \equiv A$ is a constant operator, then this definition of the mild solution for (1.1) coincides with the one given in Section 15.1. Stability definitions for (1.1) are the same as in Section 15.1.

Suppose that for some positive $r \leq \infty$, the relation

$$||F(h, t)||_X \leq l||h||_Y \ (l = const;\ h \in \Omega(Y, r);\ t \geq 0). \tag{1.4}$$

holds. Recall that $\Omega(Y, r)$ is the ball in Y with the center at zero and the radius r.

Lemma 16.1.1 *Let Assumption 15.1A (the Lipschitz property) and condition (1.4) hold. Suppose the Cauchy problem for equation (1.2) is well posed, and the evolution operator of (1.2) has the properties*

$$\chi_0 = \sup_t ||U(t, 0)||_Y < \infty \tag{1.5}$$

and

$$\theta_0 \equiv \sup_{t \geq 0} \int_0^t ||U(t, s)||_{X \to Y} ds < \infty. \tag{1.6}$$

If, in addition $l\theta_0 < 1$, then for any initial vector u_0 satisfying the inequality

$$\chi_0||u_0||_Y \leq r(1 - l\theta_0), \tag{1.7}$$

equation (1.1) has a unique mild solution $u(t)$. Moreover, the estimate

$$||u(t)||_Y \leq \chi_0||u_0||_Y(1 - l\theta_0)^{-1} \ (t \geq 0) \tag{1.8}$$

is valid, and zero solution of (1.1) is asymptotically Y-stable.

Proof: First, let us prove the existence and uniqueness of mild solutions. Define in $C([0, T], Y)$ with $T < \infty$, the mapping

$$(\Psi z)(t) \equiv U(t, 0)u_0 + \int_0^t U(t, s)F(z(s), s)ds$$

$$(t \in [0, T];\ z \in C([0, T], Y)).$$

Clearly, (1.3) can be written as $(\Psi z)(t) = z(t)$. Set $C_1 = C([0,T],Y))\cap\Omega(Y,r)$. According to Assumption 15.1A,

$$\|(\Psi z_1)(t) - (\Psi z_2)(t)\|_Y \leq$$

$$\int_0^t \|U(t,s)\|_{X\to Y}\|F(z_1(s),s) - F(z_2(s),s)\|_X ds$$

$$\leq \int_0^t w\eta(t,s)\|z_1(s) - z_2(s)\|_Y ds$$

$$(t \in [0,T];\ z_1, z_2 \in C_1). \tag{1.9}$$

where

$$\eta(t,s) = \|U(t,s)\|_{X\to Y}.$$

Furthermore, conditions (1.4) and (1.5) imply

$$\|(\Psi z)(t)\|_Y \leq \chi_0\|u_0\|_Y + l\int_0^t \eta(t,s)\|z(s)\|_Y ds\ (t \in [0,T];\ z \in C_1). \tag{1.10}$$

Thanks to (1.6), (1.7) Ψ maps C_1 into itself. Denote by V a Volterra operator in $C(0,T)$ with the kernel $K(t,s) = w\eta(t,s)$. Then relation (1.9) can be written in the form

$$\|(\Psi z_1)(t) - (\Psi z_2)(t))\|_Y \leq (V\|z_1 - z_2\|_Y)(t).$$

Simple calculations show that according to (1.6) the spectral radius of V is equal to zero, since $T < \infty$. Now the existence and uniqueness are due to Lemma 15.3.1. The estimate (1.7) almost immediately follows from (1.3) and (1.10). It gives the stability. The asymptotic stability can be proved by a small perturbation. □

The following result can be considered as the stability in the first approximation.

Corollary 16.1.2 *Under Assumption 15.1A, let function F satisfy the relation*

$$\lim_{\|h\|_Y\to 0} \frac{\|F(h,t)\|_X}{\|h\|_Y} = 0$$

uniformly in t, and let (1.5) and (1.6) hold. Then the zero solution of (1.1) asymptotically Y-stable.

This assertion almost immediately follows from Lemma 16.1.1.

16.2 Equations in Banach spaces with the Lipschitz continuous linear parts

Again $Y \supseteq D$ is a dense imbedded Banach subspace of X. In this section it is assumed that $-A(t)$ is sectorial for each $t \geq 0$, and

$$\|(A(t) - A(s))v\|_X \leq q|t - s|\|v\|_Y \quad (t, s \geq 0;\ q = const;\ v \in D). \tag{2.1}$$

In addition, let us suppose that the analytic semigroup $e^{A(\tau)t}$ generated by operator $A(\tau)$ has the properties

$$\chi \equiv \sup_{\tau \geq 0} \sup_{t \geq 0} \|e^{A(\tau)t}\|_Y < \infty, \tag{2.2}$$

and

$$\theta \equiv \int_0^\infty [l + tq] \sup_{\tau \geq 0} \|e^{A(\tau)t}\|_{X \to Y} dt < 1. \tag{2.3}$$

Theorem 16.2.1 *Under Assumption 15.1A and condition (1.4) with $\Omega(Y, r) = Y$, let conditions (2.1-2.3) be fulfilled. Then the zero solution of (1.1) is globally asymptotically Y-stable.*

Proof: Since $-A(t)$ is sectorial, the Cauchy problem for equation (1.2) under condition (2.1) is well posed due to Theorem 14.1.3. Furthermore, it is simple to show that equation (1.3) is equivalent to the following one:

$$u(t) = e^{tA(\tau)}u_0 + \int_0^t e^{(t-s)A(\tau)}[(A(s) - A(\tau))u(s) + F(u(s), s)]ds \tag{2.4}$$

with a fixed $\tau \geq 0$. Repeating the reasonings of Lemma 16.1.1 one can show that (2.4) has a unique solution $u : [0, \infty) \to Y$. Moreover, by virtue of conditions (2.1)-(2.2), we arrive at the inequality

$$\|u(t)\|_Y \leq \chi\|u_0\|_Y + \int_0^t \|e^{(t-s)A(\tau)}\|_{X \to Y}(\|[A(s) - A(\tau)]u(s)\|_X +$$

$$\|F(u(s), s)\|_X)ds \leq \chi\|u_0\|_Y + \int_0^t \eta(t - s)(q|\tau - s| + l)\|u(s)\|_Y ds,$$

where $\eta(t) = \sup_{\tau \geq 0} \|e^{A(\tau)t}\|_{X \to Y}$. With $t = \tau$ this gives,

$$\|u(\tau)\|_Y \leq \chi\|u_0\|_Y + \int_0^\tau \eta(\tau - s)(q(\tau - s) + l)\|u(s)\|_Y ds.$$

Hence, for any finite t_0,

$$\sup_{0 \leq \tau \leq t_0} \|u(\tau)\|_Y \leq \chi\|u_0\|_Y + \sup_{0 \leq \tau \leq t_0} \|u(\tau)\|_Y \int_0^{t_0} \eta(t_0 - s)(q(t_0 - s) + l)ds.$$

But

$$\int_0^{t_0} \eta(t_0 - s)(l(t_0 - s) + q)ds \le \int_0^{\infty} \eta(s)(qs + l)ds = \theta.$$

Thus,

$$\sup_{t \le t_0} \|u(t)\|_Y \le \chi \|u_0\|_Y + \theta \sup_{t \le t_0} \|u(t)\|_Y.$$

Now (2.3) implies

$$\sup_{t \le t_0} \|u(t)\|_Y \le \chi \|u_0\|_Y (1 - \theta)^{-1}.$$

Since t_0 is arbitrary, this inequality provides the stability. By a small perturbation it is easy to prove the asymptotic stability. □

16.3 Equations in a Hilbert space

Now we are going to specialize Theorem 16.2.1 in the case of systems of evolution equations in a Hilbert space.

Let $H = E^n$ be an orthogonal sum of the same Hilbert spaces $E_1 = ... = E_n = E$, and S an invertible selfadjoint operator in H commuting with all $n \times n$-matrices. Recall that for a positive number μ, H^{μ} is the set $D(S^{\mu})$ with the graph norm $\|h\|_{\mu} = \|S^{\mu} h\|_H$ $(h \in D(S^{\mu}))$. Consider in H the equation

$$\dot{u} = Q(S,t)u + F(u,t) \ (u = u(t),\ t \ge 0) \tag{3.1}$$

where F maps $[0, \infty) \times H^1$ into H,

$$Q(S,t) \equiv -S^2 + c_1(t)S + c_2(t),$$

and $c_k(t)$ are variable $n \times n$-matrices. More exactly, $c_k(t)$ are operators in H defined by matrices. It is assumed that

$$q_k \equiv \sup_{t \ge 0} \|dc_k(t)/dt\|_{C^n} < \infty \ (k = 1, 2). \tag{3.2}$$

Moreover, F satisfies the Lipschitz property as a mapping from H^1 into H and

$$\|F(w,t)\|_H \le l \|Sw\|_H \ (w \in D(S)). \tag{3.3}$$

Recall that the quantity $g(.)$ is defined in Section 2.1 Furthermore, for the matrix pencil

$$Q(s,\tau) = -s^2 I_{C^n} + c_1(\tau)s + c_2(\tau) \ \ (\tau \ge 0;\ s \in \sigma(S)) \tag{3.4}$$

put

$$\alpha(Q(s,\tau)) \equiv \max_{k=1,\dots,n} Re\lambda_k(Q(s,\tau)) = -s^2 + \max_{k=1,\dots,n} Re\lambda_k(c_1(\tau)s + c_2(\tau)).$$

It is assumed that

$$\sup_{s\in\sigma(S)} \alpha(Q(s,\tau)) \le \alpha_0 < 0 \text{ and } v(\tau) \equiv \sup_{s\in\sigma(S)} g(c_1(\tau)s + c_2(\tau)) \le v_0 < \infty$$

$$(\tau \ge 0). \tag{3.5}$$

For instance, if $c_1(t) = c_1^*(t)$, then Proposition 2.1.1 gives

$$v(\tau) \le \sqrt{1/2}N(c_2(\tau) - c_2^*(\tau)). \tag{3.6}$$

If $c_1(t) = -c_1^*(t)$, then due to Proposition 2.1.1

$$v(\tau) \le \sqrt{1/2} \le N(c_2(\tau) + c_2^*(\tau)).$$

Denote

$$\eta_1(t) \equiv \sup_{\tau\ge 0}\sup_{s\in\sigma(S)} |s|e^{t\alpha(Q(\tau,s))}\sum_{k=0}^{n-1}\frac{t^k v^k(\tau)}{(k!)^{3/2}}.$$

Theorem 16.3.1 *Let the conditions (3.2), (3.3), (3.5), and*

$$\int_0^\infty [(q_1 + q_2\|S^{-1}\|_H)t + l]\eta_1(t)dt < 1 \tag{3.7}$$

be fulfilled. Then the zero solution of equation (3.1) is globally asymptotically H^1-stable.

Proof: We need only to check that the hypothesis of Theorem 16.2.1 is fulfilled with $A(\tau) = Q(\tau, S)$, $X = H, Y = H^1$. Indeed, condition (3.2) yields

$$\|\dot{Q}(S,t)h\|_H \le (q_1\|Sh\|_H + q_2\|h\|_H) =$$

$$(q_1 + q_2\|S^{-1}\|_H)\|Sh\|_H$$

$$(h \in D(S),\ t \ge 0).$$

That is, inequality (1.4) holds with $q = q_1+q_2\|S^{-1}\|_H$. Furthermore, according to Lemma 13.10.2,

$$\|e^{Q(S,\tau)t}\|_{H^1\to H} = \|Se^{Q(\tau,S)t}\|_H \le \eta_1(t)\ (\tau,\ t \ge 0),$$

and

$$\|e^{Q(S,\tau)t}\|_{H^1} \le \chi\ (t \ge 0).$$

Now the result is due Theorem 16.2.1. □

Consider the case

$$c_1(t) \equiv 0. \tag{3.8}$$

Omitting simple calculations, we get

$$\max_{s\in\sigma(S)} e^{-s^2t}|s| = \zeta(t) \equiv \begin{cases} \exp[t\alpha(-S^2)]\sqrt{|\alpha(S^2)|} & \text{if } t \geq \frac{1}{2|\alpha(-S^2)|}, \\ \frac{exp[-1/2]}{\sqrt{2t}} & \text{if } t \leq \frac{1}{2|\alpha(-S^2)|}. \end{cases}$$

Consequently, under (3.8), the relation

$$\eta_1(t) = \zeta(t) \sup_{\tau\geq 0} e^{\alpha(c_2(\tau))t} \sum_{k=0}^{n-1} \frac{t^k v^k(\tau)}{(k!)^{3/2}}$$

holds. The integral in (3.7) is now simply calculated.

16.4 Example

Consider in $\mathbf{C}^n$ the problem

$$\frac{\partial u}{\partial t} = \frac{\partial^2 u}{\partial x^2} + \tilde{c}_1(t)\frac{\partial u}{\partial x} + \tilde{c}_2(t)u + f(t, x, u, u_x)$$

$$(t > 0;\ 0 < x < \pi;\ u = u(x, t)), \tag{4.1}$$

$$u(0, t) = u(\pi, t),\ u_x(0, t) = u_x(\pi, t)\ (t \geq 0) \tag{4.2}$$

with boundedly differentiable in t $n \times n$-matrix-valued functions $\tilde{c}_1(t), \tilde{c}_2(t)$, and a sufficiently smooth

$$f : [0, \infty) \times [0, \pi] \times (\mathbf{C}^n)^2 \to \mathbf{C}^n.$$

Note that the following system can be written in the form (4.1):

$$\frac{\partial u}{\partial t} = \frac{\partial^2 u}{\partial x^2} + b_1(t, x)\frac{\partial u}{\partial x} + b_2(t, x)u + f_1(t, x, u, u_x).$$

with bounded matrices $b_1(t, x), b_2(t, x)$, if we take

$$\tilde{c}_k(t) = \pi^{-1}\int_0^\pi b_k(t, x)dx,$$

and

$$f(t, x, u, u_x) = f_1(t, x, u, u_x) + [b_1(t, x) - \tilde{c}_1(t)]u_x + [b_2(t, x) - \tilde{c}_2(t)]u.$$

Further, assume that the following conditions hold:

$$\tilde{c}_1(t) = \tilde{c}_1^*(t)\ (t \geq 0) \text{ and } v_1 \equiv \sup_{t\geq 0} \sqrt{1/2}N(\tilde{c}_2(t) + \tilde{c}_2^*(t)) < \infty, \tag{4.3}$$

and

$$\|f(t, x, h_1, h_2)\|_{C^n} \leq l_0\|h_1\|_{C^n} + l_1\|h_2\|_{C^n}\ (l_0, l_1 = const;\ h_1, h_2 \in \mathbf{C}^n). \tag{4.4}$$

To formulate the result put

$$c_1(t) = -2I + i\tilde{c}_1(t) \text{ and } c_2(t) = -I + i\tilde{c}_1(t) + \tilde{c}_2(t). \tag{4.5}$$

Finally, denote

$$\eta_2(t) = \sum_{m=0}^{n-1} \frac{t^j v_1^m}{(m!)^{3/2}} \max_{j=0,\pm1,\pm2,\dots} |2j-1| e^{t\alpha_j},$$

where

$$\alpha_j = -(2j-1)^2 + \sup_{\tau\geq 0} \max_{k=1,\dots,n} Re\, \lambda_k(c_1(\tau)(2j-1) + c_2(\tau)).$$

Recall that q_1, q_2 are defined by (3.2). Take $H = L^2([0,\pi], \mathbf{C}^n)$, and

$$H^1 = D(S) = \{w \in L^2([0,\pi], \mathbf{C}^n),\ w' \in L^2([0,\pi], \mathbf{C}^n), w(0) = w(\pi)\}.$$

Proposition 16.4.1 *Let the conditions (4.3), and*

$$\int_0^\infty [(q_1+q_2)t + l_0 + 2l_1)]\eta_2(t)dt < 1$$

be fulfilled. Then the zero solution of system (4.1-4.2) is globally asymptotically H^1-stable.

Proof: Define S by the formula

$$Sy = -iy' - y, \text{ for } y \in D(S).$$

Then (4.1), (4.2) can be rewritten as

$$\frac{\partial u}{\partial t} = [-(S+I)^2 + i\tilde{c}_1(t)(S+I) + \tilde{c}_2(t)]u + f(t,x,u,i(S+I)u) =$$

$$[-S^2 + c_1(t)S + c_2(t)]u + f(t,x,u,i(S+I)u)$$

with $c_1(t), c_2(t)$ defined by (4.5). Consequently, the problem (4.1), (4.2) has the form (2.1) with $\tilde{F}(t,u) = f(t,x,u,i(S+I)u)$. It is clear that the eigenfunctions of S_0 are e^{2ikx}, and the eigenvalues are

$$\lambda_k(S_0) = 2k-1 \ (k = 0, \pm1, \pm2, \dots).$$

Hence, $||S^{-1}||_H = 1$. According to (4.4),

$$||f(t,x,u,i(S+I)v)||_H \leq l_0||v||_H + l_1||(I+S)v||_H \leq$$

$$[(l_0+l_1)||S^{-1}||_H + l_1]||Sv||_H \ (v \in D(S)).$$

Thus. condition (3.3) is fulfilled with $l = l_0 + 2l_1$. Furthermore, we have by Proposition 2.1.1 and (4.5)

$$g(c_1(t)s + c_2(t)) = g(i(1+s)\tilde{c}_1(t) + \tilde{c}_2(t)) =$$

$$g(s\tilde{c}_1(t) - i\tilde{c}_2(t)) \le \sqrt{1/2}N((1+s)\tilde{c}_1(t) - i\tilde{c}_2(t) - (1+s)\tilde{c}_1^*(t) - i\tilde{c}_2^*(t))).$$

Due to (4.3), this implies

$$g(c_1(t)s + c_2(t)) \le \sqrt{1/2}N(-i\tilde{c}_2(t) - i\tilde{c}_2^*(t)) = \sqrt{1/2}N(\tilde{c}_2(t) + \tilde{c}_2^*(t)) \le v_1$$

for any $t \ge 0$. We thus get

$$\eta_1(t) \le \sum_{j=0}^{n-1} \frac{t^j v_1^j}{(j!)^{3/2}} \max_{k=0,\pm 1,\pm 2,\ldots} |2k-1| e^{t\alpha_k} = \eta_2(t).$$

Now Theorem 16.3.1 yields the result. □

16.5 Notes

The contents of this chapter are based on the method of linearization and the paper

Gil', M.I. (1996). Stability of linear evolution equations in lattice normed spaces, *Zeitschrift fur Analysis, 15. No 4*, 949-959.

About some other related works see Notes to Chapter 15.

Chapter 17

APPENDIX 1

In this appendix we shall prove Theorem 2.1.5, Lemma 8.1.1 and Proposition 3.1.1.

17.1 Proof of the estimate for regular matrix-valued functions

In the sequel $\|.\|$ denotes the Euclidean norm. For the notation see Sections 1.1 and 2.1. Recall the statement of Theorem 2.1.5.

Theorem 17.1.1 *Let* $A \in \mathbf{B}(\mathbf{C}^n)$*. and let* f *be a function regular on a neighborhood of the closed convex hull* $co(A)$ *of the eigenvalues of* A*. Then the inequality*

$$\|f(A)\| \leq \sum_{k=0}^{n-1} \sup_{\lambda \in co(A)} |f^{(k)}(\lambda)| g^k(A) \frac{\gamma_{n,k}}{k!}$$

is true.

Recall that $\gamma_{n,0} = 1$,

$$\gamma_{n,p} = [\sum_{1 \leq k_1 < k_2 < \ldots < k_p \leq n-1} 1]^{1/2}(n-1)^{-p/2} = [C_{n-1}^p (n-1)^{-p}]^{1/2},$$

$(p = 1, ..., n-1)$, $\mathbf{B}(\mathbf{C}^n)$ is the set of all linear operators in $\mathbf{C}^n$.

We have divided the proof into a series of lemmata.

A subspace $M \subset \mathbf{C}^n$ is *an invariant subspace* of $A \in \mathbf{B}(\mathbf{C}^n)$ if the relation $h \in M$ implies $Ah \in M$. By Schur's theorem (Marcus and Minc, 1964, Section I.4.10.2), $A \in \mathbf{B}(\mathbf{C}^n)$ can be decomposed as

$$A = D + V \qquad (1.1)$$

with a normal (diagonal) operator D and a nilpotent (upper-triangular) operator V. In addition, all the invariant subspaces of D and V are common, and $\sigma(A) = \sigma(D)$. We will call (1.1) *the triangular representation of A*. D the *diagonal part of A* and V *the nilpotent part of A*.

Lemma 17.1.2 *For any nilpotent operator $V \in \mathbf{B}(\mathbf{C}^n)$ the inequality*

$$||V^p|| \leq \gamma_{n,p} N^p(V) \quad (p = 1, \ldots, n-1)$$

is valid.

Proof: Since V is nilpotent, we can always represent it by an upper-triangular matrix with the zero diagonal:

$$V = (a_{jk})_{j,k=1}^n \text{ with } a_{jk} = 0 \ (j \geq k).$$

Denote

$$||x||_m = (\sum_{k=m}^n |x_k|^2)^{1/2} \text{ for } m < n,$$

where x_k are coordinates of a vector x. We can write

$$||Vx||_m^2 = \sum_{j=m}^{n-1} | \sum_{k=j+1}^n a_{jk} x_k|^2 \text{ for all } m \leq n-1.$$

Now we have (by Schwartz's inequality) the relation

$$||Vx||_m^2 \leq \sum_{j=m}^{n-1} h_j ||x||_{j+1}^2, \tag{1.2}$$

where

$$h_j = \sum_{k=j+1}^n |a_{jk}|^2 \ (j < n).$$

Further, by Schwartz's inequality

$$||V^2 x||_m^2 = \sum_{j=m}^{n-1} | \sum_{k=j+1}^n a_{jk} (Vx)_k|^2 \leq \sum_{j=m}^{n-1} h_j ||Vx||_{j+1}^2.$$

Here $(Vx)_k$ are coordinates of Vx. Taking into account (1.2), we obtain

$$||V^2 x||_m^2 \leq \sum_{j=m}^{n-1} h_j \sum_{k=j+1}^{n-1} h_k ||x||_{k+1}^2 = \sum_{m \leq j < k \leq n-1} h_j h_k ||x||_{k+1}^2.$$

Hence,

$$||V^2||^2 \leq \sum_{1 \leq j < k \leq n-1} h_j h_k.$$

Repeating these arguments, we arrive at the inequality

$$\|V^p\|^2 \le \sum_{1\le k_1<k_2<\ldots<k_p\le n-1} h_{k_1}\ldots h_{k_p}. \tag{1.3}$$

Introduce the following scalar-valued function $Q_{n,p}$ of positive scalar variables $y_1, \ldots, y_{n-1}$:

$$Q_{n,p}(y_1,\ldots,y_{n-1}) = (\sum_{1\le k_1<k_2<\ldots<k_p\le n-1} y_{k_1}y_{k_2}\cdots y_{k_p})(\sum_{\kappa=1}^{n-1} y_k)^{-p}.$$

Let us check that $Q_{n,p}$ has a unique maximum when $y_1 = y_2 = \ldots = y_{n-1}$. In fact,

$$\frac{\partial Q_{n,p}(y_1,\ldots,y_{n-1})}{\partial y_1} = [(\sum_{2\le k_1<k_2<\ldots<k_{p-1}\le n-1} y_{k_1}y_{k_2}\cdots y_{k_p})(\sum_{k=1}^{n-1} y_k)^p -$$

$$p(\sum_{1\le k_1<k_2<\ldots<k_p\le n-1} y_{k_1}y_{k_2}\cdots y_{k_p})(\sum_{k=1}^{n-1} y_k)^{p-1}]L(y_1,\ldots,y_{n-1}),$$

where

$$L(y_1,\ldots,y_{n-1}) = (\sum_{k=1}^{n-1} y_k)^{-2p}.$$

Take $y_1 = y_2 = \ldots = y_{n-1} = c$, where c is an arbitrary positive constant. Then

$$\frac{\partial Q_{n,p}(c,\ldots,c)}{\partial y_1} = [(\sum_{2\le k_1<k_2<\ldots<k_{p-1}\le n-1} 1)(n-1)^p -$$

$$p(\sum_{1\le k_1<k_2<\ldots<k_p\le n-1} 1)(n-1)^{p-1}]L_0 = [C_{n-2}^{p-1}(n-1)^p - pC_{n-1}^p(n-1)^{p-1}]L_0$$

$$= (n-1)^{p-1}[\frac{(n-1)(n-2)!}{(p-1)!(n-p-1)!} - \frac{p(n-1)!}{p!(n-p-1)!}]L_0 = 0.$$

Here $L_0 = c^{2p-1}L(c,\ldots,c)$. Similarly we can prove that

$$\frac{\partial Q_{n,p}(c,\ldots,c)}{\partial y_k} = 0$$

for any index $k < n$.

Now we shall prove that $Q_{n,p}$ has only one extremum if its arguments are positive. Indeed, an extremum of $Q_{n,p}$ is equivalent to the conditional extremum of the function

$$R_{n,p}(y_1,\ldots,y_{n-1}) = \sum_{1\le k_1<k_2<\ldots<k_p\le n-1} y_{k_1}y_{k_2}\cdots y_{k_p}$$

under the condition

$$(\sum_{k=1}^{n-1} y_k)^p = a^p$$

where a is a given positive number. So the extremums are the roots of the equations

$$\frac{\partial R_{n,p}(y_1, ..., y_{n-1})}{\partial y_j} - \lambda \frac{\partial}{\partial y_j} (\sum_{k=1}^{n-1} y_k)^p = 0 \ (j = 1, ..., n-1),$$

where λ is the Lagrange factor. Therefore,

$$F_j(y_1, ..., y_{n-1}) \equiv \frac{\partial R_{n,p}(y_1, ..., y_{n-1})}{\partial y_j} = \lambda p a^{p-1}. \ (j = 1, ..., n-1).$$

Hence $F_j(y_1, ..., y_{n-1}) = F_k(y_1, ..., y_{n-1})$ for all $k \neq j$. Obviously $F_j(y_1, ..., y_{n-1})$ does not depend on y_j, and $F_k(y_1, ..., y_{n-1})$ does not depend on y_k. So the latter equality is posssible only if $y_j = y_k$, since F_k monotonically increases. So $Q_{n,p}$ has only one extremum if its arguments are positive. Obviously, it is the maximum. Thus, we have

$$\max_{y_1, ..., y_{n-1} \geq 0} Q_{n,p}(y_1, ..., y_{n-1}) = Q_{n,p}(c, ..., c).$$

Clearly,

$$Q_{n,p}(c, ..., c) = (\sum_{1 \leq k_1 < k_2 < ..< k_p \leq n-1} 1) \ (n-1)^{-p} = C_{n-1}^p (n-1)^{-p} = \gamma_{n,p}^2.$$

Therefore due to (1.3),

$$||V^p||^2 \leq Q_{n,p}(h_1, ..., h_{n-1})(\sum_{j=1}^{n-1} h_j)^p \leq \gamma_{n,p}^2 (\sum_{j=1}^{n-1} h_j)^p.$$

But

$$\sum_{j=1}^{n-1} h_j \leq N^2(V).$$

We thus have received the required result. □

For more exact estimates for powers of nilpotent operators see (Gil', 1995, Chapter 1).

Lemma 17.1.3 *Let $Q, V \in \mathbf{B}(\mathbf{C}^n)$ and let V be a nilpotent operator. Suppose that all invariant subspaces of V and of Q are common. Then VQ and QV are nilpotent operators.*

Proof: Since all invariant subspaces of V and of Q are common, these operators have the same basis of the triangular representation. Taking into account that the diagonal entries of V are equal to zero, we easily obtain that the diagonal entries of QV and VQ are equal to zero. This proves the result. □

Lemma 17.1.4 *Let $\{d_k\}$ be an orthonormal basis in $\mathbf{C}^n$, $A_1, \ldots, A_j$ be linear operators in $\mathbf{C}^n$, and $\phi(k_1, ..., k_{j+1})$ be a scalar-valued function of $j+1$ natural arguments. Define orthoprojectors $Q(k)$ by*

$$Q(k)h = (h, d_k)d_k \ (h \in \mathbf{C}^n, \ k = 1, ..., n),$$

and set

$$T = \sum_{1 \le k_1, ..., k_{j+1} \le n} \phi(k_1, ..., k_{j+1}) Q(k_1) A_1 Q(k_2) \ldots A_j Q(k_{j+1}).$$

Then the following inequality is true:

$$\|T\| \le m(\phi) \| |A_1||A_2| \ldots |A_j| \|,$$

where

$$m(\phi) = \max_{1 \le k_1, ..., k_{j+1} \le n} |\phi(k_1, ..., k_{j+1})|,$$

and $|A_k|$ are operators whose matrix elements in $\{d_k\}$ are moduli of matrix elements of A_k with respect to the basis $\{d_k\}$.

Proof: For any entry $T_{sm} = Q(s)TQ(m)$ $(s, m = 1, ..., n)$ of the operator T we have

$$T_{sm} = \sum_{1 \le k_2, ..., k_j \le n} \phi(s, k_2, ..., k_j, m) a^{(1)}_{sk_2} \ldots a^{(j)}_{k_j, m},$$

where $a^{(i)}_{jk} = (A_i d_k, d_j)$ are the entries of A_i. Hence

$$|T_{sm}| \le m(\phi) \sum_{1 \le k_2, ..., k_j \le n} |a^{(1)}_{sk_2} \ldots a^{(j)}_{k_j m}|.$$

From this and from the equality

$$\|Tx\|^2 = \sum_{j=1}^n |(Tx)_j|^2 \ (x \in \mathbf{C}^n),$$

where $(.)_j$ means the j-th coordinate, we get the result. □

Lemma 17.1.5 *Let V be the nilpotent part of $A \in B(\mathbf{C}^n)$. Then the equalities*

$$N^2(V) = N^2(A) - \sum_{k=1}^n |\lambda_k(A)|^2 = g^2(A), \tag{1.4}$$

and

$$g^2(A) = 2\,(N^2(A_I) - \sum_{k=1}^{n} |Im\,\lambda_k(A)|^2) \tag{1.5}$$

are true, where $A_I = (A - A^*)/2i$.

Proof: Due to Lemma 17.1.3 both matrices V^*D and D^*V are nilpotent. Therefore,

$$Trace(D^*V) = 0 \text{ and } Trace(V^*D) = 0. \tag{1.6}$$

It is easy to see that

$$Trace(D^*D) = \sum_{k=1}^{n} |\lambda_k(A)|^2. \tag{1.7}$$

Due to (1.1) and (1.6) we can write down

$$N^2(A) = Trace(D+V)^*(V+D) = Trace(V^*V + D^*D) =$$

$$N^2(V) + \sum_{k=1}^{n} |\lambda_k(A)|^2,$$

and relation (1.4) is proved. Furthermore, thanks to (1.1) and (1.6) we have

$$4N^2(A_I) = Tr(A - A^*)(A^* - A) = -Trace((A - A^*)^2) =$$

$$-Trace((V - V^*)^2) - Trace((D - D^*)^2).$$

Set $V_I = (V - V^*)/2i$ and $D_I = (D - D^*)/2i$. We can write

$$N^2(A_I) = N^2(V_I) + N^2(D_I).$$

It is not hard to see that

$$N^2(V_I) = \frac{1}{2}\sum_{m=1}^{n}\sum_{k=1}^{m-1} |a_{km}|^2 = \frac{1}{2}N^2(V).$$

From this and from (1.4) relation (1.5) follows. □

Furthermore, let $|V|$ be the operator whose matrix elements in the orthonormal basis of the triangular representation (the Schur basis) are the moduli of the matrix elements of the nilpotent part V of A with respect to this basis. That is,

$$|V| = \sum_{k=1}^{n}\sum_{j=1}^{k-1} |a_{jk}|(.,e_k)e_j,$$

where $\{e_k\}$ is the Schur basis and $a_{jk} = (Ae_k, e_j)$.

Lemma 17.1.6 *Under the hypothesis of Theorem 17.1.1 the estimate*

$$\|f(A)\| \le \sum_{k=0}^{n-1} \sup_{\lambda \in co(A)} |f^{(k)}(\lambda)| \frac{\| \, |V|^k \, \|}{k!}$$

is true, where V is the nilpotent part of A.

Proof: It is not hard to see that the representation (1.1) implies the equality

$$R_\lambda(A) \equiv (A - I\lambda)^{-1} = (D + V - \lambda I)^{-1} =$$

$$(I + R_\lambda(D)V)^{-1} R_\lambda(D)$$

for all regular λ. According to Lemma 17.1.3 $R_\lambda(D)V$ is a nilpotent operator because V and $R_\lambda(D)$ have common invariant subspaces. Hence,

$$(R_\lambda(D)V)^n = 0.$$

Therefore,

$$R_\lambda(A) = \sum_{k=0}^{n-1} (R_\lambda(D)V)^k (-1)^k R_\lambda(D). \tag{1.8}$$

Due to the representation for functions of matrices

$$f(A) = -\frac{1}{2\pi i} \int_\Gamma f(\lambda) R_\lambda(A) d\lambda = \sum_{k=0}^{n-1} C_k, \tag{1.9}$$

where

$$C_k = (-1)^{k+1} \frac{1}{2\pi i} \int_\Gamma f(\lambda) (R_\lambda(D)V)^k R_\lambda(D) d\lambda.$$

Here Γ is a closed contour surrounding $\sigma(A)$. Since D is a diagonal matrix with respect to the Schur basis, then

$$R_\lambda(D) = \sum_{j=1}^{n} \frac{Q_j}{a_{jj} - \lambda},$$

where $Q_k = (., e_k)e_k$. We have

$$C_k = \sum_{j_1=1}^{n} Q_{j_1} V \sum_{j_2=1}^{n} Q_{j_2} V \dots V \sum_{j_k=1}^{n} Q_{j_k} I_{j_1 \dots j_{k+1}}.$$

Here

$$I_{j_1 \dots j_{k+1}} = \frac{(-1)^{k+1}}{2\pi i} \int_\Gamma \frac{f(\lambda) d\lambda}{(a_{j_1 j_1} - \lambda) \dots (a_{j_{k+1} j_{k+1}} - \lambda)}.$$

Lemma 17.1.4 gives the estimate

$$\|C_k\| \le \max_{1 \le j_1 \le \dots \le j_{k+1} \le n} |I_{j_1 \dots j_{k+1}}| \| \, |V|^k \, \|.$$

Due to Lemma 1.10.1

$$I_{j_1 \dots j_{k+1}} \leq \sup_{\lambda \in co(A)} \frac{|f^{(k)}(\lambda)|}{k!}.$$

This inequality and (1.9) imply the result. □

Proof of Theorem 17.1.1: Lemma 17.1.2 implies

$$\| \, |V|^k \, \| \leq \gamma_{n,k} N^k(|V|) \, , k = 1, \dots, n-1.$$

But $N(|V|) = N(V)$. Now Lemma 17.1.5 yields the required result. □

17.2 An independent proof of Corollary 2.1.7

The next theorem follows from Theorem 17.1.1 and the inequality

$$\gamma_{n,p}^2 \leq \frac{1}{p!} \; (p = 1, 2, \dots, n-1).$$

Theorem 17.2.1 *Let $A \in \mathbf{B}(\mathbf{C}^n)$, and let f be a function regular on a neighborhood of the closed convex hull $co(A)$ of the eigenvalues of A. Then the inequality*

$$\|f(A)\| \leq \sum_{k=0}^{n-1} \sup_{\lambda \in co(A)} |f^{(k)}(\lambda)| \frac{g^k(A)}{(k!)^{3/2}}$$

is true.

But in this section we give the independent proof.

It is well-known that any finite-dimensional nilpotent operator is a quasinilpotent Hilbert-Schmidt operator (Gohberg and Krein, 1970). Recall that a quasinilpotent operator is an operator whose spectral radius equals zero.

Lemma 17.2.2 *Let V be a Hilbert-Schmidt completely continuous quasinilpotent (Volterra) operator acting in a separable Hilbert space H. Then the inequality*

$$\|V^k\| \leq \frac{N^k(V)}{\sqrt{k!}} \text{ for any natural } k \tag{2.1}$$

is true.

Proof: Let $L^2[0,1]$ be the Hilbert space of scalar-valued functions defined on $[0,1]$ with the scalar product $(.,.)$. Consider in $L^2[0,1]$ a Volterra Hilbert-Schmidt integral operator W defined by

$$(Wh)(t) = \int_0^t K(t,s)h(s)ds, \; h \in L^2[0,1],$$

where $K(t,s)$ is a scalar kernel. It is well known that the kernel satisfies the condition

$$\int_0^1 \int_0^t |K(t,s)|^2 ds dt = N^2(W) < \infty.$$

cf. Gohberg and Krein (1970). Here as above $N(.)$ is the Hilbert-Schmidt norm. Employing Schwarz's inequality, we have

$$||Wh||^2 = \int_0^1 |\int_0^t K(t,s)h(s)ds|^2 dt \leq$$

$$\int_0^1 \int_0^t |K(t,s)|^2 ds \int_0^t |h(s_1)|^2 ds_1 dt.$$

Setting

$$w(t) = \int_0^t |K(t,s)|^2 ds, \tag{2.2}$$

one can rewrite the latter inequality in the shape

$$||Wh||^2 \leq \int_0^1 w(t) \int_0^t |h(s_1)|^2 ds_1 dt.$$

Using this inequality, we obtain

$$||W^k h||^2 \leq \int_0^1 w(t) \int_0^t |W^{k-1} h(s_1)|^2 ds_1 dt$$

Once more apply Schwarz's inequality :

$$||W^k h||^2 \leq \int_0^1 w(t) \int_0^t w(s_1) \int_0^{s_1} |W^{k-2} h(s_2)|^2 ds_2 ds_1 dt.$$

Repeating these arguments, we arrive at the inequality

$$||W^k h||^2 \leq \int_0^1 w(t) \int_0^t w(s_1) \int_0^{s_1} w(s_2) \ldots$$

$$\int_0^{s_{k-2}} w(s_{k-1}) \int_0^{s_{k-1}} |h(s_k)|^2 ds_k \ldots ds_2 ds_1 dt.$$

Taking

$$||h||^2 = \int_0^1 |h(s)|^2 ds = 1,$$

we get

$$||W^k||^2 \leq \int_0^1 w(t) \int_0^t w(s_1) \ldots \int_0^{s_{k-2}} w(s_{k-1}) ds_{k-1} \ldots ds_2 ds_1 dt. \tag{2.3}$$

It is simple to see that

$$\int_0^1 w(s_1)\dots\int_0^{s_{k-1}} w(s_k)ds_k\dots ds_1 =$$

$$\int_0^m\int_0^{z_1}\dots\int_0^{z_{k-1}} dz_k dz_{k-1}\dots dz_1 = \frac{m^k}{k!},$$

where

$$z_k = z_k(s_k) \equiv \int_0^{s_k} w(s_{k-1})ds_{k-1}$$

and

$$m = \int_0^1 w(s)ds.$$

Thus (2.3) gives

$$\|W^k\|^2 \le \frac{(\int_0^1 w(s)ds)^k}{k!}.$$

But according to (2.2)

$$m = \int_0^1 w(s)ds = N^2(W).$$

Thus,

$$\|W^k\|^2 \le \frac{N^{2k}(W)}{k!} \tag{2.4}$$

and for W we have proved the inequality (2.1).

Due to the well-known Theorem 5.2.2 from the book (Gohberg and Krein, 1970), there exists a nonessential extension of the operator V which is unitarily equivalent to a some Volterra Hilbert-Schmidt operator V_0 acting in the space $L^2([0,1],\mathbf{C}^n)$ of vector-valued functions defined on [0,1] with the scalar product

$$(h,g) = \int_0^1 \sum_{k=1}^n h_k(s)\overline{g_k}(s)ds$$

where $h_k(s), g_k(s)$ are the coordinates of the functions $h, g \in L^2([0,1],\mathbf{C}^n)$, respectively.

Applying our arguments above to the operator V_0, we obtain the following inequality which is similar to (2.4):

$$\|V_0^k\|^2 \le \frac{N^{2k}(V_0)}{k!}.$$

Since V and V_0 are unitarily equivalent,

$$\|V^k\| = \|V_0^k\|, \text{ and } N(V) = N(V_0).$$

Thus, we get the inequality (2.1). □

Proof of Theorem 17.2.1: Lemma 17.2.2 implies

$$\| \, |V|^k \, \| \leq \frac{N^k(|V|)}{\sqrt{k!}} \text{ for any natural } k < n.$$

But $N(|V|) = N(V)$. Now Lemma 17.1.6 yields the required result. □

17.3 Proof of the estimate for the norm of resolvents

Recall that $\rho(A, \lambda) = \inf_{s \in \sigma(A)} |\lambda - s|$.

Theorem 17.3.1 *Let A be a linear operator in $\mathbf{C}^n$. Then the following estimate is true:*

$$\|R_\lambda(A)\| \leq \sum_{k=0}^{n-1} g^k(A) \gamma_{n,k} \rho^{-k-1}(A, \lambda) \textit{ for all regular } \lambda.$$

Proof: According to Lemma 17.1.3, $R_\lambda(D)V$ is a nilpotent operator. So by virtue of Lemma 17.1.2

$$\|(R_\lambda(D)V)^k\| \leq N^k(R_\lambda(D)V)\gamma_{n,k}. \tag{3.1}$$

Since D is a normal operator, we can write down $\|R_\lambda(D)\| = \rho^{-1}(D, \lambda)$. It is clear that

$$N(R_\lambda(D)V) \leq N(V)\|R_\lambda(D)\| = N(V)\rho^{-1}(D, \lambda). \tag{3.2}$$

This, and (1.8) imply the relations

$$\|R_\lambda(A)\| \leq \sum_{k=0}^{n-1} \|(R_\lambda(D)V)^k R_\lambda(D)\|. \tag{3.3}$$

Now (3.1) yields,

$$\|R_\lambda(A)\| \leq \sum_{k=0}^{n-1} N^k(V)\gamma_{n,k}\rho^{-k-1}(D, \lambda).$$

This relation and (1.4) prove the stated result, since A and D have the same eigenvalues. □

The assertion of Lemma 8.1.1 follows from the latter theorem and from inequality

$$\gamma_{n,p}^2 = \frac{(n-2)(n-3)\ldots(n-p)}{(n-1)^{p-1}p!} \leq \frac{1}{p!} \quad (p = 1, 2, \ldots, n-1).$$

The direct proof of Lemma 8.1.1: According to Lemma 17.1.3, $R_\lambda(D)V$ is a nilpotent operator. So by virtue of Lemma 17.2.2

$$||(R_\lambda(D)V)^k|| \leq N^k(R_\lambda(D)V)(k!)^{-1/2}.$$

Now (3.2) and (3.3) yield

$$||R_\lambda(A)|| \leq \sum_{k=0}^{n-1} N^k(V)(k!)^{-1/2}\rho^{-k-1}(D,\lambda).$$

This relation and (1.4) prove the stated result, since A and D have the same eigenvalues. □

17.4 Proof of Proposition 2.1.1

Inequality $g^2(A) \leq \frac{1}{2}N^2(A^* - A)$ is due Lemma 17.1.5. Furthermore, take into account that the norms of the nilpotent parts of the matrices A and $Ae^{it} + zI$ with a real number t and a complex one z coincide. Then due to Lemma 17.1.5 obtain the required relation $g(Ae^{it} + zI) = g(A)$.

References for Appendix 1

Gil', M. I. (1995). *Norm Estimations for Operator-valued functions and Applications.* Marcel Dekker, Inc, New York.

Gohberg, I. C. and Krein, M. G. (1970). *Theory and Applications of Volterra Operators in Hilbert Space*, Trans. Mathem. Monographs, vol. 24, Amer. Math. Soc., R. I.

Marcus, M. and Minc, H. (1964). *A Survey of Matrix Theory and Matrix Inequalities.* Allyn and Bacon, Boston.

LIST OF MAIN SYMBOLS

$\mathbf{C}^n$	The complex Euclidean space	
$\mathbf{R}^n$	The real Euclidean space	
$(.,.) = (.,.)_H$	The scalar product in H	
$\|\cdot\|_n$	The arbitrary norm in $\mathbf{C}^n$	
$\|A\| = \|A\|_X$	The norm of A in X	
$TrA = TraceA$	The trace of A	
$det(A)$	The determinant of A	
$\lambda_k(A)$	The eigenvalue of A	
$\sigma(A)$	The spectrum of A	
$I = I_X$	The identity operator (in the space X)	
$R_\lambda(A)$	The resolvent of A	
C_n^p	The binomial coefficient	
$\rho(\lambda, A)$	The distance between $\sigma(A)$ and λ	
$r_s(A)$	The spectral radius of A	
$\|\cdot\|_{C^n}$	The Euclidean norm	2
$N(A)$	The Hilbert-Schmidt (Frobenius) norm of A	2
$\alpha(A) \equiv \sup Re\sigma(A)$		5
$g(A)$		21
$\gamma_{n,p}$		24
$\beta(A) \equiv \inf Re\sigma(A)$		25
$\Gamma(A)$		26
$\|A\|$	The matrix whose elements are absolute values of A	
H	Hilbert space	
A^{-1}	The operator inverse to A	
A^*	The operator conjugate to A	
$\chi(A)$		26
$\Omega(r)$		81
$\|\cdot\|_{L[a,b]} = \|\cdot\|_{L^2([a,b],C^n)}$		
$L^p(\omega, \mathbf{C}^n)$		134
$C(\omega, \mathbf{C}^n)$		134
$C^1([a,b], \mathbf{C}^n)$		226
$D(A)$	The domain of A	262
$\Omega(X, r)$		316

Index

www.ingramcontent.com/pod-product-compliance
Lightning Source LLC
LaVergne TN
LVHW011907200926
848484LV00002B/553
9780792382218